Modelling Learners and Learning in Science Education

Keith S. Taber

Modelling Learners and Learning in Science Education

Developing Representations of Concepts, Conceptual Structure and Conceptual Change to Inform Teaching and Research

Springer

Keith S. Taber
Faculty of Education
University of Cambridge
Cambridge, UK

ISBN 978-94-007-7647-0   ISBN 978-94-007-7648-7 (eBook)
DOI 10.1007/978-94-007-7648-7
Springer Dordrecht Heidelberg New York London

Library of Congress Control Number: 2013956003

Printed on acid-free paper

Springer is part of Springer Science+Business Media (www.springer.com)

Essentially, all models are wrong, but some are useful.

(Box & Draper, 1987, p. 424)

Preamble

Modelling is widely recognised to be important in both the doing of science and – increasingly – in the teaching of science. Modelling is equally important in research in science education as it allows us to develop ways of thinking about the nature and structure of complex phenomena and, in particular, to devise simplifications suitable for formal testing. This is especially important in an area like learning. Learning is an absolutely fundamental concern in education (science education or otherwise), and the term refers to a complex set of processes that are not fully understood. Our research in science education often relies on assumptions about the nature of learning and learners, and research to better understand learning in science is a core area of activity. Research in these areas uses and develops constructs, representations and models that are either assumed or proposed to in some sense reflect the 'real' nature of learning. Yet the central role of modelling in research into learning in science has not always been as clearly acknowledged as – for example – the role of modelling in the teaching and learning of the subject. In this book it is suggested that this is problematic as often researchers write (and so perhaps think) as though learning and associated notions such as understanding, thinking and knowing are not problematic. Research papers (as will be illustrated in Chap. 1) commonly report research findings in terms of student knowledge, or understanding, as though the terms are well understood and unambiguous and the notions behind them are unproblematic. Yet, as will be demonstrated here, this is hardly the case.

There is a problem, then, that because notions such as learning and knowing are so familiar from everyday use, researchers often fail to treat them as technical constructs when reporting the outcome of science education research. In some cases, this lack of sophistication undermines the value of the research to the community, as results are presented in terms that treat complex phenomena as though we all understand and agree about their nature. Yet, as this book shows, this is not a warranted assumption.

This book then explores the role of modelling in research exploring learning in science. The fundamental assumption of the book is that research into learning in science necessarily involves making knowledge claims based upon building and

representing models. Reports in research papers giving accounts of student knowledge, and/or conceptual change, rely upon the models researchers adopt or develop
when interpreting their data. It follows that the modelling processes and the nature
of the models produced need to be understood by those who read the research
reports if the knowledge claims made in research are to be correctly understood and
appropriately judged.

In this volume I offer an account of the modelling processes involved in research
into student understanding and learning in science. I argue the case for the importance of acknowledging the modelling processes necessarily underpinning any
account of student thinking, knowledge or learning. The book reviews the current
state of knowledge in science education in relation to the modelling of scientific
thinking, understanding and learning in science.

A Note on Teachers as Knowers and Learners

Much work in science education focuses on the ideas of learners in formal education – school pupils or other students. In general I have written the book referring to
learners and largely have such students in mind. However, the book is equally relevant for research into informal learning (e.g. adult museum visitors who would not
consider themselves learners) and into teachers – who of course continue to learn
about both the subjects they teach and about professional matters such as pedagogy.
In recent years there has been a significant literature looking at such matters as
science teachers' beliefs about, for example, the nature of science or constructivist
pedagogy. Researchers exploring these areas are subject to the same methodological
problems and questions regarding the assumptions behind their research as those
exploring 8-year-old children's ideas about the shape of the earth or postgraduate
students' notions of types of chemical bond. The type of modelling processes inherent
in research is the same in all these cases.

Moreover, although the book is largely framed in terms of formal research as
reported in journals and other scholarly works, the core argument here is equally
important to teachers going about their work in school or college classrooms.
Teachers are not usually developing formal models of their students' learning to
present in research papers. However, the process of teaching science involves
designing instruction that takes into account a learner's current knowledge and
understanding. Teaching is a highly interactive process where the teacher seeks to
facilitate learners' sense making to shift current knowledge and understanding
towards what the curriculum sets out as target scientific knowledge and understanding. That requires a good appreciation of what students already know, where they
may have alternative conceptions and how well they understand key concepts.

The effective science teacher is therefore constantly seeking to update their own
understanding of the current states of their students' learning. That is, science teaching depends upon an ongoing, informal process of modelling the mental states of
learners. Teachers do this by constantly collecting data (by asking questions, by

reviewing students' work) and analysing it to inform the next steps of the teaching process. In effect all science teachers are involved in a process of continuous action research to inform and improve their teaching, based upon developing models of the mental states of their learners. Action research is primarily context-directed research to inform practice locally rather than the theory-directed research of academic researchers that is intended to have relevance beyond specific research sites and which is formally reported in publications. The models generated by teachers in their day-to-day work will therefore not be subjected to the scrutiny of others in the same way as the academic research discussed in this volume. Yet teachers, often even more so than researchers, tend to take for granted notions of knowledge, understanding, learning, etc., and, just as researchers, they can benefit from problematising these issues and reflecting on just how they understand – and so implicitly model – these concepts that are so central to their day-to-day work.

Acknowledgements

The author would like to thank John Gilbert for originally encouraging the preparation of this volume and Bernadette Ohmer – my publishing editor at Springer – for her patience and support during the writing of this book. I would also like to acknowledge and thank the anonymous referee who offered positive feedback with helpful suggestions on a full draft of the manuscript. As always, it is useful to have a reader's perspective on what one writes – especially for a book that emphasises how imperfect public representations of our thinking (such as the text of this book) are interpreted in other minds that impose their own sense on them.

This book is located in science education but draws strongly on cognitive science and psychological research and scholarship. Whilst reasonable attempts to offer appropriate citations to previous work have been made in accordance with academic conventions, I am very aware that the message of Chap. 5 is that human memory is not primarily a faculty for providing accurate records of past experience, but rather has evolved to support future action in the world. Memories are often hybrids of our past interpretations of events, patched up with what seems so feasible (based on our general knowledge of the world) that we can no longer see the seams between what we actually experienced and what it now just seems most likely we experienced.

Introspection suggests that my own memory certainly is quite good at recalling 'gist', but often poor at reporting specifics of past experience. My suspicion (or perhaps my hope) is that such a memory supports a strongly synthetic way of thinking that is able to make connections and recognise general patterns rather than focusing on the minutiae of specific details. That does not excuse bad scholarship, but I am aware that I inevitably own an enormous debt to the authors of many things I have read over the years that are not cited here as well as to colleagues and students for things I have heard in presentations and in conversations in both formal and informal contexts. I have sought to acknowledge those key sources I am aware have informed my thinking, and I would here like to acknowledge that I am aware that I am surely drawing on many other sources that I either no longer specifically recall or have simply not recognised as influences in writing this book.

I suspect there may even be some good ideas in here that I present as if original, but which have worked their way into my consciousness so slowly that I was unaware that their original inspiration was something I had long ago read or heard. I take some comfort in knowing that if this is indeed so, my failure is probably not so unusual, as is indicated by occasional high-profile examples such as when George Harrison was sued for a great deal of money for not acknowledging a highly popular song was very similar to an earlier hit written by someone else. At least working in the academic world, rather than 'the material world', such unconscious plagiarism is unlikely to lead to claims for vast amounts of unpaid royalties.

My only defence is that the model of cognition offered in this book suggests that my failings here are due to the effective way in which the human brain interprets and integrates experience to provide a viable knowledge base for ongoing effective action in the world. Human memories, like the models mooted in this book, need to be understood not as being true replicas of aspects of the world, but rather as useful thinking tools that are fit for purpose. Both should be judged by how they help us move forward, and I very much hope this book will provide useful ways of thinking for those working in science education and looking to develop pedagogy or advance research programmes.

Finally, I would like to dedicate this volume, with love, to my wife Philippa, who has suffered significant health issues during the preparation of the book. At times my academic work has had to be given a low priority in our lives, and the writing has been disrupted. Despite considerable challenges, however, Philippa has understood my drive to push forward in my work and has offered her love and support to help me balance family and work responsibilities. Completing this manuscript despite major distractions is an achievement of 'team Taber'.

Any errors, or other failings, are of course the author's own.

Contents

Meta-understanding and Multiple Understanding 133
Understanding Distinguished from Beliefs... 134
Alternative Interpretations of Perceived Manifold Conceptions.......... 135
Describing Student Understanding: Challenges
of the Idiographic Approach 136
The Researcher's Dilemma 137
Comprehending Language 138

7 The Learner's Thinking 141
A Study on 'Scientists and Scientific Thinking' 141
Establishing a Meaning for 'Thinking' 143
Thinking and Processing 144
The Significance of Preconscious Processing 147
Thinking as an Inclusive Term 148
Forms of Thinking Valued in Science Education 149
Scientific Thinking 149
Science-as-Logic: Logical Thinking... 150
Creative Thinking.. 152
Analogical Thinking... 154
Imagery 155
Critical Thinking 156
Problem-Solving 156
Metacognition 157
The Fallacy of 'Machine Code' 158
The Limits of Computing Analogies for Cognition....................... 158
A Different Type of Processing System 159
The Ghost in the Machine: Who Tunes Our Processing Networks?.... 161
Emergent Systems.. 161
Key Terms from the Mental Register... 162

Part III   Modelling the Science Learner's Knowledge

8 Introduction to Part III: Knowledge in a Cognitive
System Approach 167
The Cognitive System Approach 168
Linking Back to the Mental Register 170
Seeking to Understand 'Knowledge' Within the Cognitive
System Approach 170

9 The Nature of the Learner's Knowledge.................................... 173
Knowledge as a Problematic Notion... 173
Public and Personal Knowledge.. 173
What Does It Mean to Know?... 174
Knowledge as True Reasoned Belief 174
Finding a More Useful Notion of Knowledge
for Science Education 176
Use of the Term 'Knowledge' in Science Education 179

Part I
Introduction

The introductory part comprises of a single chapter on the centrality of models for knowledge claims in science education. This chapter sets the scene for the rest of the book by making a case that in the field of science education it is very common for research reports to discuss the complex phenomena of student thinking and learning as though these processes were simple and well understood. Chapter 1 contextualises the theme of this volume in terms of a well-established 'research programme' within science education and in effect sets out the agenda for the rest of the book. The various terms often used unproblematically in research reports, as shown in Chap. 1, are central to the phenomena being investigated and potentially explained in our research. Parts II, III and IV of the book will explore these notions, and consider to what extent we do understand these phenomena, and how we go about modelling them in the way we approach and report research. It is not too much of a 'spoiler' to note here that we do not fully understand student thinking and learning, and that forming representations and models of these processes is far from straightforward.

Chapter 1
The Centrality of Models for Knowledge Claims in Science Education

Our knowledge of others, in short, is essentially no different from our knowledge of the world. Because it is the result of our own perceiving and conceiving, it cannot be a true representation of independently existing entities; but insofar as we can use it as a basis for further acting and thinking it constitutes a viable model of these very special elements of our experiential world. (Glasersfeld, 1988, p. 6)

Science education is an active field of research and scholarship (Fensham, 2004), concerned with the teaching and learning of science. Although a broad field of research, a key focus has been on learners' knowledge and understanding of aspects of science and how this changes – that is, science learning. Such enquiry has been undertaken with a view to informing better pedagogy, to support teachers in their role in facilitating learning.

Studying learning from a science education frame clearly has significant potential to overlap studies of learning undertaken as part of psychology and often carried out in science contexts. Indeed there need be no absolute distinction here, and certainly some published studies can contribute to both disciplines: however, psychological research is likely to be motivated primarily by general questions about the nature of human learning, with science learning providing contexts seen as suitable for particular studies, whereas work undertaken in science education will tend be undertaken with a view to being directly relevant to informing more effective science teaching.

Research into what students 'think', 'know' and 'understand' about science – and as will be discussed these words are often taken for granted but deserving careful specification when used as technical terms in a field of scholarship – serves a number of purposes.

From the perspective of effectiveness, if science teaching is about facilitating student learning, then we need to find out whether students have learnt. For that we need to find out what students know or understand, both before and after teaching, to judge if there has been any learning. In principle this could be down to individual teachers using classroom assessment. However, in planning curriculum (e.g. at a National level) it is important to have a fairly good overview of what students generally know and understand and are likely to be able to learn next, at particular ages.

K.S. Taber, *Modelling Learners and Learning in Science Education: Developing Representations of Concepts, Conceptual Structure and Conceptual Change to Inform Teaching and Research*, DOI 10.1007/978-94-007-7648-7_1,

Much thinking about such issues within science education has been informed by a constructivist perspective (Bodner, 1986; Gilbert & Watts, 1983; Glasersfeld, 1989), which can be characterised as appreciating how future learning is highly contingent upon the current state of an individual's knowledge/understanding (Taber, 2009b). This will be considered in more depth later in the book, but for the moment it is important to note that as a result of the widespread influence of a con-structivist perspective, *a good deal of research in science education has made claims about what learners know and understand or how their knowledge and understanding have changed* – that is, claims about learning. Some examples of such claims are considered below.

The central argument of this book is that these knowledge claims are *inevitably based on models*, and so the claims made in research reports can only be fully appreciated by readers who both recognise the models for what they are and understand something of the modelling processes used to derive them. The motivation for this book derives from concerns that this is not always made explicit in research reports nor fully appreciated by those who use them. This lack of appreciation of the status or research findings and the processes that produce them undermines the potential of the research to inform more effective classroom work.

Some Examples of Knowledge Claims Made in Studies

It is useful to provide readers with a few examples from the literature of what I mean by 'knowledge claims' in studies of student thinking, understanding and learning in science. I have selected a range of examples that in this form are necessarily stripped of their context within the original authors' accounts to illustrate something of *the range* of kinds of claims made in this area of research (see Table 1.1). Anyone familiar with the literature in this area will recognise these types of knowledge claims as being very common in science education research. I have italicised some key terms to highlight the kinds of entities being referred to in these claims. Clearly for a reader of a study to fully understand its conclusions, there needs to be a 'shared' (as far as this may be possible) understanding of what is meant by these terms in the contexts of these claims.

Most of the examples in Table 1.1 refers to what later in the book will be described as the mental register: terms such as ideas, thinking and understanding. Some of the terms used are less familiar from everyday life (p-prim) and so are likely to strike the reader as technical terms. However, when research results make claims about learners' ideas or beliefs, then these words ('ideas', 'beliefs', 'understanding', etc.) are being used as technical terms, even though they are everyday words and readers may therefore take for granted a shared meaning for such terms with the report authors. This book argues that it is unwise to assume that we do all share common understandings of such terms and seeks to explore how such words should be understood when recognised as technical terms in science education research.

Table 1.1 Examples of knowledge claims in science education research

1. 'about one-third of the pupils at the compulsory school *have little understanding* of chemical change' (Ahtee & Varjola, 1998, p. 310)

2. 'there is *a common core* to the pupils' explanations and predictions in such widely differing areas as temperature and heat, electricity, optics and mechanics' (Andersson, 1986, p. 155)

3. 'A large percentage of teachers (76 %) and students (46 %) *believe* that, for the same concentration the pH of acetic acid will be less than or equal to that of hydrochloric acid solution in water' (Banerjee, 1991, p. 491)

4. Data suggest that many students begin post-16 studies with a wide range of *misunderstandings* about chemical reactions. However, students' *understanding improves* steadily as the course progresses (Barker & Millar, 1999, p. 645)

5. 'Everyone recognizes the phenomenon that earthly motion essentially always dies away… dying away is often taken intuitively as a *primitive*. This p-prim is essentially the stipulation that a certain pattern of amplitude (gradual diminuendo) is natural for a particular class of amplitudes (actions by inanimate objects that are not subject to continuous influence). Novice adults often treat dying away as a relative primitive. That is, they will often be satisfied with an explanation that does not have any particular cause for the dying away'. (diSessa, 1993, p. 133)

6. '…many people have striking *misconceptions* about the motion of objects in apparently simple circumstances. The misconceptions appear to be grounded in a systematic, *intuitive theory* of motion that is inconsistent with fundamental principles of Newtonian mechanics'. (McCloskey, 1983, p. 114)

7. 'Students' *conceptual understanding* of photosynthesis and respiration in plants was *measured* … The conceptual change instruction, which explicitly dealt with students' *misconceptions*, produced significantly greater achievement in *understanding* of photosynthesis and respiration in plant *concepts*'. (Yenilmez & Tekkaya, 2006, p. 81)

8. 'for some participants personal *beliefs* (including religious beliefs) appear to override their scientific training and the norms of their profession; for others personal beliefs are paramount; and, for some personal beliefs and *scientific thinking* are compartmentalized'. (Coll, Lay, & Taylor, 2008, p. 211)

9. '[Sister Gertrude Hennessey] pursued a structured approach to science instruction that made students' *thinking* visible and therefore accessible to her observation'. (Lehrer & Schauble, 2006, p. 167)

10. '…pre-Galilean *ideas* about force and movement are not only prevalent among school children, but also in certain cases do persist even after years of formal exposure to physics teaching. There is also evidence to suggest that, at least when projectile motion (vertical or composite) is considered, the *conceptions* are closer to the mediaeval impetus theories than to the older Aristotelian conceptions'. (Gilbert & Zylbersztajn, 1985, p. 117)

Knowledge Claims in Research

Science, in the broad sense of the word, is about the furthering of human knowledge, that is, *public* knowledge. In philosophy, 'knowledge' is sometimes defined as justified, true belief (Matthews, 2002). By such a definition, we can only consider something as knowledge if it is true, *and* we have had it demonstrated as being true. This is considered further later in the book, in the context of what we mean by a learner's knowledge; see Part III. Philosophers of science have argued that science is not able to produce any general abstract knowledge which can logically be demonstrated to be true in this sense, that is, absolutely true for all time. Science deals with conjectural knowledge, which – to be accepted into the canon of scientific knowledge – must have strongly supported grounds *but* remains provisional, at least in principle. Scientific knowledge is fallible, and the scientific attitude is to always consider that in principle even the best established ideas are open to challenge and should be revisited if strong evidence comes to light that undermines their authority (Popper, 1934/1959).

The process by which knowledge becomes widely accepted in science is somewhat organic but starts with the presentation of a new knowledge claim, and the evidence for its support, for peer review within the scientific community, that is, the submission of a report of research to a recognised peer-reviewed scientific journal. The publication of research reports in such journals is used as a criterion that the research is considered to be sound, and so that the knowledge claims made are well supported. This does not mean that the conclusions are accepted as 'proven' knowledge but, rather that the claims are seen to have sufficient support to deserve to be taken seriously. The processes by which the claims made in individual papers are variously challenged, elaborated, ignored, forgotten, built upon or come to be seen as seminal are certainly important but have been somewhat open to dispute in different accounts of the scientific enterprise. Suffice here to say that *an accumulation of evidence from programmes of research* establishes major new ideas as accepted components of scientific knowledge, even when it seems clear that a particular 'seminal' study plays a major role in stimulating a particular research direction (Lakatos, 1970).

Locating This Work Within a Research Programme

I have elsewhere (Taber, 2006a) considered how research into student understanding and learning of science may be understood as a scientific research programme (SRP), in the sense proposed by the philosopher of science Imre Lakatos (1970). In particular, I have argued that conceptualising research associated with the 'alternative conceptions movement' or 'constructivism in science education' as a research programme (RP) is necessary both as (a) a basis for clarification of a diffuse and diverse body of work to defend this area of work from its critics (Taber, 2006c) and (b) to identify the 'progressive' elements which should direct continued research (Taber, 2009b). That is, the process of characterising an RP is important both in 'external' terms (in establishing its identity, location in a field and relationship with other RP in that field) and in 'internal' terms (because a key feature of a RP is that it offers heuristic guidance to those working within the programme).

The Constructivist Research Programme

In that previous work I argued for (1) the existence of this RP as an identifiable component of the field of science education (something that had been broadly accepted), (2) a particular characterisation of the RP (somewhat different from some earlier characterisations) and, most importantly, (3) its status as a *scientific* RP (SRP) in the sense in which Lakatos (1970) sets out demarcation criteria – that is, that the RP was theoretically and empirically 'progressive'. Each of these points is open to challenge, and in the latter case there could be an argument for considering the current status of the RP as 'somewhat' progressive as developments in the area in the past decade have not been as frequent or as coherent as might be expected in an active international programme of enquiry. Nonetheless I consider the general case to be very strong, and here I largely assume the existence of the RP rather than conceptualise this area of work as a number of discrete RP or a disparate set of researchers/research groups working largely independently.

In my previous work there was emphasis on how the identification of a common core of ontological and epistemological key commitments – the programme's hard core in Lakatos's terms – provides a basis for demarcation: for judging which work falls within the tradition of a particular RP. I suggested (Taber, 2009b, p. 123), based on analysis of much-cited key studies, that the hard-core commitments of the research programme into the contingent nature of learning in science ('constructivism') are:

- Premise 1. Learning science is an active process of constructing personal knowledge.
- Premise 2. Learners come to science learning with existing ideas about many natural phenomena.
- Premise 3. The learner's existing ideas have consequences for the learning of science.
- Premise 4. It is possible to teach science more effectively if account is taken of the learner's existing ideas.
- Premise 5. Knowledge is represented in the brain as a conceptual structure.
- Premise 6. Learners' conceptual structures exhibit both commonalities and idiosyncratic features.
- Premise 7. It is possible to meaningfully model learners' conceptual structures.

The assumptions are carried into the present study. It is important to note that although this particular account of the essence of constructivism in science education is my own formulation, all of these principles are long established in the science education literature (cf. Sjøberg, 2010). I have simply drawn out, reformulated and re-presented the key ideas proposed, modified and largely accepted by many other researchers. I refer readers interested in the original sources of these ideas to previous work (Taber, 2006a, 2009b) and here look to build upon and develop aspects of that earlier analysis. The assumption of an SRP provides a focus for the justification for the present work and offers a coherent context or range of application for the ideas discussed here. However, the arguments made in this volume do not *depend* upon accepting the notion of the SRP in science education: the issues explored here are fundamental to a wide range of research studies, however those studies are collectively conceptualised.

The research context for the present volume is shown in Fig. 1.1. A key feature of this representation is that the relationship between learning and thinking is two

Fig. 1.1 The research
context of the present study

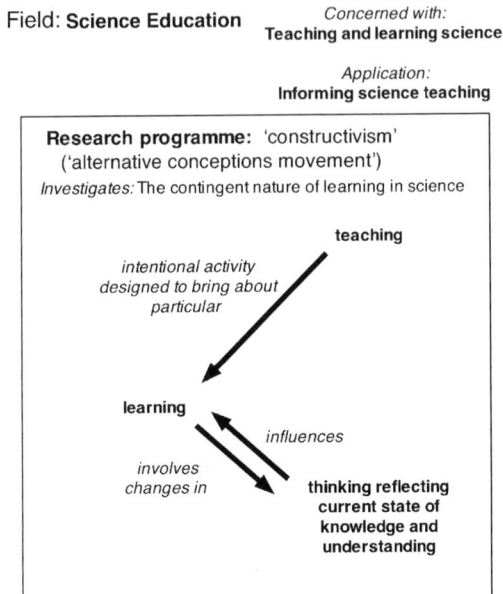

Field: **Science Education**

Concerned with:
Teaching and learning science

Application:
Informing science teaching

> **Research programme:** 'constructivism'
> ('alternative conceptions movement')
> *Investigates:* The contingent nature of learning in science
>
> **teaching**
>
> *intentional activity*
> *designed to bring about*
> *particular*
>
> **learning**
>
> *influences*
>
> *involves*
> *changes in*
>
> **thinking reflecting**
> **current state of**
> **knowledge and**
> **understanding**

way: that is, not only can learning facilitate changes in the individual's thinking but thinking (which depends upon the current state of the learner's knowledge) influences learning.

The reader should note therefore that the arrow from 'teaching' to 'learning' is labelled in terms of *the intentions* informing teaching. An individual will learn from interactions with those around them, whether they are conscious of having a teaching role or not (Vygotsky, 1978): teaching is the term we use to describe behaviour which is deliberately undertaken to facilitate learning, and educational research has this type of behaviour as a key focus (Pring, 2000). There is no direct causal relationship between a teacher's intentions and the student's learning: teaching *behaviour* is likely to have consequences, but not always the intended or anticipated ones – as all experienced teachers will likely acknowledge. For the constructivist, teaching can certainly facilitate learning, but not in an unmediated way.

Progressing the Research Programme

The analysis of this area of work in terms of a RP serves the heuristic role of helping identify priorities for research – drawing upon what Lakatos called the positive heuristic of a RP (Taber, 2009b) – and the present volume follows up on one of the key areas that my earlier analysis suggested could impede progress in this area of work. It was argued in my previous work that the premises of the RP lead to broad research questions (this is what Lakatos, 1970, referred to as the positive heuristic of a programme: the way it suggests the direction for research). In particular, the premise 'It is possible to meaningfully model learners' conceptual structures' leads us to ask 'What are the most appropriate models and representations?' (Taber, 2006a).

That is, work in the constructivist tradition in science education assumes that it is possible to make meaningful claims about the knowledge in learners' minds – its extent, organisation, match to target knowledge and so forth – as in the examples quoted above (in Table 1.1). Yet, in a field which notoriously has failed to develop an agreed terminological canon (Abimbola, 1988), there is little agreement on how best to understand and describe the nature of personal knowledge – as again is illustrated in the range of the examples quoted.

The focus in the present volume then is on the nature of studies which contribute to the RP and, in particular, the way that key concepts are used and how this effects how data (e.g. student utterances, such as replies to a teacher's or researcher's questions) are understood, and results are conceptualised (e.g. being considered as alternative frameworks, mental models and proportions of samples reported to have acquired concepts or to hold particular conceptions) and reported. These are central issues in interpreting research reports within the RP.

It was clear from work reviewed previously (Taber, 2009b) that even within what could be considered the 'same' overall programme, there was not only limited common agreement on the meanings given to key terms but often also a lack of clarity in the precise nature of the phenomena discussed and the theoretical entities inferred or posited. Such issues clearly impede effective communication *between* researchers and *with* other 'users' of research, such as teachers and curriculum planners, and undermine the smooth development of a research programme.

Put succinctly: *in many research papers it may not be entirely clear to readers what the descriptors used in knowledge claims are really understood to refer to*. It is that concern which provides strong motivation for the present book.

Assumptions Informing the Research Process May Not Be Explicit

It will be argued in this book that to some extent the confusion, ambiguity and vagueness that can be found in the research literature in this field can be understood in terms of

(a) the uncertain nature of
(b) the inaccessibility to direct observation of

the objects of research.

Referring back to Fig. 1.1, it is clear that the central foci of research cannot be readily observed. I have suggested (Taber, 2009b) that learning is best understood in terms of a change in the behavioural *repertoire* of an individual, as all that can be observed is the behaviours of the individual that are produced in particular contexts. What can be observed, and so recorded to form research data, are such behaviours as utterances (what the individual says), inscriptions (such as written answers to questions) and 'body language' such as nods, shrugs and various gestures. The same is true in regard to the teacher seeking information to support effective teaching. The teacher may use diagnostic probes to test background knowledge or ask questions in class to check on student understanding but then has to interpret the learners' behaviours – their written or spoken responses – to infer what they currently know and understand. Thinking (the focus of Chap. 7) cannot be observed directly, nor can 'understanding' (the focus of Chap. 6) or 'knowledge' (considered in Part III). It will be suggested in Chap. 2 on what I label the 'mental register' that these common-sense notions of knowledge and understanding are actually quite problematic in a research context.

It should be noted, however, that these references to *the central significance of behaviour* to the research field certainly do not suggest adherence to a *behaviourist* position (J. B. Watson, 1967). The behaviourist (or behaviorist) perspective considered that research in psychology should not concern itself with non-observables, and, for example, Watson not only argued that the notion of consciousness was neither definable nor usable but claimed the terms was simply an alternative to 'soul' and so had inherent religious (and so superstitious) connotations (J. B. Watson, 1924/1998). The stance taken in the present book shares with the behaviourists a concern that not-directly-observable foci such as thinking, understanding, knowledge and the various descriptors for aspects of learners' conceptual structures are inherently problematic constructs for research; however, I certainly do not share the behaviourists' response to this problem in excluding these constructs from consideration (J. B. Watson, 1924/1998). Rather, I certainly welcome how 'information processing and constructivist models of learning have supplanted behaviourism as the dominant theory. They encompass a much wider set of variables, including content, perception of context, abilities, prior knowledge, attitudes, and purposes' (White, 1998, p. 61). Many variables of interest are *not directly observable* and so need to be inferred indirectly, but this does not mean we must exclude 'mental' terms such as understanding from our academic and professional discourse. Rather we have to keep in mind that such terms are often used without careful definition, and that they refer to what we infer rather than what we can directly observe.

Arguably *even teaching cannot readily be observed directly*. Behaviours of teachers in teaching contexts can certainly be observed, but to the extent that teaching involves *intentional* acts directed at bringing about learning, observed behaviour needs to be interpreted before it can be classed as teaching. So a stare which is a behaviour which can be observed could be intended as a behavioural prompt to a particular student, or could be a gestural illustration or analogy to make some teaching point, or could just be an unintentional by-product when the teacher has paused for thought – or has just noticed a drastic change in a student's hairstyle. Similarly with verbal behaviour, a question about whether anyone in the class watched a certain television programme the previous evening could be a lack of engagement in the lesson, an attempt to develop rapport to support a teaching relationship or the opening move in drawing a teaching analogy from something in shared experience. The present volume does not focus on teaching *behaviours* in any detail, but the

Fig. 1.2 Observable
correlates of teaching
and learning

Field: **Science Education**

Concerned with:
Teaching and learning science

Application:
Informing science teaching

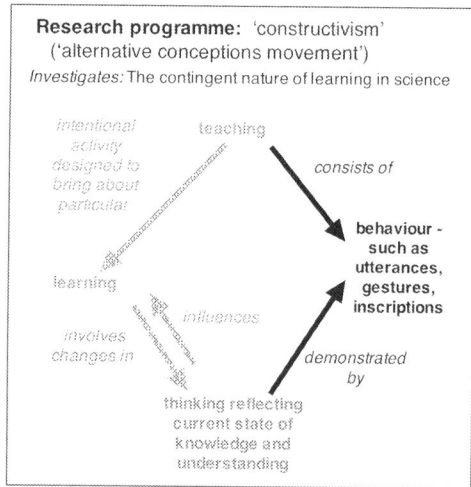

Research programme: 'constructivism'
('alternative conceptions movement')
Investigates: The contingent nature of learning in science

*intentional
activity
designed to
bring about
particular*

teaching

consists of

**behaviour -
such as
utterances,
gestures,
inscriptions**

learning

*involves
changes in*

influences

*demonstrated
by*

*thinking reflecting
current state of
knowledge and
understanding*

fundamental problems for research in exploring learner thinking discussed here have their counterparts in research looking at *teacher thinking*.

This situation is reflected in Fig. 1.2, where within the RP the only available direct observables are indirect evidence of the key concerns of researchers. Arguably, then, teaching and learning are not phenomena in a strict sense: classroom phenomena such as talk have to be interpreted in terms of theoretical notions relating to teaching and learning.

Alternative conceptions, mental models, conceptual ecologies (discussed later in the book) and the various other notions introduced to discuss this area of research are certainly *not* phenomena in the usual sense of that which can be observed and needs to be explained. What is to be explained is what students say and write etc. (in normal classroom situations, in formal assessment and in research investigations), and notions of student knowledge and understanding, etc. are theoretical constructs that have been used to help develop explanatory accounts for patterns in *those* phenomena.

So a classroom teacher may observe learner behaviour when asked, for example, to explain the significance of the periodic table. As a result of considering the learner response, the teacher may then undertake certain teaching behaviours intended to facilitate a different response to the question on a future occasion. Some time later the teacher may ask the learner the same question, and again listen to the response, and so evaluate whether teaching has had the desired effect. In making decisions about what and how to teach, and judgements about whether the teaching has been successful, the teacher will be conceptualising the learner responses in terms of a mental model of the learner knowing and understanding certain things and having learnt (or not) as a result of the teaching intervention. The teacher's thinking is in terms of the learners' knowledge, understanding and learning – but all the teacher actually experiences is what the learner says when asked questions, and so the teacher is working with a theoretical interpretation of this 'data' to infer mental properties and events that can only be inferred.

The same limitations apply to researchers, with the important caveat that the teacher usually has ongoing opportunities to test and correct their interpretations of learner behaviour as they interact with the same learners over extensive periods. The researcher, however, often has a much more limited window to collect data from any particular learner and so limited opportunities to test their interpretation of that data. Moreover, whilst the teacher is involved in a process of modelling the mental states of learners, the purpose of the modelling is to inform practice within the present teaching context, whereas the researcher needs to produce interpretations that have relevance to (even if they cannot be said to directly apply to) other teaching and learning contexts: that is, researchers are charged with producing generalisable theoretical knowledge, where teachers need to work with fit-for-purpose understandings of their own classrooms. As the researcher usually lacks the myriad opportunities for self-correction of interpretations of learner behaviour available in teaching and is expected to produce a public account of work which is theorised in formal terms suitable for communication to the wider science education community, it becomes much more important that the researcher is aware of, and clear about, how they interpret and conceptualise their data.

The Centrality of Models in Research

Given these very real impediments to research, it becomes important for the researcher, and readers of the research, to be very clear about *the indirect nature* of much of the research and *the status of the various entities* discussed. In particular, it should be recognised, and made explicit, that the descriptions researchers offer of aspects of learners' knowledge structures are inevitably of the form *of models*. Research reports that fail to make this clear can be misleading, and this can have unfortunate consequences. Firstly such reports can give false impressions to practitioners wishing to learn from and apply the findings of research in educational practice. Secondly, such imprecision means that knowledge claims in research reports can be readily misunderstood by other researchers – contributing to some of the less helpful examples of claim and counterclaim in the research literature (Taber, 2009b). When teachers not intimately involved in research processes themselves are users of such reports, there is a considerable scope for the overliteral reading of reports (i.e. reifying terms carelessly used as nouns and assigning inappropriate status to conjectural, 'first-approximation' and overgeneralised notions).

It is unlikely researchers commonly deliberately mislead their readers in this regard. Rather these flaws in many research reports are likely to be due to a combination of the following: (a) deliberate attempts to keep reports clear and concise (an admirable enough aim in itself), (b) failures to explicitly specify important qualifications that researchers may take for granted in their own work due to overfamiliarity and (c) failures to fully appreciate, or at least make explicit, the indirect nature of their own enquiries into the phenomena of interest – that is, insufficient attention to the ontology of the phenomena they study. The aim in this book is not to diagnose which of these factors is likely to have contributed to insufficient clarity in particular research reports. Rather the purpose here is to clarify the nature of the research process in the field of research exploring student understanding, knowledge, thinking and learning so that readers of research reports can better 'read between the lines' – and to contribute to scholarship in the field, with the aspiration that authors, journal referees and editors might come to expect greater explicit clarity about the status of the entities referred to in research reports.

From my own reading of this literature, I have come to suspect that factor (c) may be quite significant in many cases – that is, many researchers are not sufficiently problematising the research process by being explicit about the assumptions underpinning their work. It will be suggested in the next part that a major factor is the way that research draws upon everyday 'lifeworld' terminology – in that there is a register of terms such as thinking, knowing, understanding and learning that are widely used in everyday discourse and readily understood at a non-technical level, but which lack operational definitions when adopted for research purposes.

For example, consider the following quotation from a research report published in the journal *Research in Science Education* in 1986:

> Various approaches have been used to identify students' understanding and misconceptions of science phenomena. Of these approaches, interview methodologies have acquired strong support as a viable approach… Although interviews with students have been successful in ascertaining students' understanding of science phenomena the interview methodology has possible limitations if it is to be used by classroom teachers. (Peterson, Treagust, & Garnett, 1986, p. 40)

The message here is clear. The extract suggests that interviewing is in principle a 'successful' approach to identify misconceptions and ascertain students' understanding in research – 'interviews with students have been successful in ascertaining students' understanding' – although the practicalities of classroom work make it problematic as an assessment tool for the teacher. I do not disagree with Peterson and colleagues: I consider that interviewing students is often the best technique to find out about their ideas and understanding of topics. However, there is a real issue here: How can we know if interviewing is 'successful in ascertaining students' understanding of science phenomena'? We could in principle know this if we already had independent access to students' understanding of science phenomena, when we could compare the outcomes of the analysis of research interviews with what we know about student understanding. If there was a good match, however defined and calculated, then we could be confident in the research technique. However, if we already had that knowledge, we would not need the research technique!

Shared Community Commitments

This is not a problem specific to science education of course but is a general issue in research. How does a scientist know that a particular technique can accurately date rocks, or find the energy of an x-ray emission, or identify a metabolic pathway? Ultimately all such techniques rely upon providing results that are consistent with a wide range of other measurements and evidence that we are reasonably confident in: usually in large part because they, in turn, are generally also consistent with what we feel is secure knowledge.

It is in this context that the ideas of Thomas Kuhn may be relevant. Although much of Kuhn's description of science and how it progresses has been criticised (Lakatos, 1970; Popper, 1970), his notion of scientists being inducted into a paradigm (T. S. Kuhn, 1970) or disciplinary matrix (T. S. Kuhn, 1974/1977) remains helpful. Part of Kuhn's thesis was that during what he termed 'normal science', the scientists working within a field hold a shared set of commitments (i.e. the disciplinary matrix). So in particular fields of scientific research, the existence and nature of electrons, the evolutionary relationship between particular groups of organisms, the mode of operation and interpretation of results from a mass spectrometer, the components of the human immune system, the means of denoting muons, the appropriate level of precision for citing the age of fossils, the appropriate way of interpreting temperature in terms of molecular motion, etc. come to be accepted by a research community and may be taken for granted.

In principle, at least, in science all such matters are provisional and open to revisiting in the face of new evidence – but for the purposes of normal scientific business, they can be considered as taken as given and *need not be argued from first principles in research reports*. In contrast, what cannot be assumed to be a shared commitment in the field needs to be justified in a research paper. So, for example, when ideas from one area of science are adopted in research within a different field, then journal editors and referees are likely to ask for justifications that would not be seen as needed in the host field.

One purpose of science education, especially at the highest (postgraduate) levels, is to train up new scientists, and this involves the induction of new researchers into the traditions and commitments that make up the norms and what is taken for granted in a particular scientific field:

> Kuhn's model of normal science education centres on the principle that the student is initiated into the dominant scientific paradigm of the day. A primary aim of science education, therefore, is to produce competent researchers, and research can only occur in line with the methods and concepts of the paradigm that define the puzzles being researched. (Bailey, 2006, p. 15)

Research in science education has sometimes been discussed in paradigmatic terms – for example, that constructivism forms the basis of the paradigm, 'as if a period of Kuhnian normal-science has descended upon the science and mathematics education communities' (Matthews, 1992) – but has never completely reflected Kuhn's ideas in terms of a set of shared community commitments at the ontological and epistemological levels. For example, Gilbert (1995, p. 181) noted that 'whilst the ethnographic/naturalistic research paradigm was developing, research within the older 'normative' tradition was continuing'.

Even though the premises of the constructivist programme as listed above (Taber, 2009b) would be shared by a good many researchers in the field (Sjøberg, 2010), there is no widespread agreement on how to operationalise the fundamental ideas underpinning these tenets in terms of the concepts most useful for carrying out and reporting research. To draw on analogies from the natural sciences, science education might be considered to be like the biological sciences after Darwin's ideas had been widely influential, but before anyone had a clear notion of where to look for the mechanism of genetic inheritance. Or, taking the physical sciences, we might compare the situation in science education to the state of affairs after Dalton had suggested the basis for modern atomic theory, but before there was agreement on the meaning of terms like atom and molecule, or anyone had clear notions of how such entities might be identified, or what kinds of interactions and structure they might have. Such 'ignorance' was still an advance, as the commitment to submicroscopic particles at least allowed the questions to be posed and so provided the impetus for research.

So in science education there was an 'explosion' of interest and activity in the field in the last quarter of the twentieth century that coincided with the development of widespread shared commitments to a constructivist notion of learning in science. However, that establishment of a central focus has not yet led to consensus models and constructs to describe, explain and explore the central concerns of the field encompassing student thinking, understanding and learning about science. In this sense, explorations of student thinking in science better reflect areas of enquiry such as personality or motivation in psychology (where there are widely discussed models and theories, and commonly used instruments, but no strong consensus), than many areas of the natural sciences where concepts are tightly described and standard instrumentation is well established. This is of course not a coincidence: the kinds of phenomena studied in the field discussed here (mental models, conceptions, understanding, etc.) are quite similar to many constructs studied in the behavioural sciences.

Being Explicit About the Frameworks Underpinning Educational Research

When new graduate researchers in education and other social sciences are taught to approach their research, they are commonly told that in setting out their work, they will need to present both a conceptual framework and a theoretical perspective for their study (Taber, In press). That is, they not only have to motivate their research questions by reviewing previous literature about what is already known, and where there may be 'gaps' in existing public knowledge, but they also have to justify how their research design will enable the production of knowledge of a suitable form to answer their research questions.

This is in effect *not* the methodology of a study in a specific sense (in terms of which research techniques are being used and how they fit into an overall design) but a step back to consider the paradigmatic grounds upon which a particular methodology (survey, experiment, grounded theory, case study, etc.) stands. There is a process whereby in approaching a study, the researcher is expected to move through at least three levels of thinking about what they are going to do: these may be considered (Taber, 2007, 2013a) the levels of philosophy (metaphorically, the executive level, setting out the paradigmatic vision), strategy (the managerial level of methodology) and tactics (the technical level of specific techniques).

Of course research journals cannot publish the level of detailed discussion of such matters expected in a graduate thesis. However, in many research papers published in science education, there is *little* explicit information for how a methodology was chosen in terms of the nature of the entities being studied and the nature of the kind of knowledge that the research might be able to produce. As well as limitations of journal space, this may often reflect the natural science background of many researchers in science education, where research training has traditionally been somewhat different to the social science model outlined above and where within an established paradigm the choice of methodology to approach a standard type of problem may often be seen as generally unproblematic (T. S. Kuhn, 1970). Perhaps it may not occur to some researchers that it is important to examine and present methodological justifications in these terms.

Whatever the reasons, many research reports on student understanding and learning in science education leave a great deal unstated about the fundamental nature of the entities they discuss and the status of the claims they make. (An example of the use of the term 'misconception' is discussed below.) For this reason this present chapter sets out an account of the overall process by which research in this field is carried out.

Claims About Technical and Common-Sense Notions

We think we understand a word, such as 'cause', and as long as we keep going all is well. If we stop to analyse it, however, all is lost. In daily life, this odd phenomenon may not matter, but there are occasions in which it is important to know what we mean. (Johnson-Laird, 2003b, p. 41)

It seems that claims made in research reports can to a first approximation be considered to be of two kinds. Some reports make claims in technical language, for example, refer to such entities as 'alternative conceptions', 'p-prims', 'conceptual frameworks', 'mental models', …. By using such terminology the paper makes a claim that is explicit about at least some aspects of the way the researchers are thinking about learners' cognition or knowledge structures. Of course, there may still be issues about how such terms are defined, understood and used and the extent to which they are shared as useful constructs in the research community. There may also be issues of whether teachers can readily make sense of research reports using such technical terms. The latter point is important, for although most research reports are written primarily for the research community, most researchers in education are at least hopeful that their work can impact upon practice. There are good reasons to argue that research with direct classroom implications should be written up both for research journals and for practitioner journals, with appropriate conventions and writing styles according to the intended audiences (British Educational Research Association, 2000). So editors of research journals expect an account of methodology and appropriate referencing to the relevant literature that informed a study, where editors of practitioner journals often look for a focus on classroom relevance and application and often prefer a limited bibliography of useful further reading rather than a formal reference list. Often the different genres also involve distinct expectations about nomenclature – with technical vocabulary being more suitable for the research journal than the practitioner journal.

However, it is also clear from the examples presented above (see Table 1.1) that some research reports although published in the academic literature make claims using what seems everyday, non-technical language. These reports claim to tell us what learners (or teachers) 'think', 'understand' or 'believe'. Such writing certainly seems more reader-friendly: any reader of a journal with a good grasp of the English language will understand [sic] a statement such as 'students' *understanding improves* steadily as the course progresses' (Barker & Millar, 1999, p. 645), when not all will be familiar with specific technical constructs (e.g. such as p-prims or alternative conceptions; see Chap. 11) used in other reports.

Accounts that make claims in terms of learners' understanding or knowledge, or thinking, or beliefs may then be considered as more 'reader-friendly', in the sense of being more readily and widely appreciated, than those which describe findings in more specific technical terms. However, there is also an argument that such reports are open to more ready misinterpretation. If we all 'know' what is meant by a student understanding something, or knowing something, because in everyday life notions such terms as 'understand', 'know', 'believe' and 'think' are taken for granted, then claims phrased in these terms can also seem unproblematic. If we claim that a learner knew nuclei were positively charged or understood how acids reacted with carbonates, or believed that plants only respire during the hours of darkness, or thought that a continuous force was needed to maintain an object's motion, then we seem to be saying something very clear and definitive.

This would be fine if finding out what people know, understand, believe and think was straightforward (and if indeed knowing, understanding, thinking and believing were simple matters, open to pithy descriptions). Yet, taking an overview of the last few decades of research in science education, it is clear both that:

(a) These processes (knowing, understanding, thinking and believing) are often not simple matters than can be authentically described in simple statements.
(b) There are genuine methodological difficulties in finding out what someone thinks, believes, understands or knows in any definitive sense (as will be detailed later in the book).

The Value of Clarity in Language

This is not an argument for excluding everyday language from research reports. There is a strong case for 'headline' statements (e.g. in paper titles and abstracts) offering a very clear and straightforward statement of what a paper is about and what the researchers think they have found. Indeed given the lack of consensus on the technical terms used in this field (conceptions, frameworks, etc.), there is a strong argument for ensuring that all researchers can quickly identify papers likely to be relevant to their research regardless of the specific conceptualisations and approaches adopted in different studies. Pithy statements referring to such everyday notions as students' knowledge and understanding can be very valuable in this regard.

However, it is argued here that the same clarity that can offer a quick impression of what a paper is about can also lead to researchers with different understandings and assumptions about what is involved in understanding or knowing, for example, misinterpreting what a study actually offers in terms of new knowledge. However, this should not happen if the 'headline' claims are underpinned by more technical explanations of the research. Research papers should of course make it clear just what is being claimed and how the research undertaken supports those claims. However, it is suggested here that research in science education often falls somewhat short of being fully explicit about such matters. This may be because researchers sometimes assume others working in the field will share what they see as obvious assumptions and commitments, or it may sometimes be because the researchers are working with a good deal of tacit knowledge (Polanyi, 1962) – that is, drawing upon assumptions which are so well established in their thinking, they are implicit in the researcher's work and are not 'brought to mind' when writing reports.

Making the Research Process Explicit

Inevitably we all operate with a great deal of tacit knowledge, and indeed we could not function in any sphere of life if we had to stop and analyse everything we do (every keystroke I am making now – what am I doing, and what do I expect the outcome to be?) Humans only operate at the higher levels of cognitive function because our cognitive apparatus allows us to automate so much (see Chap. 7).

However, when it comes to research, there are some things that need to be made explicit for a research report to provide a sufficiently detailed and clear account of our work. Unsurprisingly, these relate to ontology and epistemology. If we wish to investigate, for example, student understanding, then we need a clear idea of what kind of thing 'understanding' is, that is, we need to operationalise it as part of our 'conceptual framework' setting out the background to the study – what existing research already suggests. We also need a good idea of the kind of knowledge it might be possible to develop about another person's 'understanding', informed by the 'theoretical perspective' that with the conceptual framework justifies the methodological choices made during the research (Taber, In press). What is being argued here then is that in our field of research, these aspects of the research report are often limited and inadequate.

An Example of a Study Reporting Student Misconceptions

To illustrate the nature of my general argument here, I wish to briefly consider a paper from an international research journal in the field, which discussed an aspect of learners' 'misconceptions' in a science topic (Banerjee, 1991). I have not identi- fied this paper as being especially problematic, and indeed I feel it makes a useful contribution to the field. Rather, I suggest that it is somewhat typical of many papers published in the research literature about aspects of students' ideas in science. Additionally, it usefully – for present purposes – uses the key term (here, miscon- ception) throughout the text and is consistent in using *that* term (rather than precon- ceptions, alternative conceptions, alternative frameworks or other related alternative terms). This allows the preparation of a useful concordance of each time the term is used in the paper (excluding the reference list), which is presented in Table 1.2:

Reading of Table 1.2 obviously only gives a flavour of the full paper, but it dem- onstrates that the notion of a misconception is not explained in any detail: there is not a part where the author feels the need to explain to the reader what is meant by the term 'misconception' in the context of this paper. By the time of this study, the term misconception was in widespread use in science education, and the author presumably felt that it was well enough established that the readership of research journals in the field would know what was meant. The Banerjee paper is some years old now, but at the time of writing, papers in top journals continue to use terms such as misconception in a taken-for-granted way (Bivall, Ainsworth, & Tibell, 2011; Jaakkola, Nurmi, & Veermans, 2011; Ratinen, 2011).

We might expect to see a similar approach in mature scientific fields: for exam- ple, papers in chemistry research journals that refer to something being a 'com- pound' do not usually explain what is meant by the term. It can be taken for granted that the readership of research literature will share a common understanding of the term: something that would not have been the case when the term 'compound' was first being mooted and had not been widely accepted in the discipline. For com- pound, we can substitute any number of now accepted terms: gene, energy, neutrino, tectonic plate, brown dwarf, etc.

My argument here is that where within the natural sciences, there are mature fields where it is reasonable to assume other workers share a fairly close under- standing of what is denoted by common terms, but science education does not yet have this level of maturity (Fensham, 2004), and many terms are used in looser ways. As well as science education being less 'mature' than fields in the natural sciences, it also deals with subject matter of an inherently different nature, due to the complexity of the phenomena studied in behavioural and social sciences (Taber, In press).

Table 1.2 Concordance for the term 'misconception' found in Banerjee (1991)

Paper part	Occurrence of term 'misconception'
Title	'Misconceptions of students and teachers in chemical equilibrium'
Abstract	'A written test was developed and administered to diagnose misconceptions in different areas of chemical equilibrium…'
	'Analysis of the responses reveals widespread misconceptions among both students and teachers…'
Introduction	'There is a large body of research on the misconceptions of students in a variety of science subjects…'
	'The usual method for obtaining information about students' misconceptions has been through individual student interviews'
	'… used interview techniques to study misconceptions of students in stoichiometry and … in chemical equilibrium'
	'Another line of research on misconceptions uses multiple-choice tests'
	'… and … developed and used tests to identify misconceptions of year 11 and year 12 students about covalent bonding and chemical structure'
	'The present study covers broad aspects of chemical equilibrium including gaseous, ionic, solubility and acid-base equilibria, and diagnoses misconceptions among 162 undergraduate chemistry students'
	'To obtain information on the question of whether misconceptions are removed with increased content knowledge and experience, the diagnostic test was also given to 69 secondary and senior secondary school chemistry teachers. Apart from knowing whether teachers also have misconceptions in the areas of equilibrium, the study would indicate whether misconceptions among students may have originated from the misconceptions of the teachers'
Development of the test	'An analysis of the responses indicated widespread misconceptions'
Administration of the test	'However, conceptual difficulties and misconceptions in the different areas of equilibrium were not specifically covered in these lectures'
	'In this paper, the discussion is concentrated on 12 test items (listed in the appendix) which were used to diagnose conceptual difficulties and misconceptions…'
Analysis and discussion	'Misconceptions were mostly identified from the explanation given in support of the answer by the student and teacher'
	'These responses, in general, indicate widespread misconceptions among both teachers and students in topics relating to…'
	'A sizeable percentage of teachers and students have the misconception that…'
	'Similar student misconceptions were reported by…'
	'There are widespread misconceptions in the areas relating rate with equilibrium'
	'They have the misconception that a large value of equilibrium constant implies a very fast reaction'
	'Similar misconceptions have been reported by…'
	'However, the present study clearly indicates that the rate approach should be used with caution and should not be overemphasized, in order to avoid the possible development of misconceptions'
	'Students and teachers show a high rate of misconceptions in acid-base and ionic equilibria'

(continued)

Table 1.2 (continued)

Paper part	Occurrence of term 'misconception'
	'A comparative study of the responses given by students and teachers reveals that the extent of misconceptions is equally high among both groups. One possibility is that teachers might have developed these misconceptions during their student days. The misconceptions are retained, despite professional experience over the years'
	'According to the general constructivism theory of knowledge … it is very difficult to remove misconceptions from the minds of learners. The findings of this study on misconceptions among students and teachers should not be treated as specific to this sample. Many misconceptions of a similar nature about chemical equilibrium have been reported with students from Australia and the United Kingdom'

A hypothetical scholar from a distant time or place who did not know what a twentieth century science educator might mean by 'misconception', and finding that Banerjee did not define this term, could look for clues in the text and attempt some kind of hermeneutic exercise to tease out what a misconception might be. We can find in Banerjee's paper a number of knowledge claims about the nature of misconception deriving from the study:

- Some misconceptions are widespread among students and teachers.
- Some misconceptions occur at an equally high rate among students and teachers.
- Particular misconceptions may be had [held] by substantial proportions of the sample.
- There can be degrees of similarity between different reported misconceptions.
- The development of misconceptions may be facilitated by teaching approaches.
- Misconceptions can be retained for extended periods.
- Misconceptions can be very difficult to remove from minds.

These claims derive from a study that is set up in a particular way because of the researcher's assumptions about the nature of misconceptions (i.e. ontology) and how one could investigate them (i.e. epistemology). Epistemological assumptions informing the research would seem to be that:

- Misconceptions may be diagnosed/identified by written tests – for example, from justifications of respondent answers.
- Misconceptions may be explored through student interviews.
- Misconceptions may be diagnosed with multiple-choice tests.

Ontological assumptions (the researcher's assumptions about the type of entity misconceptions are, i.e. their nature), which support these epistemological assumptions, would seem to be:

- Misconceptions can be widespread.
- Misconceptions can occur among students and teachers.
- (And more particularly) misconceptions are found among undergraduate students.
- Misconceptions can relate to a variety of science subjects.

- Misconceptions occur in the minds of learners.
- Misconceptions are the kind of things that in principle could be removed.
- Misconceptions may be passed from individual to individual.
- Misconceptions could potentially be 'covered' in lectures.

These qualities of misconceptions are largely assumed by the author and are implicit in what is written, and, I would suggest, only one of the points here (namely, misconceptions occur in the minds of learners) reflects the essential quality of misconceptions that is the central focus of the paper. It could be argued that research reports of this type suggest the reported studies may themselves be under-theorised, as rather well-defined technical procedures are used to investigate foci that are themselves only vaguely characterised, and so the technical procedures are themselves largely operationalised without explicit rationale. Whilst I have examined one study in some detail here, similar analyses could be obtained for many of the papers reporting empirical results in this field.

Knowledge Claims Need to Be Understood as Being About Models

So a central argument of this book is that research reports need to be more explicit about the processes by which we feel we can make claims about aspects of a person's knowledge and understanding in science. In particular, I will argue that such research involves a series of modelling stages. Consider, for example, the question what do 15-year-old students (in some educational context) know about photosynthesis (or atomic structure, or the electromagnetic spectrum, or plate tectonics, etc). If our focal topic were part of the school curriculum, we might expect that a key source of any knowledge they may have would be expected to derive from teaching. Figure 1.3 sets out in schematic form the key modelling steps both in the teaching process and in the research process.

The left-hand side of the figure illustrates something of the processes by which scientific knowledge is transformed in the curriculum and classroom and then interpreted by the individual learner in forming and developing their own mental models of scientific concepts (Gilbert, Osborne, & Fensham, 1982). A key point here is that the 'standard' by which student knowledge and understanding is usually judged in educational contexts is not scientific knowledge itself, but a specified target knowledge in terms of a prescribed curriculum. That curriculum will include models of the scientific knowledge (Taber, 2008b). In part, this will be a deliberate modelling process, designing appropriate simplifications for learners of a certain age and expected background knowledge; in part, it will be the inevitable limitations of curriculum developers themselves in knowing and understanding the latest scientific knowledge.

Moreover, the curriculum models are generally moderated by the presentation in class and in textbooks (Chevallard, 2007). The teaching models presented in class will be based upon the curriculum models as understood by the teacher but may

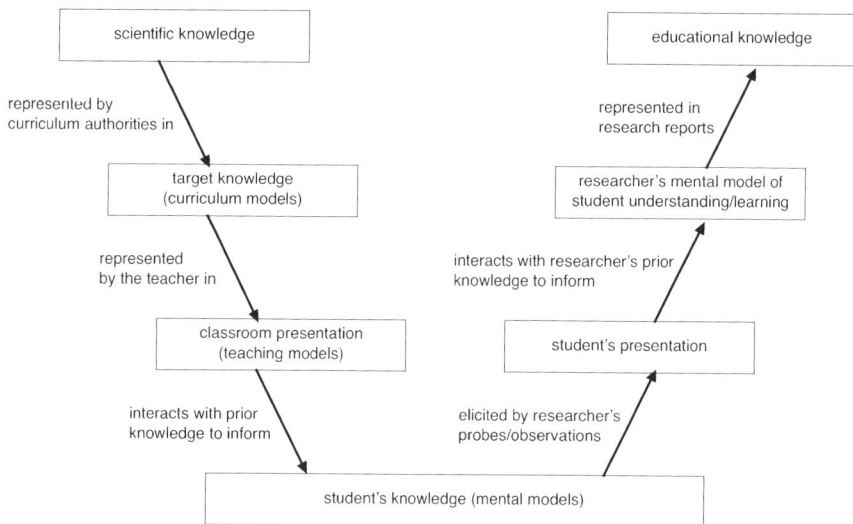

Fig. 1.3 Overview of the process of producing educational models of student understanding of scientific knowledge

include further simplifications, taking into account the specific class of students, or distortions (if the teacher does not fully understand the target knowledge). Even if there is limited deliberate modification of target knowledge, there will always be potential for distortion of curriculum models as the process of teaching inevitably involves processes of re-representation (Taber, 2009b, p. 46, fig. 4.1): the curriculum documents are read by the teacher who forms some kind of mental representations of them so the knowledge represented in the curriculum documents is re-represented in a different form (this is discussed in Chap. 4) and then presents an account of his or her understanding in talk, gesture, written inscriptions, etc. in the classroom, so re-representing the mental representation in public communication.

This set of processes is well described in literature (Gilbert et al., 1982), and Fig. 1.3 shows how a similar set of processes are usually involved in research exploring aspects of student knowledge and understanding. The right-hand side (rhs) of the figure offers an overview of the processes by which researchers develop representations of student knowledge to report in the literature. The result of this process is formal public knowledge (a notion explored in Chap. 10).

Knowledge claims made in literature rely upon the cases – argument chains, supported by warrants (Toulmin, 1972) – made by the authors for the interpretation of evidence collected in research, informed by the researchers' own conceptual and theoretical frameworks. In the case of scientific knowledge, the conceptual frameworks will normally be based upon widely accepted principles (e.g. natural selection; molecular orbital theory, etc.), and the theoretical principles underpinning data collection and analysis (e.g. C-14 dating; PCR analysis of genetic material, etc) will also be widely accepted (T. S. Kuhn, 1996). Whilst the production of educational knowledge (i.e. the rhs of Fig. 1.3) parallels that in the natural sciences, the assumptions made by researchers are less likely to be so widely agreed within the research community (Black & Lucas, 1993b). Therefore, it is important that researchers' accounts are explicit about the assumptions underpinning their work to allow others to make considered judgements about their claims (Taber, 2007).

This volume then explores the processes by which knowledge claims about students' knowledge and developing understanding are produced: processes that at their core involve researchers collecting and analysing evidence *to build models* (1) of learners' knowledge and thinking and (2) of the shifts in that knowledge associated with learning and conceptual development.

Part II
Modelling Mental Processes in the Science Learner

Chapter 2
Introduction to Part II: The Mental Register

> Sister Gertrude Hennessey was the sole science teacher for grades one through seven in a small parochial school in Wisconsin. As a result, she had the unusual opportunity to think about the goals and trajectory for students' scientific reasoning across all those grades of schooling…Hennessey not only planned the course of instruction and taught her students from day to day; she also kept detailed records and videotapes of her students' learning and conducted regular interviews of individuals, small groups, and intact classes. She pursued a structured approach to science instruction *that made students' thinking visible and therefore accessible to her observation*. (Lehrer & Schauble, 2006, p. 167, present author's emphasis)

The quotation above is taken from a chapter on 'Scientific thinking and science literacy' in an academic book. I have highlighted a statement about a teacher's classroom work that I would consider to be a 'knowledge claim', *something asserted as being so* as often found in formal academic writing. It is this type of knowledge claim that I feel needs to be examined more critically if we are to move this area of research forward and ensure that the research literature in science education is precise about the concepts it uses to describe and investigate student learning.

This particular quote is taken from a handbook in psychology, rather than a research paper in science education (cf. the examples given in the previous chapter, Table 1.1), and the statement highlighted is not intended as a 'result' of enquiry, but rather is presented more as a justification for having confidence in research findings. Nonetheless, it stands as an interesting example of the kind of writing which is commonplace when discussing aspects of student cognition and learning, and it represents what I feel is a dilemma for the research community. The claim here is that Hennessey's approach to teaching science 'made students' thinking visible and therefore accessible to her observation'. I would suggest this could be understood in two ways.

The first interpretation is a literal, perhaps pedantic, one. Lehrer and Schauble are claiming that some classroom approaches allow us to observe student thinking. Now I understand the term 'thinking' to refer to mental activity and therefore to be part of a way of talking about the mind, a construct we use to explain aspects of our own experience (i.e. related to consciousness), and by analogy as a means of interpreting the behaviour of others. As infants we develop a 'theory of mind' (Knight, Sousa, Barrett, & Atran, 2004) which posits that some other objects in our environment (e.g. mummy, daddy, doggy, perhaps my teddy bear, perhaps even the wind) also experience wants and needs like I do and are capable of planning action to bring about desired states.

However, the mind is just a theoretical construct, albeit a very useful, and widely accepted one, in the sense that no one has even seen or weighed or produced an infrared spectrum of a mind. Rather it is an explanatory device to make sense of much human (and sometimes other) behaviour. This is not to deny that mind exists – in a sense it is the subject matter for this part of the book – but to seek to be clear about its ontological status. The key point is that the mind is the realm of personal subjective experience. I can *describe* my experience in terms of thoughts, ideas, emotions and so forth, but these descriptions will only ever be just that: accounts of something that can only ever be directly available to introspection and not *made available* to others. Of course, there are many science *fiction* stories where devices are used that allow people to directly experience the mental life of others – but at the moment this idea remains just that, fiction.

So from a technical perspective, I can never observe anyone else's thinking but only activity that I perceive as behaviour and which – because like other normally developing humans I have a well-established and much drawn upon theory of mind – I then further interpret as evidence that these others are thinking certain things. In other words, I use the only resources I have available to conceptualise mental activity (i.e. my own thoughts, beliefs), to model the mental activity I assume is going on in the minds I presume other people have. A reader may feel this is an overcautious account, as others can tell us what they think, but as I will discuss in later chapters, that is still not *direct* evidence of thinking.

So if, technically, no one can observe another's thinking, why would such a claim be made in a handbook, supposedly a technical book intended as a standard work of reference in an academic area. Clearly there must be another interpretation of what is meant by making 'students' thinking visible and therefore accessible to … observation'.

The other possibility is that this is not meant in a literal, technical sense, but rather is more a kind of figure of speech. Clearly, Hennessey could *not* observe her students' thinking, but because of the way she taught, and her extended familiarity with the students, she was able to make observations that *seem to unproblematically* indicate what the students were thinking. This is still quite an ambitious claim, but one which seems less incredible. In terms of the conceptualisation being adopted in this book, I might paraphrase this along the lines:

> Sister Gertrude Hennessey was the sole science teacher for grades one through seven in a small parochial school in Wisconsin. As a result, she had the unusual opportunity to think about the goals and trajectory for students' scientific reasoning across all those grades of schooling…Hennessey not only planned the course of instruction and taught her students from day to day; she also kept detailed records and videotapes of her students' learning and conducted regular interviews of individuals, small groups, and intact classes. She pursued a

structured approach to science instruction that [gave her myriad opportunities during her extended engagement with the same learners to formulate, test and modify hypotheses about students thinking, allowing her over time *to develop models of student thinking that offered strong explanatory and predictive power*]. (Lehrer & Schauble, 2006, p. 167, present author's reformulation)

My reformulation is not as concise as the original and perhaps not as poetic. I imagine it would not have quite the same impact on the reader – however, I would argue that is a good thing because in making a claim like this, I would wish the necessary provisos to be explicit – in the way that claiming that a teacher can make 'students' thinking visible and therefore accessible to her observation' does not.

The dilemma here is why, given the formal academic context of the quote, Lehrer and Schauble choose to claim something that is clearly technically impossible, rather than make a more measured claim that might be supportable. I suggest the answer to this question relates to the dominant role that the mind concept has in our lives: what we come to call thoughts, beliefs, perceptions, suspicions, opinions, etc. are the very stuff of our human experience, and having a common language to talk about such matters is so central to communicating with others that we come to take those lifeworld notions, and our mutually reinforced but sometimes colourful ways of discussing them, for granted and forget they are not technical terms with clearly operationalised definitions.

There is then a kind of folk psychology of mind that permeates our own thinking (sic) and dialogue and which functions perfectly adequately in normal conversation, but which lacks the precision expected in technical communication. We might refer to this as *the mental register*, where key terms would include thinking, ideas, understanding, knowledge and beliefs. The mental register does effective work for us in everyday discourse in many communicative contexts, but when we need to specify more precisely what some of these terms refer to, we may soon run into difficulties.

The Problem of Natural Language in Technical Studies

This part of the book contains five chapters exploring what we understand by, and how we might investigate, such matters as learners' ideas, learners' thinking and learners' understanding. All of these themes have been commonly discussed in science education, although they suffer from focusing on terms that are commonly used in everyday life without clearly defined meanings. This is illustrated by an extract from a research paper, reporting a study of the 'relationships between students' thinking and use of language when reasoning about a problem' (Anderberg, 2000).

The extract is taken from an interview where a teacher (denoted R) was asked (by the interviewer, I) about 'different ways of managing fundamental mathematics instruction' (p. 93). During the interview, the teacher was asked what 'understanding involves':

R: **Yes, it is of course a learning process or …**
I: A learning process. When you say a learning process, what are you thinking about?

R: **Well, I'm thinking of a time when you learn to measure and then work out these connections and that, all of that is ... a learning process** (lowers voice).

I: A learning process. If you say understanding, what do you feel that you are describing then?

R: **... Well, I think that when I've gone through this learning process then I have acquired an understanding ... of the whole thing.**

I: So the understanding is, sort of, the learning is the process, then, or ...

R: **Yes it is, and then comes the understanding, after I've gone through that part. ...Well, then you have a, you know ... you can have an aha experience that it's ... that's how it is.**

I: That's how it is?

R: **Yes. And if you've experienced that then I think it stays with you for the rest of your life.**

I: Yes. And what have you got then?

R: **I've got knowledge.**

I: You've got knowledge, yes. And what is that, having knowledge?

R: (PAUSE) **Well ...**

I: It's not easy. But when you say it, what idea do you want to express?

R: **Well, it's that you really have learned something ...**

I: ... You really have learned something. ... And what is that?

R: ... **Yeah, that's a good question (laughter) ... you mean, what happens inside the head here?** (p. 4) (Anderberg, 2000, pp. 100–101)

Anderberg's teacher interviewee 'R' clearly has concepts of understanding, learning and knowledge that are linked and which support thinking about teaching and learning. Understanding is a state – in relation to knowledge – which is an outcome of a process, learning, which leads to an experience recognised as significant: because of the type of knowledge it produces (i.e. permanent knowledge, reflecting 'real' learning). All of this is understood as related to happenings in the head. Whilst R does not find it easy to pin down precise meanings for terms, a clear relationship between key ideas is presented, and there are even some ontological and epistemological characteristics implied, that is:

- Learning is a process, whereas knowledge is something acquired.
- Understanding (as an outcome of the learning process) can be recognised by a certain type of subjective experience, but the learning process itself is only inferred to be something going on in the head.

Figure 2.1 is a representation of one reader (i.e. the present author)'s interpretation of the data presented by Anderberg: an interpretation based on a small snippet of secondary data. As such, its validity as a model of R's thinking about these terms must be open to challenge. However, the scheme in Fig. 2.1 offers some indication of the way that many people, including professional educators such as R, operate with key ideas that are central to the whole field of teaching and learning. Learning, thinking, understanding and knowing are core foci of education, but tend to be operated with by many practitioners as 'fuzzy' concepts that lack clear operational

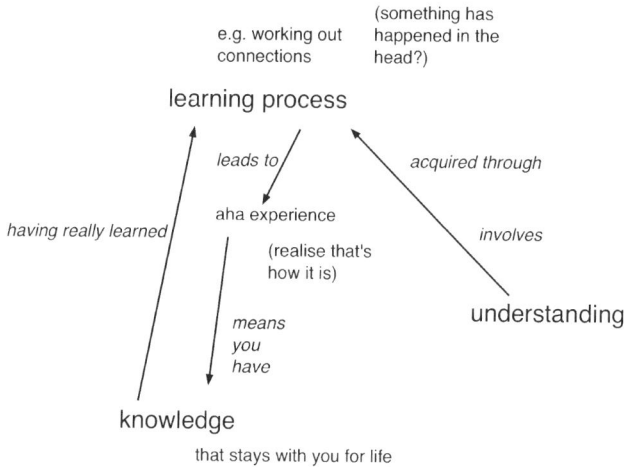

Fig. 2.1 Key ideas presented by teacher 'R' as reported in Anderberg (2000)

definitions. Whether or not that is the case because of the difficulty of pinning down these concepts, or because there is some practical advantage for teachers to work in this way, it certainly seems inappropriate for educational researchers studying teaching and learning to show the same lack of precision in their use of these terms.

Most likely, teachers talk about these notions in such imprecise ways simply because that is the way language is generally used. Whilst science may aspire to deal with tightly defined terms, which allow statements to be judged as clearly correct or not, this is not how 'natural' language tends to operate,

> natural language concepts have vague boundaries and fuzzy edges and that, consequently, natural language sentences will very often be neither true, nor false, nor nonsensical, but rather true to a certain extent and false to a certain extent, true in certain respects and false in other respects. (Lakoff, 1973, p. 45)

This then is a key issue with much writing in science education: whilst there is an academic community of science educators who seek to undertake formal research into aspects of teaching and learning in science, much of this work is reported in terms of natural language where key terms are 'fuzzy' and understood in a 'vague' way, making such reports accessible to non-specialists such as teachers but lacking in the precision expected in a technical field.

Folk Psychology as a Lifeworld Structure

The mental register, then, comprises of a set of terms used to refer to fuzzy concepts such as learning, thinking and understanding, which are part of natural language and perform adequately for communication during most everyday situations. Yet these terms lack the technical definition and operationalisation of scientific concepts.

This situation somewhat parallels that reported by Solomon, who explored secondary students talking about energy in the classroom. She proposed that students have everyday knowledge of energy which could be considered to be in a different domain to academic knowledge (Solomon, 1992). In making these claims she drew upon the notion of the 'lifeworld', where

> In the normal or 'natural attitude' we all tend to categorize our experiences rather loosely – to 'typify' them – so that they can be absorbed into 'meaning structures'. These are then reinforced by communication with others and by language itself, which gives this 'life-world' knowledge both social value and great persistence. Since each practical situation is only in limited need of explanation, such meaning structures will be fragmentary, not logically integrated with one another and tied to the particular type of experience which prompted them. (Solomon, 1983, p. 50)

Solomon found that her students' talk about energy fitted this pattern of lifeworld knowledge, with its fragmentary, 'local' nature (cf. Claxton, 1993). It is rather important that energy is well characterised in science, but in the lifeworld energy is a vague and more flexible explanatory device, perhaps so variable as to not really be coherent enough to be labelled a concept.

The mental register, the notions of thinking, believing, understanding, learning and so forth, seems to reflect Solomon's characterisation of thinking about lifeworld knowledge: terms used loosely, reinforced by communication with others but somewhat fragmentary, not tightly, logically integrated with one another. Arguably, due to the development of theory of mind, and then social induction into how others describe their mental experiences in terms of phrases like 'thoughts', 'ideas', 'understand' and 'believe', people acquire a 'folk psychology' which enables them to make sense of, and discuss, their own subjective experiences and – by inference – to explain the behaviour of others in terms of various components of the mental register.

Mental Life

The starting point for this set of chapters is a consideration of the general nature of mental experience. It is suggested that there is a very real problem in understanding the nature of ideas and thinking and in considering how these can relate to the kind of observables that tend to be used as data in science education research: such as recordings of speech, writing and drawings. Indeed the same issue arises in neuroscience, where the data from various types of 'brain scans' stand in a similar mysterious relationship to thought as much of the data collected in science education research.

The first chapter here then considers issues that could be considered to be related to the philosophy of mind, rather than science education as normally understood. This is not however an aside or diversion: for if in our research we wish to discuss mental constructs such as thinking and understanding, then we need to move beyond the lifeworld register of folk psychology and be clear about the ontology of mind (e.g. what kind of thing is thought, and how does it relate to the kind of things we can collect in our research), which is a logical prerequisite to considering the epistemological questions (e.g. how can we know about the 'contents' of minds, i.e. ideas, thinking) that in turn need to be considered before selecting methodology – that is, before deciding how to collect data to answer such questions as 'what are students' ideas about…' and 'do learners think scientifically about…'.

This first chapter in this part (Chap. 3) therefore identifies key issues and explains why they are fundamental to the research programme and then sets out how mental constructs will be understood in the treatment in this book. This provides an important part of the background that will be drawn upon in the rest of the book.

This will then lead to a series of chapters exploring what we understand by the key terms for mental processes: namely, ideas (Chap. 4), memory (Chap. 5), understanding (Chap. 6) and thinking (Chap. 7).

Chapter 3
Modelling Mental Activity

...since cognitive scientists aim to understand the human mind, they, too must construct a working model. It happens to be of a device for constructing working models. (Johnson-Laird, 1983, p. 10)

It was argued in the opening chapter that it is important to strip away the assumptions underpinning research, to problematise the research process. Only by doing that can be sure that we are aware of the limitations of our work, and so identify areas where it is most tentative, or would benefit from further development. This is especially important in an area where researchers working in the same field only share commitments to a limited extent. So, in science education, it is widely accepted that students' ideas about science topics are important for their further learning (and so for how they should best be taught), but there are different views on how to understand the status and nature of those ideas and so how best to elicit and represent them. As pointed out in Chap. 1, and in the introduction to this part, a key issue is that much of the professional discourse refers to everyday ('fuzzy') concepts such as knowledge, thinking, ideas, understanding and learning, which are used to make sense of observables (i.e. behaviours).

Everyday Notions Related to Conceptual Learning

So in addition to technical notions such as conceptions and conceptual frameworks, the research literature uses a range of terms that are drawn from everyday discourse and are *in that context* generally seen to be unproblematic: knowledge, belief, thinking and understanding. Some of these common terms are included in the left hand column of Table 3.1.

To the extent that when such terms are used in everyday discourse, they generally seem to support effective communication (i.e. communication that seems satisfactory to those involved in the conversations); they are not problematic. It is widely accepted

K.S. Taber, *Modelling Learners and Learning in Science Education: Developing Representations of Concepts, Conceptual Structure and Conceptual Change to Inform Teaching and Research*, DOI 10.1007/978-94-007-7648-7_3,

Table 3.1 Some descriptors used to describe research foci

Everyday terms: descriptors used in research drawn from everyday discourse	Technical terms: descriptors used drawn from technical literature
Belief	Cognitive ecology
Ideas/thoughts	Cognitive structure
Imagining	Conceptions
Learning	Concepts
Knowledge	Conceptual frameworks
Reasoning	Conceptual structure
Thinking	Intuitive theories
Understanding	Mental models
	Misconception
	Personal constructs
	P-prims
	Preconception

that in normal discourse, individual words do not take their meanings from technical definitions, but in part from the specific context of use (Vygotsky, 1934/1986), and in normal communication such as in conversations, including classroom discourse, the interactive nature of the communication, its dialogic aspect, affords opportunities to detect and correct misunderstandings (Bruner, 1987).

However, research reports are not designed to be conversational, but rather have to make explicit what is being discussed so that they can be understood remotely, at different places and times, without relying on interactive clarifications. The genre of a research report in effect eschews any deliberate explicit dialogic intent (i.e. to write in an open way which invites the reader to bring a personal meaning to the text) and is intended to make an argument for specific knowledge claims and to support that specificity by closing down as far as possible opportunities for readers to interpret the text very differently from the meanings intended. From a constructivist perspective, the meaning that a reader takes from a text is always constructed by the reader using the idiosyncratic conceptual resources available to that individual. So the writer always faces a challenge in guiding a reader towards a specific meaning through a text. This becomes more difficult when 'fuzzy' lay terms are used, as readers will interpret such reports according to their own understanding of the terms used. In the context of the area of research discussed in this book, this raises the issue of what is meant, for example, by an individual's *knowledge* or what is meant by reporting that someone *understands* something.

There are publications within the wider subject area of education that define themselves as research journals, but which would not expect the kind of explicit writing described here, and that publish some materials that are less explicit and deliberately invite readers to make their own interpretations and draw their own conclusions. For example, the *International Journal of Qualitative Studies in Education* (QSE) will publish material in the form of fiction (Behar, 2001) or poetry (Norum, 2000). Education spans the humanities and the social sciences, and such forms of scholarship and research are admitted in some traditions encompassed within education. Such work can certainly have merits.

A short poem (entitled 'Poem') published in QSE using the metaphor of plant propagation by rhizomes invites reflection on the ownership of data and ideas and the question of who owns the ideas developed by researchers (e.g. the kind of modelling discussed in this book), given that they build their writings through reading the work of other scholars and using data gifted by study participants (Trainor, 2003). At least, that was the meaning constructed for the poem by this reader – but it is in the nature of such artistic works that they can have multiple readings. By contrast, research journals that locate themselves within the more specific field of science education are likely to have specific expectations about the genres of writing accepted and what should be included in a submission (Taber, 2012). They also expect very clear explicit arguments rather than allusion and metaphor.

The Mind

As Descartes famously pointed out, the knowing subject can be sure of one thing – his or her own thoughts, the ideas that are the basis of conscious experience. I have thoughts or ideas, and – *assuming* that other people are like me in that regard – I expect others to also have ideas. This is certainly an assumption for as Descartes had deduced we cannot *know* whether our thoughts represent anything outside ourselves (an external world beyond our own selves), and even if we do accept the existence of the world and others inhabiting it, they could – as some philosophers have commented – be automata lacking mental experiences like our own.

Although such grounds for adopting solipsism may be strictly true in a logical sense, we generally accept that there is an external reality, and that what we perceive as an external world, and the people in it, are more than figments of our own imagination. It would be hard to imagine how life could be lived otherwise. So for the purpose of writing this book, and indeed, an essential assumption to make it worth my while setting out to write a book, I imagine my readers do have mental experiences, and that *you* will share my assumption that all humans experience thoughts and have ideas just as we do. So, here are some assumptions that are not *absolutely* logically secure, but ones I imagine most science educators will have no difficulty accepting. So, here at least are some common commitments among colleagues working in the research programme! Most of us include such assumptions in our worldview, as taken-for-granted assumptions.

Figure 3.1 presents a very simple representation of the individual, considered as a thinker – having conscious thoughts – located in and acting in an external world. The thinker, who I will call Jean, is an embodied person, inhabiting a body that is physically demarcated from the external world, that is, there is a fairly clear boundary between Jean and Jean's surroundings. The space outside Jean has been labelled as a public space, as other thinking subjects (not shown in this figure) will also inhabit this same physical world.

Fig. 3.1 The individual in
the world

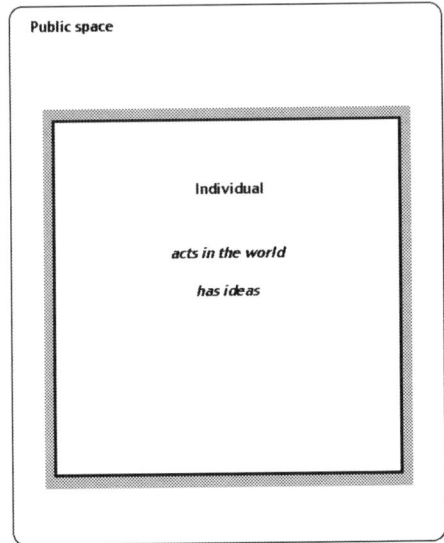

Mind and Matter

The simple structure of Fig. 3.1 ignores a major problem that will be of considerable importance in exploring research into student thinking. Conscious thought is considered to be in some sense located 'within' Jean as an individual. Thoughts are 'internal' mental experiences, so clearly they must occur 'inside' us. However, if we ask *where* these thoughts occur, we find two common suggestions.

One is that mental activity takes place in the brain. There is certainly extremely strong scientific evidence to consider that our thoughts *are related to* electrical activity in parts of the brain, and that in terms of physical processes such as chemicals diffusing across synapses and electrical signals passing through networks of neurons, this is the location of thinking in the body. Arguably there is no absolute distinction between the brain and the rest of the nervous system, but it at least seems reasonable to consider that thinking is due to activity in the nervous system.

However, although it seems fairly clear that our thoughts are *correlates* of electrical activity in the nervous system, the thoughts themselves are subjective experiences that cannot be *objectively* (i.e. by any neutral observer) observed. There are a number of techniques that allow electrical activity in the brain to be investigated, some with good temporal resolution and some with good spatial resolution. However, unlike in some science-fiction scenarios, there is as yet no device that allows anyone to experience the thoughts of another. Indeed there are good reasons to suspect this could never happen.

The second suggestion for locating thoughts is inside our *minds*. Unlike the brain or central nervous system, the mind is not a physical entity. Rather the notion of the mind is a theoretical construct that has been found useful in a range of explanatory schemes (Claxton, 2005). People are minded to do certain things and are commonly

Table 3.2 The Mind-brain distinction

Domain	Mental	Physical
Location	Mind	Brain
Activity	Thinking – having ideas	Processing (chemical/electrical)
'Substance'	Thought	Matter

said to have beautiful (or troubled, or diseased, or dirty, etc.) minds. In effect, the mind is the hypothetical 'place' where our mental lives occur, that is, 'where' we have subjective experiences (thoughts). Table 3.2 sets out a simple demarcation between mind and brain. The key point is that the mind, unlike the brain, is not a physical entity; it does not have physical form or material substance; it is immaterial.

Arguably *the whole notion of mind is a kind of analogical model*. In the physical world, objects have substance and are located somewhere. So we can understand our thoughts (by analogy with experience of the physical world) as being the (metaphorical) substance of our mental worlds, and so the mind may be considered as the (metaphorical) container where we have ideas, or perhaps even the (metaphorical) factory where we produce them or the (metaphorical) home where they live, or the (metaphorical) theatre where they are played out, etc. Even referring to a mental 'world' would seem to be adopting an *analogy with the physical world*. According to Lakoff and Johnson (1980a), all our concepts must ultimately be grounded in our experience of the material world, but this means that again the metaphors we adopt can come to be taken for granted: of course the mind is a place, what else could it be?

One of the key issues in philosophy, and indeed one source of the 'big' questions in science, is how mind and brain are related, that is:

- How can physical reality be experienced subjectively?
- How can our minds interact with matter to bring about actions in the physical world?

The first question is clearly a genuine and potentially scientific question. For anyone who accepts the existence of an external physical reality, the question of how we can experience and know about that reality is a 'big' question, quite fundamental to understanding ourselves in the world.

Some would suggest that the second question is based on a false premise. We consider that our thoughts are somehow able to bring about changes in the world: I would like some tea, so I intentionally go to the kettle. Certainly we perceive that we carry out physical actions such as filling the kettle with water and turning on the switch that seem to be necessary to bring about our plans to make a cup of tea.

Thought as an Epiphenomenon

However, it is logically quite possible that our thoughts are better understood as *by-products* of the physical processes occurring in the brain. That is, feeling thirsty is a perception of the body's homeostasis system being triggered to bring about water intake. In principle it could be possible for the biological mechanisms to trigger processing in the brain that leads to my making a cup of tea without any need for conscious thought. This need not imply that making a cup of tea is not intelligent or learned, for my brain has built up neural circuits to allow my body to operate in my environment and carry out crucial actions – such as preparing tea. However, this physical system (based on chemical sensors and electrical signals) might well operate perfectly without my *needing to* have conscious thoughts about it. My being aware of thirst and 'deciding' to make a cup of tea could well be just a by-product of the process.

If this seems fanciful, it is worth noting that we often carry out many complex procedures without consciously directing them. It is common experience that we sometimes make mistakes because of this. We go straight home, or directly to work, instead of pausing at the postbox where we intended to post a letter, or heading off to a meeting in a less familiar location, because our minds were thinking about something else, or our minds 'were some*where* else' as we sometimes say, whilst our body carried out our familiar patterns of action. Sleepwalkers seem capable of complex actions without conscious awareness: 'some individuals have been known to drive cars for great distances whilst sleepwalking' (Mahowald & Schenck, 2000, p. 323).

This is a point that may seem more significant to philosophers than science educators, as for most everyday purposes, we talk and behave *as though* people are making conscious decisions and behaving accordingly. At least since Freud, it has been widely accepted that we may not always be aware of our motives for certain actions (or inactions), but this tends to generally be considered as a secondary pathological effect impairing the normal state of things where we are in conscious control of our actions in the world.

However, this is in effect a 'folk model' based on common-sense understanding rather than scientific knowledge. The common acceptance of a folk model of mind that informs people's ways of talking no more makes it scientifically correct than the common adoption of impetus-like notions (alternative conceptions) of force and motion. Most students believe that an object will only continue to move if it is subject to a continuous applied force; and similarly most teachers probably believe that the conscious mind controls behaviour.

However, what will be important for our analysis of what is going on in research into student thinking is the key point that mental life and the thoughts and subjective experiences people have are quite distinct in nature and 'substance' from what can be observed and measured by scientific observers, and in that sense, Fig. 3.1 may be misleading:

> Mental events and externally observable (physical) events constitute two categorically separate kinds of phenomena. They are mutually irreducible categories in the sense that one cannot, *a priori*, be described in terms of the other. (Libet, 1996, p. 98)

Figure 3.2 represents this by locating conscious thinking in a distinct 'place' within the individual, but in a part of the individual that is not within the public space.

The border around conscious thought in Fig. 3.2 is meant to represent how the mind is a distinct aspect of the individual from the physical body, although in some sense embodied. Figure 3.2 is only a schematic representation, and it is possible to

Fig. 3.2 A dualist representation: conscious thought as a distinct, non-physical, aspect of the individual (cf. Fig. 3.1)

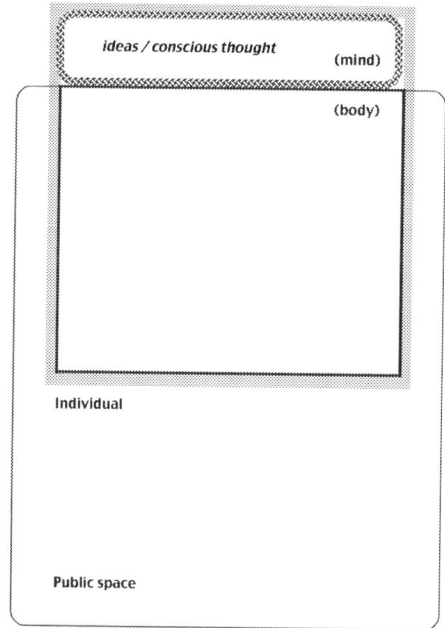

read it a number of ways. It suggests the existence of the mind as a real entity, which is part of the person, but *not* part of the physical world. An obvious implication here would be the dualist one, that mind is of a different sort of stuff to matter, and somehow communicates with it.

This is somewhat problematic as if mind is thought of as something separate from the material world, then any consideration of its form is outside of the realm of science, and mind must interact with matter by mechanisms that are totally unknown to science. However, it will be suggested below that it is possible to retain the notion of mind as a useful concept, without adopting dualism.

The Problem with a Dualist Model of Mind

A key difficulty in modelling a student's idea, knowledge, learning, etc. from a dualist perspective is that one has to take one of two positions: either that mental thought can influence matter so that ideas in the nonmaterial mind are somehow then coded into physical form in the brain to be acted upon, or that conscious thinking is an epiphenomenon of brain processes, and that the subjective perception that we have some control over our actions through conscious thought is an illusion.

The former position fits better with the way people commonly understand and speak about the mind, but there is currently no known mechanism drawing upon established scientific principles which could explain how a nonmaterial mind could control our bodies. The second position, at least when expressed in the form above, seems to suggest that we are just automatons responding to programming with an illusion of responsibility and control over our actions. This could be true but would likely be unwelcome. That it would be unwelcome would of course make it no less likely to be true! However, there is a different way of thinking about the brain-mind relationship.

Consciousness as an Emergent Property of Processing

There is an alternative perspective, which sees mind as an emergent property of brain complexity and processes, and so avoids the question of how mind interacts with matter, without necessarily giving up on personal control and responsibility. In this perspective, the brain has evolved through natural selection to process information at certain 'levels' (another spatial metaphor) – with various degrees of automation – presumably because there was an advantage to dealing with most issues quickly but some in a more 'deliberate' manner. We are aware in conscious experience of that processing that occurs at a specifically 'high' enough level in the brain, and we experience thoughts. At this level the brain is applying specific explicit thinking strategies to make sense of events and objects in the world and solve other problems, and as we are aware of this level of processing, we have a sense of an 'I' directing operations that seems other than part of the body. From this perspective we certainly do make decisions, try out ideas and do have control, but the conscious aspect is in a sense a by-product. However, all the processing occurs within the apparatus of the brain, and consciousness (whilst subjectively very important) is more a marker for the level of deliberation (or complexity of analysis being applied) in the processing than an indication of a level of control *beyond* the brain.

Table 3.3 develops this distinction between two different perspectives: a dualist perspective, where it is assumed that mind interacts with brain, and a 'unitary' perspective, which accepts the existence of mind as something real but considers it as an emergent property of brain:

In the dualist model there are two types of 'thing' or 'substance', mind and matter, which somehow influence each other. The alternative approach does see mind as *in a sense* separate from the brain but does not see it as a different type of substance, but rather *a way of conceptualising phenomena at a different level of analysis*. Mind is real, but is an emergent property of the brain, and gives us an alternative way of thinking about thinking, knowing, understanding, etc. From this perspective, the mind-body problem (of how mind can interact with matter) is seen to be an argument based on an ontological confusion that mind is some*thing* other than matter, rather than an emergent *property* of certain forms of organisation of matter. As one commentator has suggested, 'the debate on the 'mind-body' problem can exist only insofar as one denies that the functional organisation of the nervous system corresponds to its neural organization' (Changeux, 1983/1997, p. 275). In general, then, cognition is due to activity within the nervous system, which can be understood to involve processing at different levels in the system, only some of which lead to conscious awareness.

Table 3.3 Two approaches to thinking about mind in relation to the brain

Perspective	Dualist	Unitary
Consciousness	Represents our minds, as something separate from our bodies	Represents awareness of a high level of processing where deliberation is useful
Control	Our mind controls conscious actions through thought, but much mental processing is automatic and not directly under the mind's control. 'I' control my mind, but my body does not refer everything to my mind	The brain is the apparatus that controls (most of, cf. reflexes) our actions through processing information in the form of electrical activity. We are aware of only some of this processing
Thinking	Conscious thinking (undertaken in the mind) is different in kind to subconscious processes (which occur in the nervous system)	Although we only tend to use the term 'thinking' to refer to some aspects of the processing of information in the brain, distinctions between conscious and unconscious (preconscious) processing are a matter of degree, not kind

Different Levels of Analysis of Cognition

This type of distinction is familiar in the cognitive sciences where it has been suggested that cognitive systems can be described and analysed at (at least) *three* complementary levels. These levels are complementary because it is assumed that all three are valid ways of describing cognitive processes; however, the different levels are appropriate for discussing different types of questions. This is set out in Table 3.4:

Neuroscience

The first level relates to the physical basis of cognition. The brain is a physical object, connected to the rest of the nervous system and composed of units such as neurons, with their synapses, organised into a complex structure. It is widely accepted that aspects of brain activity are strong correlates of conscious thought and in particular that electrical activity in parts of the cortex (the outer layer of the brain) tends to correlate with conscious experiences such as perception, imagining and planning (Rees, 2007). The cortex can be considered the 'highest' part of the central nervous system in the sense of its being the most recent development in evolutionary terms, and in it being physically furthest from the peripheral nervous system (sensory and motor nerves).

Table 3.4 Three levels for thinking about thinking, learning, personal knowledge, etc.

Level	Physical	System – computational (functional)	Mental
Focus	Brain	Processing system	Mind
Contents	Neurons, synapses	System components, processing apparatus	Ideas, thought
Processes	Circuit activity (ultimately chemical)	Identifying input, calculating, problem-solving, storing, retrieving, executing action	Perceiving, knowing, thinking, remembering, deciding
Development	Formation/pruning of connections; modifying connection strengths	Changes in processing; increasing contents of store; changing linkage in store	Learning, conceptual development, cognitive development

Locating Consciousness in the Brain

The problem of identifying the precise location(s) of brain activity which is associated with consciousness, the 'the neural correlates of consciousness' (Crick & Koch, 1990, p. 265), is a complex one – if indeed such a question can have a meaningful, definitive answer,

> It is … likely that the operations corresponding to consciousness occur mainly (though not exclusively) in the neocortex and probably also in the paleocortex, associated with the olfactory system… it may be important to consider in detail the inputs and outputs of the hippocampal system. Structures in the midbrain or hindbrain, such as the cerebellum, are probably not essential for consciousness. It remains to be seen whether certain other structures, such as the thalamus, the basal ganglia and the claustrum, all intimately associated with the neocortex, are closely involved in consciousness. We shall include these structures together with the cerebral cortex as 'the cortical system'. (Crick & Koch, 1990, pp. 265–266)

In practice, explanations at this level are (currently, at least) only of limited *direct* relevance for much of our research in science education. Despite this, it is considered important to emphasise that in this volume, the anatomy and physiology of the brain is assumed *to underpin all cognitive processes* – such as thinking and remembering – and can *in principle* (at some point in the future) offer a basic level of explanation for such processes.

This is important because quite naturally when we talk and think about such matters as learning, understanding and knowing, we tend to focus on conscious experience, but the processes underpinning such mental activities are not only those giving rise to conscious awareness that learners are aware of and can offer reports of:

> When we try to understand conscious experience we aim to explain the differences between these two conditions: between the events in your nervous system that you can report, act upon, distinguish and acknowledge as your own, *and a great multitude of sophisticated and intelligent processes which are unconscious and do not allow these operations.* (Baars & McGovern, 1996, p. 63, emphasis added)

For present purposes, in the context of research in science education, precise details of brain functioning even when understood are not usually relevant. However, to be consistent with the assumption that cognition is underpinned by physical processes in the brain, it is considered important that any models proposed in science education – for example, linking to the 'computational' systems level of analysis (Table 3.4) – should *in principle* be *capable of being explained at the physical level* and so should be consistent with the findings of neuroscience.

Systems Analysis

The next level of analysis does not deal with the specifics of neurons and synapses but rather considers cognition in functional terms of a system that processes information – what I will refer to as the cognitive system:

> the mind can be studied independently from the brain. Psychology (the study of the programs) can be pursued independently from neurophysiology (the study of the machine and the machine code). The neurophysiological substrate must provide a physical basis for the processes of mind, but granted that the substrate offers the computational power of recursive functions, its physical nature places no constraints on the patterns of thought. This doctrine of *functionalism*, which can be traced back to Craik, and even perhaps ultimately to Aristotle, has become commonplace in cognitive science. (Johnson-Laird, 1983, p. 9)

At this level, the discussion is in terms of what happens to information in the cognitive system and what kind of systems components are needed. For example, human cognition involves memory, which means that there has to be some means of 'storing' information within the human cognitive system. (It will be argued later that it is more appropriate to consider memory a process of *representation* rather than storage, as the latter term implies fidelity that memory may not provide.) This is presumed to be based upon the neural apparatus, and this has been commonly demonstrated even if we do not understand exactly how this works. However, for the purposes of research in science education, it will be important to know something about the nature of human memory in functional terms – capacity, fidelity of representation, ease of access of 'stored' information, etc. – and it is less important if we do not know exactly how such processing occurs in terms of physiology.

Indeed, from a systems perspective, the same processing patterns could in principle be executed in different physical substrates: 'in principle', because in practice the physical properties of the system will influence its operating characteristics. My own memory and my computer hard drive both act as memory stores and so have the same basic function within different types of information processing systems, but they have rather different properties and so different strengths and weaknesses because of the different physical systems involved.

In considering cognition in systems terms, it will make sense to designate specific components to act as sensory interface, memory, central processors, etc. If the description at this level is accurate, each component will refer to some specific parts of the brain or nervous system (the retina, areas of the cortex, etc.), although nominal components of cognitive systems may be spatially distributed in the brain. In some cases there are well-established anatomical regions identified with the system component, but in other cases, it may currently just be assumed such regions exist.

Mental Activity

Mental activity, such as thinking and remembering, is according to this model (i.e. Table 3.4) at a different level of description to either the physiological or the system accounts. However, mental activity will in principle have parallel descriptions at the other two levels. So, for example, recalling something (a subjective experience) can be understood in terms of the systems level as information that had previously been 'stored' in memory being accessed and brought into some central processing unit, which is part of the system which gives rise to conscious experience. The apparatus for accessing the information, and then processing it, along with the actual physical basis for the memory store itself, will be comprised of neural circuits that are activated by other neural circuits.

Mental Apparatus and Mental Resources

In considering a systems view of mental activity, then, it is useful to distinguish between the apparatus in terms of systems components and their functions and the resources available to support processing – which we might consider the knowledge content of the system, that is, what has been 'stored' (or is represented) within the system. In an organic system such as a human being, this is not an absolute distinction as both the apparatus and the resources are embodied as configurations of the nervous systems, and development and learning can be considered to modify the apparatus itself (this is discussed later in the book). This is different to a computer system, for example, where the configuration of the hardware that processes information is not changed by changes to the information being stored. Despite this caveat, there is a systems level at which the basic architecture of the cognitive apparatus is common in all normal people and is constant through the life span. This allows the development of simple systems-level models of the learner, as discussed later in the book.

Explanatory Power of the Three Levels of Description

The value of acknowledging these three levels of description in discussing cognitive processes in learners (or teachers or researchers) is that collectively they offer a good deal of explanatory power. It is possible to make a comparison here with the

Fig. 3.3 The systems level of description for cognition offers a useful intermediate between the subjective mental level and the objective but incompletely understood physiological level of description

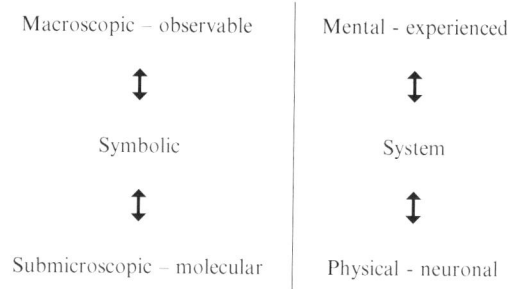

Macroscopic – observable	Mental - experienced
$\updownarrow$	$\updownarrow$
Symbolic	System
$\updownarrow$	$\updownarrow$
Submicroscopic – molecular	Physical - neuronal

way teachers can use different levels of description in chemistry lessons (Gilbert & Treagust, 2009; Johnstone, 2000; Taber, 2013c). The phenomena of chemistry are macroscopic events that can be perceived directly. So when a strip of magnesium metal is added to a beaker of hydrochloric acid, for example, bubbles are produced and rise to the surface and the metal itself appears to be broken down and (at a phenomenal level) disappears. If sufficient of the reactants are used, and the gas in the bubbles is collected, it will ignite with a 'squeaky pop' when a lighted splint is applied.

These phenomena are described in theoretical terms at the macroscopic level: metal reacts with acid to produce a salt solution and hydrogen. The hydrogen is a gas that undergoes combustion with the oxygen in the air. However, much *explanation* in chemistry occurs at a very different level – at the submicroscopic level of hypothetical entities which are conjectured to form the structure of matter at small enough scale: 'quanticles' such as atoms, electrons, ions and molecules. The properties and interactions of these entities are used to explain the properties and reactions of substances.

These two different levels are bridged by the use of a third 'level', the symbolic level. Here language, formulae, images, etc. are used to represent the macroscopic descriptions of phenomena and the submicroscopic models that provide explanations. This level can act as a bridge as the verbal and formulaic labels used in chemistry (hydrochloric acid; zinc; Cu; CH_3COOH; $2H_2 + O_2 \rightarrow 2H_2O\ldots$) are able to refer to both real substances and the hypothetical quanticles that populate the explanatory models (Taber, 2009a, 2013c). This 'bridging' function offers much potential for confusion if students are not clear when the symbols are being used to refer to the macroscopic level, and when to the submicroscopic level, but nonetheless provides a very powerful tool for conceptualising and explaining chemistry.

Although the analogy is not perfect, there is a similar relationship in considering the three levels for discussing cognition. In this comparison (see Fig. 3.3), the system level analysis offers a bridge between the way cognitive activity is experienced (directly available to introspection, but only giving glimpses of cognition) and the underlying physical basis (which is currently poorly understood, is highly complex and offers limited direct insights to inform educators).

Consequences for Science Education

The reason for focusing on this issue is to provide a basis for setting out objectively what different research studies are concerned with. When research reports discuss learners' beliefs or ideas, or conceptual frameworks, and so forth, the reader should be given a clear indication of how the researcher intends the term to be understood and methodological details that provide a convincing case that data collection and analysis techniques applied are suitable.

In the present volume, an attempt will be made to consider what is usually meant by these various referents and how they link to, and can be modelled in terms of, the different descriptive levels considered above. This will allow a consideration of the kind of data collection and analysis which is possible in different cases and so the nature of the results that can be offered. Inevitably, it will be argued, research results that discuss these foci of enquiry are necessarily presented in relation to models of some kind, and the present work sets out to offer indications of the nature and status that such models can have.

Representing the Learner

Given that knowledge of cognition at the physical level is limited, and often somewhat indirectly related to the foci of research in science education, much of the discussion in this volume will look to relate the mental and systems level of description. However, an effort will be made to make a clear distinction between these different levels of description. A representation such as Fig. 3.2, presented earlier in this chapter, would be considered flawed in this respect as it suggests that thought is somehow *separate from*, rather than at a different level of description to, the thinker as a physical person or a processing system. A schematic representation that acknowledges that limitation might be Fig. 3.4, which shows that there is a component of the system (an executive processing module) which gives rise to conscious experience. This representation is meant to suggest that subjective conscious experience is not to be considered some*where* else, but as a different level of description for what is going on inside the person.

The notion here of the 'executive processing module' should be seen as a theoretical entity, which is part of a model of the *individual as a system* that processes information and which can give rise to mental experience. Later in the book this general notion will be developed. For some purposes such a model might be useful regardless of whether there was any anatomic evidence that such a discrete component could be identified in the brain. However, for such a model to be considered as a realistic one, rather than a purely instrumental one, it would be expected that ultimately the system model could be mapped upon actual physical features of brains.

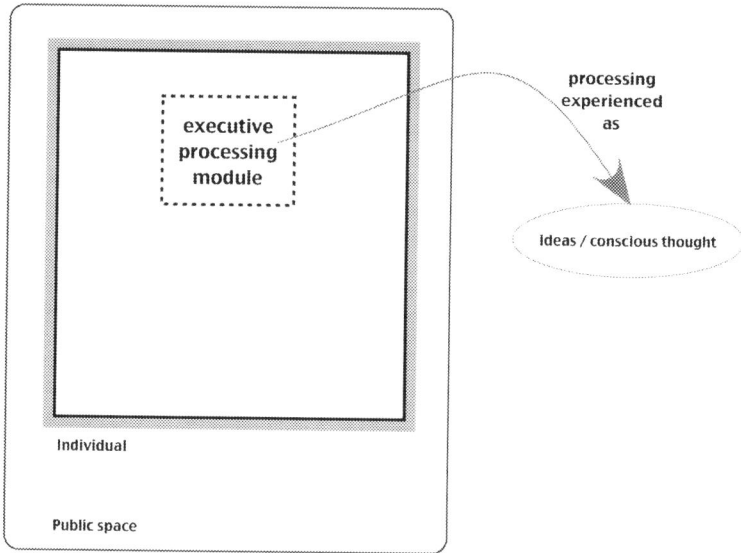

Fig. 3.4 Relationship between the system and mental levels of description

Principles Informing the Account in This Book

So cognition can be described at three complementary levels, relating to the physical nature of the brain, the cognitive system at a functional level and in terms of mental constructs. It is important to be clear about which level is being discussed at any time. In the discussion that follows, an attempt will be made to consider how research into learners' thinking, knowledge and learning can be understood primarily in terms of ideas at both the mental and systems levels. In the latter case, features of models of people as information processing systems will be based upon established ideas from the cognitive sciences – that is, what research and scholarships suggest about perception, memory and so forth. A sensible principle to adopt would seem to be that *any models proposed for research in science education should be consistent with what is understood about human cognition.*

Chapter 4
The Learner's Ideas

The paradox of science education is that its goal is to impart new schemata to replace the student's extant ideas, which differ from the scientific theories being taught. (Carey, 1986, p. 1123)

Researchers looking at student thinking and learning in science have used a wide range of terms to describe what they are exploring (Abimbola, 1988), but when Black and Lucas edited a 1993 volume about the topic they 'settled on the suggestion of *Children's Informal Ideas in Science* as a relatively neutral expression' (Black & Lucas, 1993a, p. xii). Presumably, part of this neutrality comes from 'idea' being an everyday term that each of us understands in terms of our subjective experience.

The Idea of Ideas

The *Oxford Companion to the Mind* notes that ideas 'might be called "the sentences of thought". They are expressed by language, but underlie language – for the idea comes before its expression' (Gregory, 1987, p. 337). Although we may all feel we are comfortable with what an idea is, another reference work warns that in philosophy 'idea' has been 'a term that has had a variety of technical usages' and notes that 'modern philosophers prefer more specific terms like "sense datum", "image", and "concept"' (Brockhampton, 1997, p. 262). However, the same source notes that 'an innate idea [sic] is a concept not derived from experience' (*ibid*).

Another term that is commonly used is 'thoughts' in the sense of a thought being 'an idea or a pattern of ideas' (Watson, 1968, p. 1150). I will therefore use the terms 'idea' and 'thought' interchangeably.

K.S. Taber, *Modelling Learners and Learning in Science Education: Developing Representations of Concepts, Conceptual Structure and Conceptual Change to Inform Teaching and Research*, DOI 10.1007/978-94-007-7648-7_4,

The Source of Thoughts and Ideas

Given that individuals such as Jean (our hypothetical representative science learner, from the previous chapter) have ideas, this raises the question of how thoughts arise: How do people have ideas, and what are they based upon? An obvious answer is that ideas derive from sensing the world – but clearly not all ideas can be derived *directly* from sensory experience. Ideas such as the notion of the electron, the second law of thermodynamics and the alternative conception that exercising gives you energy are not based upon direct observations in any straightforward and simple way.

Sensation and Perceiving the World

One key part of our subjective experience is that we are aware of our surroundings, that is, we form mental 'images' of the external physical world. The existence of a physical world in which we collectively exist was one of the shared commitments of researchers in science education that was assumed earlier. Indeed, this would seem to be a necessary ontological commitment of science – scientific enquiry would seem to presuppose an objective and relatively stable physical world. I can see my computer and the rain on my study window, and I can hear music playing and, now that I am paying attention, the previously unnoticed, but now irritating, sound of my fingers depressing keys on my computer keyboard.

So one aspect of mental experience is sensory: the body's sensory system allows us to experience and think about our immediate environment. So the ideas a person has at least in part derive from their experiences of an external world. So Fig. 4.1 shows that somehow objects and events in the world can be sensed and presented to consciousness. There are clearly two large areas of simplification here, relating to how our body is able to sense the environment and how those sensory inputs lead to conscious experiences.

The Apparatus of Sensation

At the functional/systems level of description (see Chap. 3), we are able to experience the world because we have systems components which act as a 'sensory interface': that is, we are equipped with apparatus which converts stimuli in the external world into a form of information that can be processed within the cognitive system. This apparatus is our eyes, ears and other sensory organs. However, using the less familiar systems language of a 'sensory interface' (see Fig. 4.2) is useful because it can remind us of something that we may otherwise take for granted. Sensation involves *representing* – or coding – one kind of thing (a pattern of illumination of the retinal cells, say) in another form (electrical signals leading to sequential activation of neurons in the nervous system).

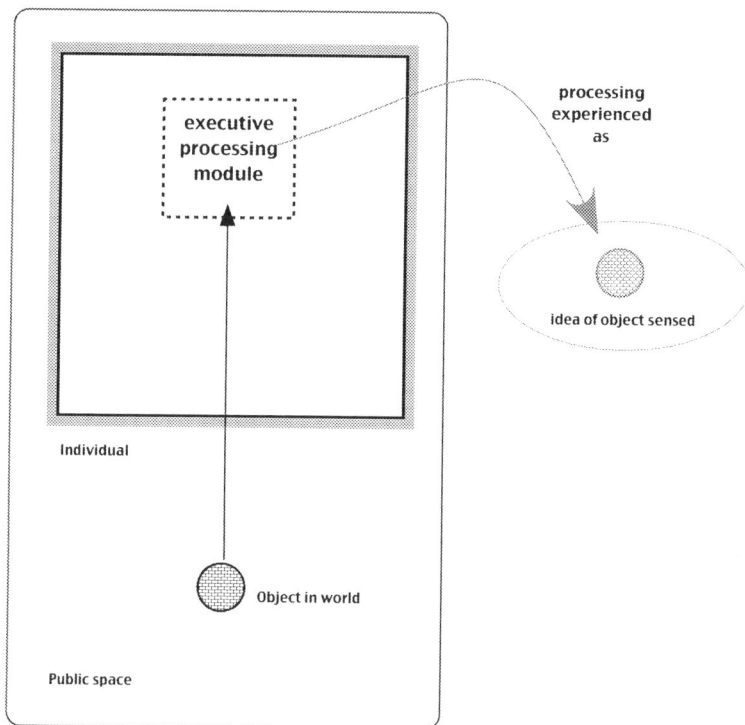

Fig. 4.1 Sensing the world represented as an unproblematic process

This is important, because any such translation from one form into another can only offer a limited fidelity to the original stimulus. This is familiar from technological applications. Telephone lines tend to lead to audio signals that lose much of the frequency range of the input. Pictures presented on video links may break up into an unrealistic staccato series of images. Photographs or screen images have limited resolution. Portable music players offer a range of file formats, and those which support the storage of the most tunes (such as mp3 files) do so by sampling original so-called lossless sources in ways that lose some of the original information.

A story I once heard that amused me, although quite probably apocryphal, concerned a man who is reputed to have approached the artist Pablo Picasso in an art gallery and complained that his pictures of women looked nothing like women. The man is supposed to have reached for his wallet and pulled out a photograph of his own wife. Showing this to Picasso, the critic claimed that this was what a real woman looked like. Picasso is said to have then asked if the man's wife was really like the picture, and – on having this confirmed – commented that the man's wife was very small and rather flat.

Representations are, of course, just that. However, if we are very familiar with the representation, it may become easy to forget this – after all we commonly use representations to stand for, and to operate on and so think about, other things as if

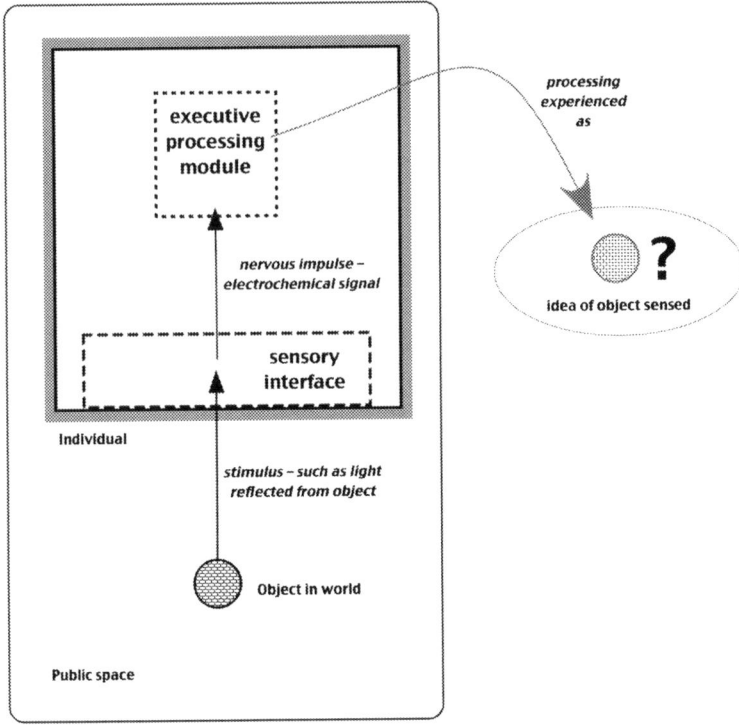

Fig. 4.2 Sensing the word necessarily involves representing external stimuli by coding the signal into a form that can be processed in the nervous system

they were those things. It is often appropriate and more productive to come to see the flickering pattern on the television screen *as* the newsreader, to see the icon on the computer screen *as* the hard drive and to see the chemical equation *as* the reaction.

If asked to reflect we might well say these symbols only stand for other things, but we *learn* to react to them and act on them *as if* they are the things. When I manipulate my computer mouse in certain ways, I conceptualise what I am doing as physically 'opening' my hard drive and imagine I am doing this at the point where I see the screen cursor overlaying the desktop icon on the screen. If I stop to think, I know that the drive is not located there, and that the images on the screen merely represent operations elsewhere in the computer system, but because the screen acts as part of the interface for interacting with my computer, it is more effective to operate with the screen images *as if* they are drives, applications, files, etc.

It is tempting to suggest that there is a parallel here with the previous chapter: that the computer screen with its visual imagery is the analogy of consciousness, reflecting particular representations of underlying processes in the computer. However, it is probably not productive to delve too deeply into such a comparison, as the observer of the screen is outside the computer system.

Representations May Seem Realistic

Most readers will have seen that with current technology it is possible to produce images of solids at the level of individual atoms. Many magazines and books carry reproductions of such images. They look, to all intents and purposes, like photographs of atoms. However, they are not. By scanning across a surface with a suitable probe, it is possible to detect the electric field that we interpret as due to the submicroscopic entities, such as ions or molecules, conjectured to make up the structure. By adjusting the position of the probe to keep the measured field constant, it is possible to build up a contour map of the surface – at least if we assume the field represents the 'position' of the surface, of course. (Whether the notion of a surface is actually appropriate at the submicroscopic level is a moot point.) This information can be coded using a false-colour scale – and, voila, we have an image, which looks like a photograph of the surface at the atomic level. The process is in some ways similar to taking sonar readings of a riverbed that can be built up into a contour map of river depth.

Similarly, most of us are familiar with seeing television images from infrared cameras – where radiation that is not energetic enough to be detected by our retinas is detected by cameras and then represented as a visible image on the display (with a scale of greys or a spectrum of colours representing the temperature of emitting objects from different parts of the scene) that can activate our retinal cells. Similar techniques are used in astronomy. For example, satellites that image the earth at different frequencies can be used to build up coloured photographs of the surface. However, the frequencies detected by the sensors do not respond to the three primary colours of light of the human visual system. This can lead to very impressive and sometimes quite beautiful false-colour images highlighting different features to those which would be salient to the human optical system (Sheffield, 1981).

In the same way, planetary probes sent out into the solar system can use radiation in the radio region of the electromagnetic spectrum to produce 'false-colour' images of other planets and of their moons. This can of course be seen as a more sophisticated version of radar, where objects can be detected using reflections of radio signals that are converted to visible blips and audible bleeps for the human operator.

The term 'false colour' however seems to imply that, by contrast, there is a standard for translating radiation signals into 'true' colour. Most humans have similar visual systems. Although we cannot know if we experience colours the same subjectively (i.e. as qualia), the apparatus we each have for detecting light is generally physiologically similar, and most people are sensitive to much the same range of 'visible' frequencies of radiation. However, there are exceptions. Some people do not have the normal pattern of cones firing in the retina and suffer from 'colour blindness'. This does not actually mean *not seeing any* colours, but missing some of the usual discriminations. Arguably, people who are colour-blind see in 'false colour' compared to most people, but this would be a somewhat arbitrary judgement based upon taking 'normal' human vision as the standard.

This would be arbitrary because there is nothing absolute about the specific nature of human vision: presumably it is an optimisation of match between the general organic properties of humans and the environment in which we live, subject to whatever constraints operated during evolution such as available genes and proteins. If, hypothetically, humans had evolved on Mars, it seems likely the visual system would have been better matched to discriminate in that environment and would not have operated so well here on earth.

Some insects have eyes with quite different frequency responses to human eyes and in particular can detect (i.e. what *for them* is visible) radiation in (what for humans is) the ultraviolet region of the electromagnetic spectrum (Peitsch et al., 1992). Although there are variations from species to species, it has been found that Lepidoptera species (e.g. butterflies) may have receptors with peak sensitivity in the UV as well as other receptors with peak sensitivity in the blue, green and red regions of the spectrum. Hymenoptera species such as bees may lack 'red' receptors and so have three main types of colour receptors, like humans, but detecting primarily green, blue and ultraviolet frequencies. By human standards it would seem that bees have 'false-colour' vision, but if eschewing an anthropocentric perspective, it would seem just as fair to suggest that, by bee standards, humans have false-colour vision.

The point here is that as vision involves the conversion of one kind of entity (a pattern of radiation) into a completely different material form (electrical activity in nerve cells), there is no 'correct' standard for the conversion process, so *bee vision is no more or less intrinsically faithful to the world sensed than human vision*. Similar arguments apply to the other senses. From a constructivist perspective, it is not meaningful to ask which system of representation offers the best fidelity to reality: rather the focus should be on the extent to which systems for representing information visually in cognitive systems facilitate the construction of models of the external world which support intelligent action in/on that external world. Having a visual system supports the development of effective mental models (see Chap. 11) that allow us to act in and upon the world in desired ways (i.e. models that show good 'fit' to the external world in terms of our experiences, Glasersfeld, 1989) is as near as we get to 'seeing the world as it is' – because seeing an object is not something that can be inherent in the object itself but is always the outcome of an interaction between the world and a particular sensory system.

Sensory Representation Involves a Coding System

Our senses are then based on transducers, units that convert energy from one form to another, so sensory information is necessarily represented in the nervous system using a coding process that has evolved to support us in effectively modelling the external world. This coding process allows different patterns of stimulation to lead to different patterns of neural activity. Presumably the coding process has evolved over a very long period of time and has been subject to natural selection. Therefore, the way our sensory organs code stimuli is not only non-arbitrary but has been proved in the field to be a very effective system for surviving in our environment.

That is reassuring, but the bee has evolved a somewhat different sensory interface which is coding data somewhat differently when in the 'same' environment and which has also proved fit for survival. Bees and humans are rather different – scale, form of locomotion, reproductive genetics, etc. – and perhaps have different priorities in their environments. For example, the vision of bees reveals patterns on flowers that are not visible to humans, patterns that have been construed as guiding the bees to the nectar and so pollen. So it should not be surprising if effective solutions for sensing the environment are different in the two cases. However, this does underline that sensing is inherently a process of representing – of modelling stimuli in electrical activity.

Sensory Information Is Selectively Filtered Before Reaching Awareness

However, this is only the first of a number of complications in building a mental model of our surroundings.

> … most sensory signals probably do not reach conscious awareness but many of them lead to modified responses and behaviours, as in the tactile and proprioceptive signals that influence simple everyday postural and walking activities, which have therefore clearly been detected and utilized in complex brain functions. (Libet, 1996, p. 97)

Sense organs do not send nerve fibres directly to the cortex but rather to subcortical areas such as the thalamus, which then have connections into the cortex (Changeux, 1983/1997). Neural 'nuclei' in the brainstem are considered to control pathways into the cortical areas that undertake the 'higher-level' analysis of information.

When concentrating, we may become oblivious to what is around us (Csikszentmihalyi, 1988), and even when we are aware of our surroundings, we are usually only aware of a fraction of the available sensory information. I had not noticed the sound of my typing, until I reached the point above where I wished to give an example of what I was sensing. We might notice a change in engine tone where we had not been consciously aware of an engine, or spot movement in a hedge that we had not previously noticed was in our field of vision. So there is a (considerable) degree of filtering of sensory input, and what we perceive is only a part of what we sense.

This relates to the body's internal sensory monitoring as well. Propriosensory nerves provide information to the brain on the position of the parts of the body, but we are usually only aware of this input when involved in coordinating movement requiring particular attention. In terms of the earlier discussion of the relationship between consciousness and brain function, an interpretation is that most sensory information can be effectively processed within the brain without reaching the 'higher levels' of processing that are associated with consciousness. Indeed, Changeux (1983/1997) reports that a baby born without a cortex and so presumably with no conscious awareness will show many expected behaviours such as to sleep and wake, to suck and to cry.

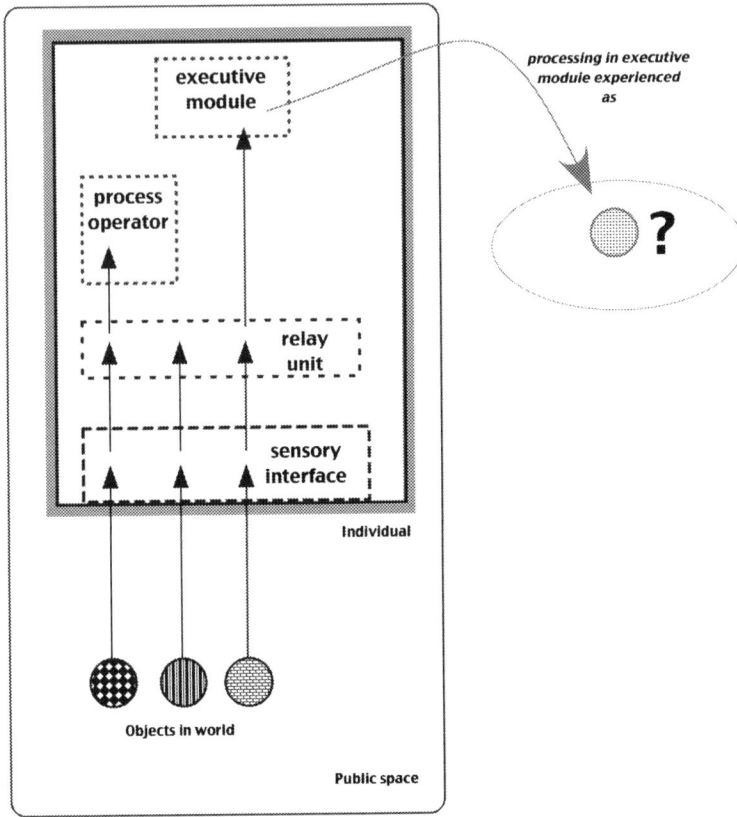

Fig. 4.3 Sensory information is filtered and is channelled selectively to different parts of the system

This leads to a slightly more sophisticated model (see Fig. 4.3) of what occurs as a result of sensory input than that presented earlier in the chapter. Sensory input is filtered within a part of the system, which acts as a kind of 'post-in' department. Some sensory information is directed to the (executive) processing module that leads to conscious awareness, and some will be directed to other parts of the system where the information will be processed and possibly acted upon through motor responses but without involvement of conscious thought.

Executive and Non-executive Processing Modules

In Fig. 4.3, the part of the cognitive system where processing is associated with consciousness has been labelled as the 'executive' module (Parkin, 1993). The term 'executive' is used in a number of contexts to refer to certain key cognitive functions and/or to the systems that undertake those functions in the system. The term will be met later in the book in the context of discussing thinking and memory.

To indicate that sensory signals as representations of sensory information may be processed and acted upon within the system without the involvement of this executive module, Fig. 4.3 includes a different processing unit that operates at a subconscious, automatic level. This has been labelled simply as a 'process operator' to indicate that it processes information without having executive function. The importance of preconscious processing will be discussed further below.

Sensory Information Is Patched-Up Before Being Presented to Consciousness: 'Filling-In'

The human sensory apparatus has various limitations due to its physical form. Some of these are obvious – we do not have eyes in the back of our heads, so our visual fields are limited but we are not really aware of the edges of the visual field. Questions about what we experience just outside the visual field (nothing, blackness, grey) do not usually occur to us and are akin, given the currently widely accepted model of a finite but unbounded universe that originated in a singularity, to questions of what there was before the formation of time or what lies outside our universe.

More interesting are features like the 'blind spot'. The mammalian eye has evolved with the retinal cell axons, which send the signals from the cell to the brain, on the 'outside' of the retinal layer – that is, in the vitreous humour – so that they have to pass through the retina. This happens for all the retinal receptor cells at the same area of the retina. So there is an area of each retina where the fibres from the receptors pass through the retina to form the optic nerve. Although not physically a large area, this is still a significant patch within the retina. Moreover, it is not near the periphery of the light sensitive area, but relatively central. So, on closing one eye, our visual field should have a 'hole' where nothing can be seen, although we do not notice any such gap. However, the blind spot can readily be demonstrated by locating a dot in the correct position so that it will become invisible when it falls within the blind spot of the eye's visual field because light reflected from the dot falls on the blind spot and not the surrounding receptors.

Our subjective experience (i.e. at the mental level) is of a continuous view of the scene, where actually there is a part of the scene that we are not seeing – where there is a dot that is now invisible. This is due to a feature of the cognitive system known as filling-in. The term can be explained by a simple analogy. Consider a science educator who was part of a team researching learners' understanding of the phases of the moon and who was given by another member of the team a set of scanned images of drawings collected from students participating as informants in the research. Consider that one of the images looked like the left-hand part of Fig. 4.4.

The researcher might well interpret the large shape at the top of the image, labelled as 's' as representing the sun (with 'm' as the moon and 'e' as the earth). The researcher might also interpret the shapes as meant to be circles that act as two-dimensional representations of the approximately spherical bodies being represented. However, the representation of the sun is slightly odd – it shows a round body but with a big

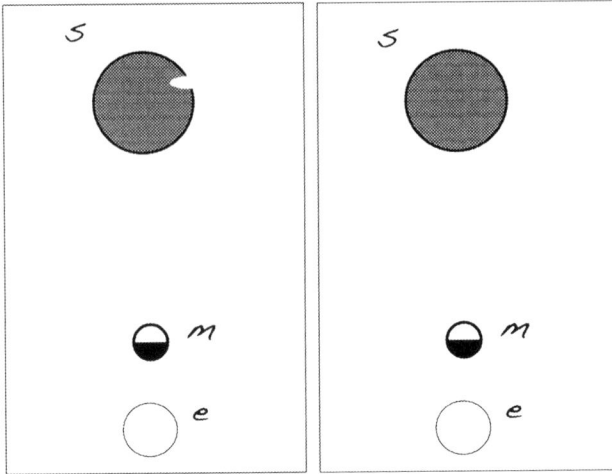

Fig. 4.4 Filling-in a copy of a student's diagram with an apparent flaw

dip, as if a part has been removed. Perhaps the student thinks that the sun is almost round, but with a chunk missing – but that seems unlikely.

The reader will probably agree that it is not likely that the student thinks the sun has this deformed shape – but it is worth noting that this is a judgement made totally interpedently of the data, or any other information about the individual student, and so is based on background knowledge deriving from previous sources that we judge should apply to this student's ideas. Clearly if we are interested in understanding the ideas of individuals, there will always be a danger of interpreting them in terms of what we have learnt from *other* individuals. However, having noted the potential for misinterpretation, and as this is a hypothetical case, we will allow the assumption.

The researcher might surmise that this is a flaw in the representation – perhaps during the scanning process something on the scanner glass, or stuck to the original drawing, obscured part of the image. The researcher may therefore decide that the image needs to be amended so it is a better representation of the original and uses a graphics programme to modify the image to give the right-hand version. If the amended figure is included in a research report, the authors might omit to mention that the image had been 'touched up' to correct an apparent flaw. If you were to read the report, then as far as you are aware, the published figure (the right-hand version) is a faithful representation of the student's original drawing.

Now this is simply meant as an analogy, because something similar happens when the cognitive system fills in the blind spot. The sensory interface (the retina) sends signals relating to the stimulus (light falling on the retina) with no information coming from the blind spot where there are no receptor cells. A separate part of the cognitive system processes ('analyses') the signals and in effect decides there will not be a hole in the external world being represented, so it fills it in by extrapolating from the available information. In the part of the cortex which processes visual signals, there is 'a patch of cortex corresponding to each eye's blind spot that receives input from the other eye as well as from the region surrounding the blind spot in the same eye' (Ramachandran, 2003, p. xv).

That processed signal is then sent to other parts of the cognitive system, so that the component of the system that gives rise to conscious experience, where vision is experienced, is not aware that the information received has been amended to correct an apparent flaw in the original data from the retina – just as the reader of the hypothetical research report was not told that the representation of the student's drawing has been altered.

Most of the time this works well, as we have two eyes, and even when we keep one eye closed, the part of our surroundings that is not registered because of the blind spot would usually appear as a continuation of adjacent areas – so filling-in can produce an accurate rendition. However, the demonstration with the disappearing spot shows that when the part of the scene missed by the blind spot is not continuous with its surroundings, the subconscious processing of the sensory signal interpolates inappropriately. An analogy here is with the student taking readings of water being heated in beaker and extrapolating the temperature series 14, 34, 56, 75 and 94 °C by assuming that the next reading will be approximately 115 °C.

Under these conditions, the processed signal gives false information about part of our surroundings, reporting what is not there and missing what is:

> What I mean by *filling-in* is simply this: that one quite literally sees visual stimuli (e.g., patterns or colours) as arising from a region of the visual field where there is actually no visual input. … In neural terms, this means that a set of neurons is being activated in such a way that a visual stimulus is perceived as arising from a location in the visual field where there is, in fact, no visual stimulus. (Ramachandran, 2003, p. xv, italics in original)

Filling-in has presumably evolved, and been retained, as part of the way our cognitive systems function because it generally works – most of the time it corrects a defect in sensory input and supports accurate perception. However, clearly there are conditions under which it can mislead. In my analogy of the student drawing, the patching-up of the apparently faulty image would seem an appropriate thing for the researcher to do – but it was based on assumptions. Perhaps had the researcher asked the student about this before amending the diagram, it might have transpired that the unusual representation of the sun was not an artefact of copying but part of the deliberate design. Perhaps the student was meaning to show how a cloud can obscure part of the face of the sun: in which case the filling-in process could have denied the reader some interesting information about the student's ideas.

Perception Is the Outcome of Active Processing of Sensory Information

Furthermore, we are usually aware of objects and events, when what we *sense* are patches of colour, tones and so forth. This is what is meant by 'perception': 'the process of recognizing or identifying something; Usually employed of sense perception, when the thing which we recognize or identify is the object affecting a sense organ'

(Drever & Wallerstein, 1964, p. 206). Roth suggests that 'the term *perception* refers to the means by which information acquired from the environment via the sense organs is transformed into experiences of objects, events, sounds, tastes etc' (Roth, 1986, p. 81). A key term here is 'transformed'. This again is something we tend to take for granted. The percept, the 'mental product of the act of perceiving' (Drever & Wallerstein, p. 206), is a conscious image of a bus or building or tree, etc., but of course our perception is not of an internal representation but rather that such objects are present in our surroundings.

By definition what is 'in' the nervous system when we perceive objects in our surroundings cannot be those objects – we do not have buses and building and trees in our central nervous systems, only representations of them, coded into electrical patterns of activation. This is obvious but reminds us that we have learnt to 'see' buses and building and trees and so forth without any independent standard of what they are 'meant' to look like. In terms of the system level of description (see Chap. 3, Table 3.4), stimuli (such as light reflecting from an object) are detected at the sensory interface (the retina in the case of light) and coded into electrical activity. The electrical representations of the stimuli are subject to processing (i.e. in circuits in the brain) in components that interpret the signals as indicating the presence of certain familiar features of our surroundings, and then the outputs of that level of processing are further signals that are sent to the 'higher' processing centre that is accessible to consciousness. Deese (1963, p. 400) has commented that 'the world presented through our senses is a vast jumbled confusion of different sensations. We are able to deal with it only by cutting it down to the size of our mental processes'.

There are then there at least three system components involved in perception: the sensory interface, one or more subconscious processing components which analyse sensory signals and interpret them and the higher-level 'executive' component which receives the interpretations (see Fig. 4.5).

Perception May Involve Over-Interpretation

There is then a considerable level of interpretation of sensory information prior to that information being accessible to consciousness, or, put another way, 'it is well known that making sense of our perceptual inputs is an 'ill-posed' problem and much 'computation' must be done to produce veridical solutions' (Crick & Koch, 1990, p. 272). We are often most aware of this when that interpretation proves to be wrong: when on investigation something we see or hear or feel turns out *not* to be present. The whisper transpires to just be the wind; the person moving in the shadows is just a 'trick' of the light; the man in the moon is just a random pattern of craters illuminated by the sun. We recognise a friend, only to then realise we are waving to a stranger. So percepts can be mis*interpretations* of the external world.

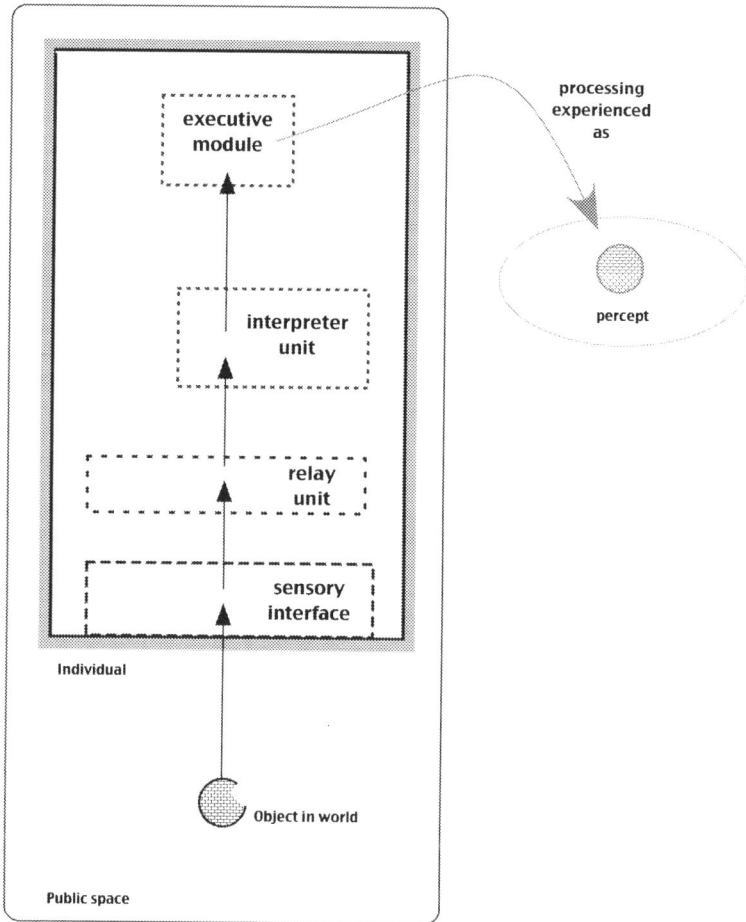

Fig. 4.5 Perception involves interpretation of sensory signals

Innate Bias in Perception

The human cognitive systems also seem to have innate, that is, built-in, biases. That is, we seem to be born with a tendency to interpret sensory information in particular ways. One example of this is ease at which we interpret patterns as faces. Recognising and identifying faces is clearly a very important part of operating as a social being, and when this aspect of cognition goes wrong, the deficit can make life difficult (Sacks, 1986). Babies seem to be able to recognise faces very early in life, and it is possible to produce quite simple iconic presentations of the face that are readily recognised as such: the emoticon, for example, ☺, originally typed simply as three punctuation marks:

:-)

This seems to be achieved due to a bias in the cognitive system to interpret quite minimal cues as faces, and the by-product of this is that people have seen faces in

Fig. 4.6 Sometimes subconscious levels of processing present us with interpretations we know cannot be correct (Image first published in Taber & García Franco, 2009)

the moon or in images of the surface of Mars, in butter melting on toast, in geological formations and so forth. Figure 4.6 is a reproduction of a photograph I took at the English coast. In the image, I can see a man's face looking down out of the clouds. I am perfectly aware that when I took the photograph, there was no giant man in the clouds and that such a thing would not be feasible. However, even though I *know* that the impression is produced by the over-interpretation of arbitrary features of the cloud formation, I still *see* the face.

Perhaps when you first looked at the image printed here, you could not make out any face? If so, and you persevered, I suspect you too (re)cognised the face, even though you also *know* there is no face there.

Learnt Bias in Perception

Moreover, once you have recognised the face, once it has become a percept, it is difficult to avoid 'seeing', no matter how convinced you are that it is merely an artefact of the imaging process. Even when our cognitive systems are not innately biased, our experiences certainly bias their future operation. It has been suggested that 'there is no such thing as an entirely new experience or memory. All that is apparently new to us happens in the context of old and well-established memory' (Fuster, 1995, p. 4).

The part of the cognitive system that consciously processes the perception does not seem to have a means of providing feedback to the subconscious component that interprets the sensory information to modify its analysis. The face will continue to be perceived regardless of our executive judgement that it is not present in the environment.

Key points to note from this discussion so far are that:

1. Attention to sensory information is selective.
2. Our senses only provide a representation of what is in the external world.
3. Perception involves subconscious analysis and interpretation.

Perceiving Communication from Others

The discussion above considers an object or event in the environment that is being sensed and perceived and provides a source for the conscious ideas of the learner. Human beings have developed communication systems based on signals and signs. These signals and signs are an important part of the individual's environment and are an important source of their sensory input. In the context of research in science education, the importance of thinking about student learning in terms of human social interaction has been increasingly recognised in recent years (Roth & Tobin, 2006; Scott, 1998; Solomon, 1993). This would include listening to the teacher, reading textbooks and discussing schoolwork with peers.

In an important sense, perceiving communication from others can be considered to be subsumed under the discussion above – Fig. 4.5 can still offer a general scheme to cover these cases. However, there are features of these situations that mark them out as particular. For one thing, language appears to be supported by areas of the brain that have evolved to specialise in language processing.

Moreover, communication is of special interest because it involves leaving traces in the environment that are intended to *represent particular meanings*. Perception of communication therefore raises the question of *to what extent the communicatee is able to reconstruct the intended meanings*. This issue will be explored later in the book, when considering the nature of understanding (Chap. 6).

Paying Attention: Distinguishing Subliminal and Preconscious Processing

It has been proposed that there are two distinct criteria for whether sensory signals, the electrical impulses from the 'sensory interface', result in conscious awareness (Kouider & Dehaene, 2007). One concerns the strength of the signal – that below a

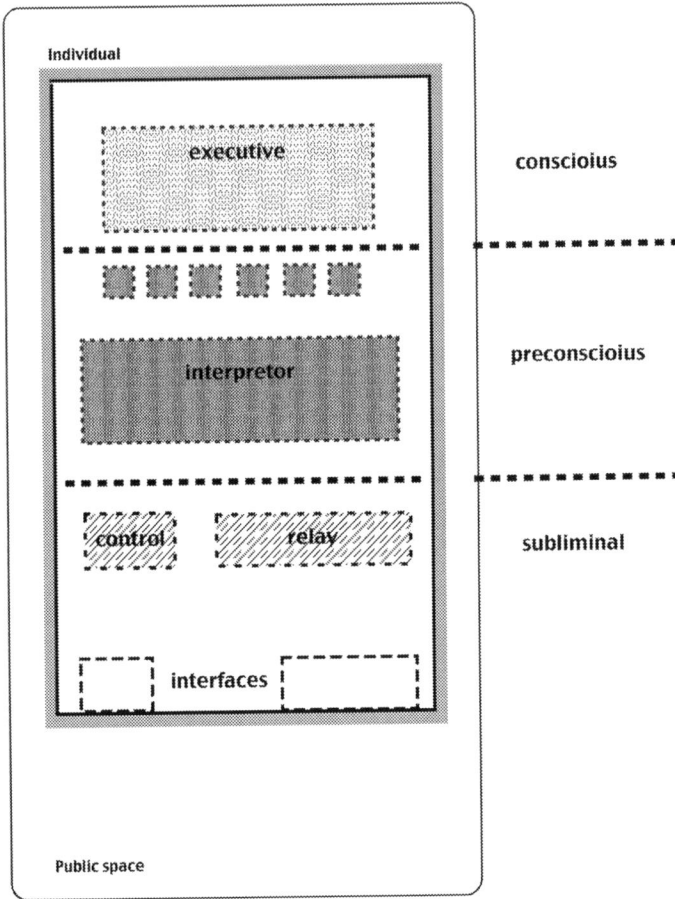

Fig. 4.7 Different levels of processing in the cognitive system

certain threshold a signal will not lead to processing at high enough levels to lead
to conscious awareness. It has been suggested that this type of subconscious
processing be referred to as *subliminal*. Subliminal processes may be attended to by
the system, but not consciously (see Fig. 4.7).

However, signals that meet the threshold *do not necessarily* lead to conscious
awareness. Rather, Kouider and Dehaene suggest such signals are stored in buffers
available to the executive processing module, where they may be accessed if attended
to, but that they may not be accessed if attention is elsewhere, in which case they
will be lost when the buffer is effectively refreshed with a new signal. Kouider and
Dehaene suggest this type of processing should be referred to as *preconscious*:

> Preconscious processing occurs when processing is limited by top-down access rather than
> bottom-up strength. According to the theory, preconscious processes potentially carry
> enough activation for conscious access, but are temporarily buffered in a non-conscious
> store owing to a lack of top-down attentional amplification (for instance owing to transient
> occupancy of the central workplace system). (Kouider & Dehaene, 2007, p. 176)

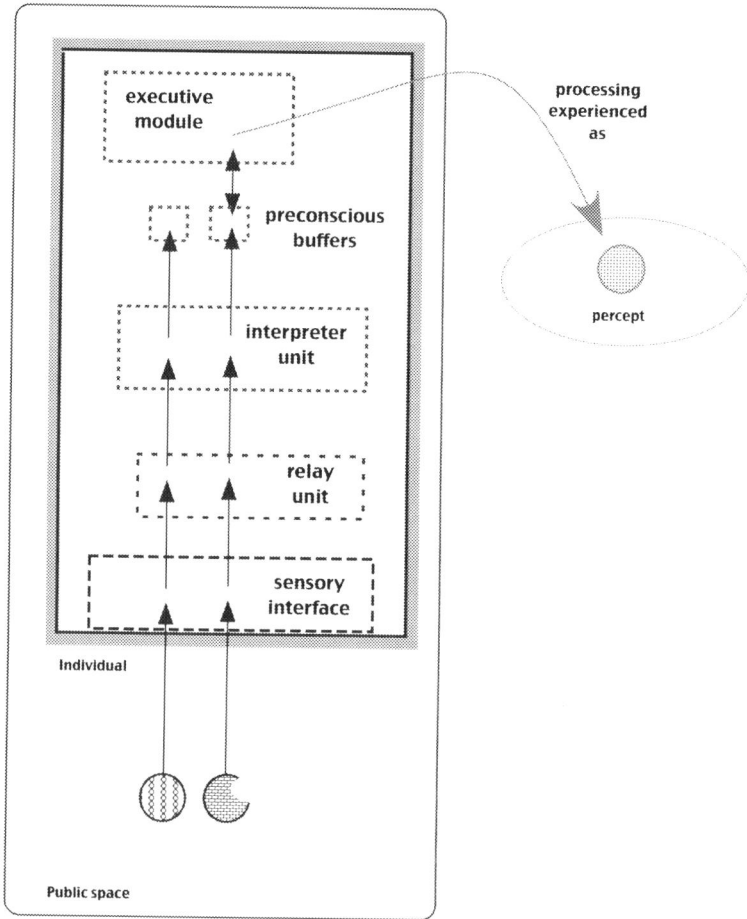

Fig. 4.8 The executive can select preconscious signals for conscious attention

So a signal that would be strong enough to get our conscious attention in some situations may not be noticed consciously at other times. We may not notice the pain from an accident in the immediate aftermath as our focus is on responding to the crisis. We may not notice the smell of food burning in the kitchen, if we are absorbed in reading an engaging book.

This is a potentially important distinction as it shows that there are both (1) cognitive processes that are not accessible to consciousness and (2) others that are potentially accessible but require the executive to in effect 'call' for the data (Fig. 4.8). Part of the role of the 'executive' module is to direct attention and thus moderate awareness of sensory information:

> Thus neural activity throughout the brain that is generated by input from the outside world may be differentially enhanced or suppressed, presumably from top-down signals emanating from integrative brain regions .... (D'Esposito, 2007, p. 17)

Recalling Experiences

As sensory information is filtered, interpreted and tidied up before reaching the executive levels of processing and so consciousness, there is a sense in which ideas based on current perceptions not only derive from current information from the external world but also draw upon features of the processing systems operating subliminally. To the extent that some of these features have themselves been shaped by previous experience, they can be considered to in a sense be *a form of memory*, albeit an automatic type of memory that operates without our awareness.

However, it is also the case that we have access to more explicit forms of memory. We are able to think about previous experiences and so have *current* mental experiences that do not directly relate to *current* perceptions of the external world. Such recollections can be triggered by something we currently perceive. We can also access memory without reference to current perceptions. So just as perception can act as a source of ideas, so can memory. The concept of memory is highly significant for a number of the key ideas in this book such as personal knowledge and learning and so is considered in more detail in the next chapter (Chap. 5).

Imagining Possibilities

It is also our experience that our thoughts are more than just perceptions and memories: as well as what seems to be now, and what we recollect was before, we can also think about the impossible, or the feared or wished-for, 'as of one who in a vision sees what is to be, but is not' (as Longfellow expresses it in his poem 'The Song of Hiawatha'). So there can be a *creative* aspect to thinking, and ideas need not be direct reflections of our experiences.

Imagination is a human faculty and (just as perception and remembering) represents a processing facility of the human brain. Imagination works with/on existing resources: our perceptions of the world and our memories of previous experiences.

> Philosophers have traditionally distinguished between 'simple' ideas and 'complex' ideas. Simple ideas were supposed to be directly derived from sensation. When combined, they can produce complex and abstract ideas far removed from sensory experience and expressed in shared language. Simple ideas are the 'atoms' of associationist accounts of the mind. (Gregory, 1987, p. 337)

Our perceptions are representations – of how the world seems to be – in the mind. Our memories are recollections of how aspects of the world previously seemed to be – so also representations in the mind. Although experience may seem holistic, it is based on a combination of both analytical and synthetic processes: analytical, as the sensory information detected by the eyes, ears, etc. are analysed and presented to the mind as the sound of an oboe, a moving car, the taste of spring onion, etc., not the patches of colour, coincidence of tones, etc. actually triggering the sensory cells; and synthetic, because we have a 'multimedia' experience of the world, visual and auditory information, for example, enter the brain by different 'channels' but lead to a unitary experience of a world where the sights, sounds, smells, etc. are associated.

Indeed, for rare individuals, synaesthesics, there are strong associations between sensory impressions that go far beyond what most people experience (e.g. particular numbers being associated with certain colours or sounds that evoke tastes) and which would seem to be suboptimal in forming mental models of the world because the association can seem to non-synaesthetes as arbitrary in relation to the objects stimulating the perceptions.

This blending of different kinds of input into a coherent perception of our surroundings clearly involves a lot of processing and suggests that the human nervous system has quite sophisticated apparatus for processing sensory information and constructing a multimedia impression of our immediate environment. Our ability to imagine things we have not experienced and even could not experience (because they do not reflect actual states of the external world and may even be physically impossible, i.e. they may not even reflect possible states of the external world) suggests that either the same or at least comparable processing apparatus is able to use the resources from which we construct our experience of the here and now and build alternative possibilities. The current view is that it seems likely that apparatus that evolved primarily to support perception has been recruited to allow imaginative thinking as well. This of course is highly relevant to science and science education, when considering the nature of thought experiments, for example (Brown, 1991). The processes by which imagination can produce novel ideas will be considered in more depth later in the book in the chapter on the learner's thinking (Chap. 7).

Expressing Ideas

As the Oxford Companion to Philosophy points out, ideas 'are entities that only exist as contents of some mind' (Brown, 1995). That in no way suggests they are not important: in a very real sense, humans are just 'such stuff as dreams are made on' (as Shakespeare's Prospero suggests in *The Tempest*). However, it is highly significant for research exploring student thinking that *the ideas learners have are only directly available to them and have to be inferred indirectly by researchers.*

Representing Ideas

If our ideas are purely personal, subjective experiences, then we cannot show them to anyone else. Rather we have to represent them in a public space if we wish to 'share' them with others. This representation can take a number of forms: we can express our ideas in prose or poems; we can draw out our ideas, act our ideas out or make films, compose music, compile photoessays or dance them out. The key point is that, whatever form representation takes, the representation is *different from the original idea – the thought – and so *the representation can never be a replica of the idea*.

There is symmetry here with the situation described in terms of perception (above). The representations of the external world in the nervous system are necessarily different in form to the information reaching the senses: equally, the representations we make in the external world *of* our ideas are necessarily different in form to those ideas.

I have referred elsewhere to Popper's use of the notion of three worlds: the physical world, the world of subjective experience and the world of ideas in the abstract, or Worlds 1, 2 and 3 (Popper, 1979; Taber, 2009b). These 'Worlds' contain fundamentally different types of things, and we can never *move* something from one of these worlds to another. So the abstract notion of a Platonic Form such as an ideal sphere only exists in World 3. A physical (World 1) object such as the sun or a ball bearing may be considered to approximate it, but even if it is considered 'spherical', it cannot be the same as the abstract notion of a sphere.

The human mind is capable of thinking about both actual objects that may be designated as spherical and about the geometrical form of the sphere in abstract, but in both cases these are mental representations (World 2 objects), so 'the image of a Platonic Form that occurs in a person's mind would be an idea in our sense' (Brown, 1995, p. 389). This might be seen as a parallel to the distinction made in discussing language with metalanguage (e.g. between the word 'banana' and a banana): a word is not what it represents, and a sentence is not its meaning.

In this book I have not emphasised the notion of the different worlds but rather have stressed the assumption that conscious experience such as 'having ideas' is associated at the cognitive systems level with processing of information, which in turn is based on the properties of the physical (electrochemical) structure of the human brain. In the terms set out in the previous chapter, an account of the conscious experience such as 'having ideas' is a description at the mental level; an account of the processing of information is a description at the computational or system level; and an account in terms of the electrochemical structure of the human brain would be a description at the physical level. However, there is a close parallel here with the three Worlds formalism, with the mental description concerning World 2, the systems description representing an idealised account in World 3 and the physical description being about World 1.

The mental experiences we have are clearly different in form to the electrical patterns of activation in the brain that are associated with them. Although those patterns 'translate' (give rise) to ideas in terms of conscious experience, they are actually in the form of a 'code'. That code also has to be translated into behaviour to represent what is coded in the external world, in the public space available to others. The representation formed in the external world is thus at least two stages removed from the idea experienced.

This is represented in Fig. 4.9, where the left-hand image offers a dualist interpretation (see Chap. 3), where ideas are translated into brain processing which leads to actions that can be examined by others in the external word (e.g. writing,

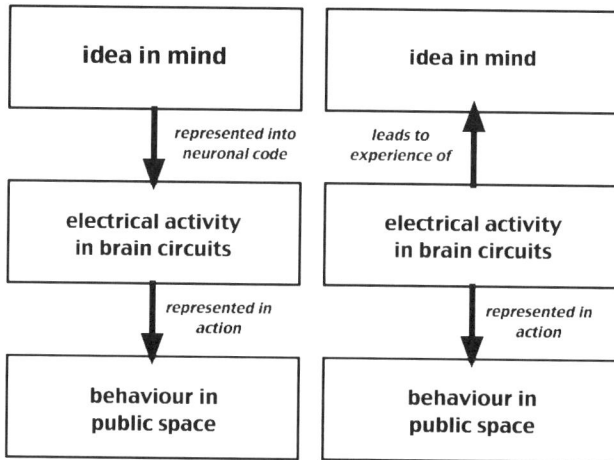

Fig. 4.9 Representations available to public scrutiny are indirect indicators of ideas

speaking). In this volume it is considered more appropriate to consider mental notions such as ideas as being an emergent property of the brain. In this conceptualisation (see the right-hand image in Fig. 4.9) electrical activity in the brain is considered as the basis for mental experience: nonetheless, ideas are still two representational stages away from behaviour that reports those ideas.

I stress this point as it is a fundamental assumption underpinning this book that because ideas can only be expressed by being represented in some other form than 'thought', *we cannot assume that representations of ideas have true fidelity to the ideas being represented.* Indeed, in principle, the representations cannot be 'the same as' what is being represented. This is indicated in Fig. 4.10, where the representation is in the external world and takes a physical form such as sound, inscriptions, bodily movement, etc., quite different from the nature of thought itself.

Figure 4.10 shows that the processing of information in the 'executive' processing module that correlates with consciousness (having an idea) can lead to action in the external world to represent our thinking. Signals from the executive processor can activate areas of the brain concerned with the voluntary control of movement (the motor cortex), which then sends signals to the muscles to bring about speech, writing, gesturing, etc. There is in effect a 'motor interface' to represent electrical activity into action in the external world, just as there is a sensory interface to represent stimuli into electrical activity. As with the sensory interface (which includes the retinas, the mechanism of the ear, the olfactory membranes, etc.), the motor interface is not physically a single entity, but rather consists of various spatially distributed components. There are many places in the body where motor nerves stimulate different muscles.

So in representing an idea, the motor system is used to produce talk or to write, etc., and so the brain needs to bring about coordinated physical activity. This is something we generally take for granted, but clearly for some individuals, especially the

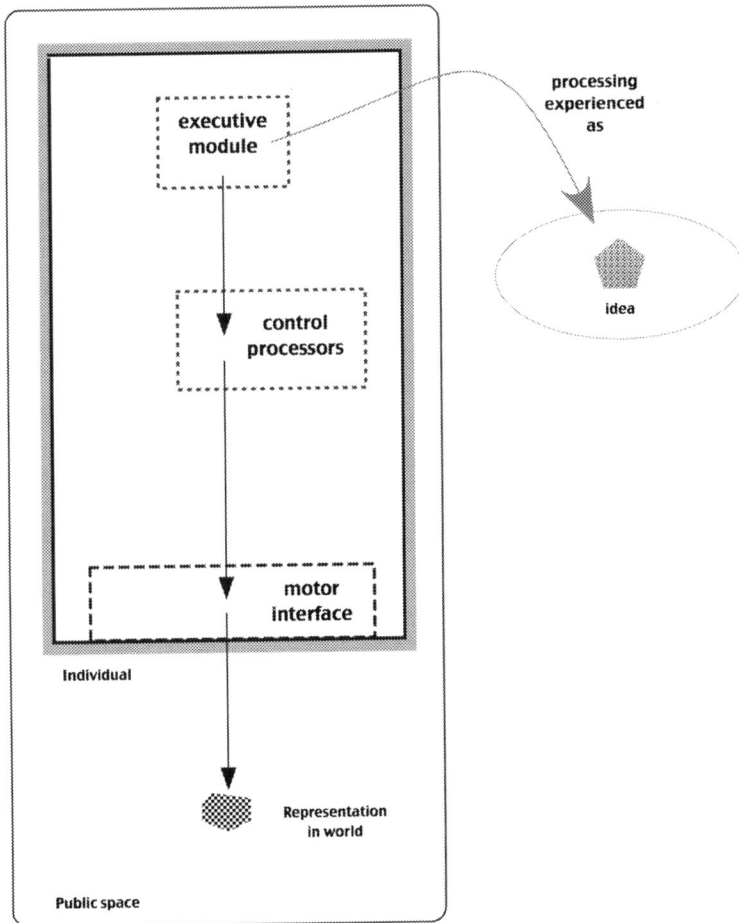

Fig. 4.10 Representing ideas in the external world

young or those with particular disorders, producing coherent speech or readable handwriting is non-trivial. For many of us, expressing our ideas in some forms may be challenging. If I was asked to dance or paint a representation of electromagnetic induction or photosynthesis, then even assuming I was able to conceptualise how I wished to express my idea of the concept in that form, I would not be confident that I could execute the representation as I would intend. I might feel more confident in using speech or writing, but that may not apply to all young people we interrogate in our research. Clearly this is important in research if we assume that the representations that people make of their ideas are a good guide to their ideas. This is often recognised when the literacy skills of young children may limit their ability to read research instructions/questions or produce written answers, but in principle this is always an issue that limits what we can achieve in research into students' ideas and so their understanding and their learning.

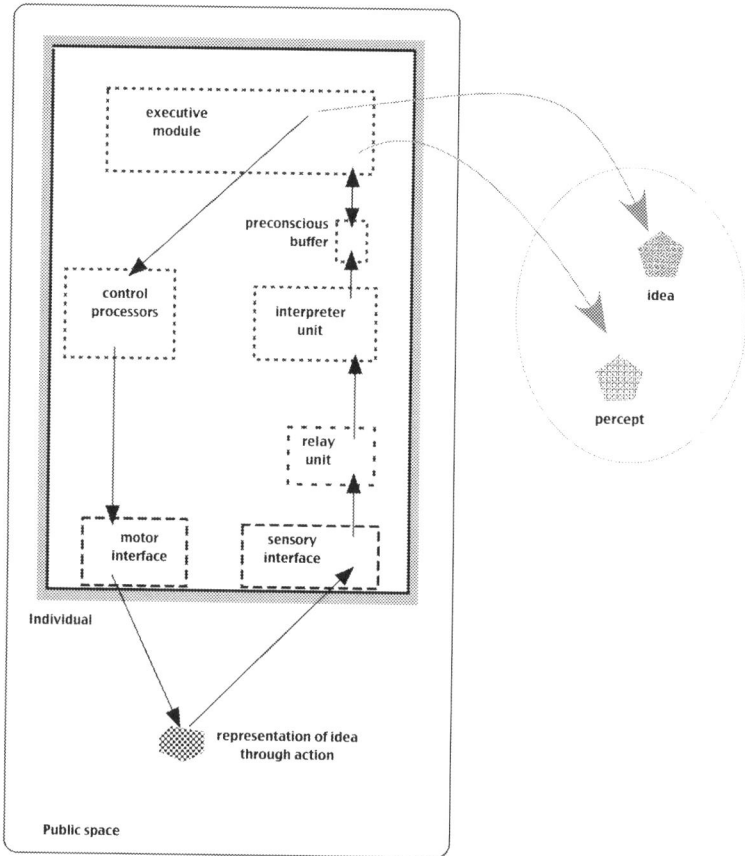

Fig. 4.11 Feedback to check on representation made in external world: does the percept of our representation match the idea we were trying to represent?

To some extent individuals can check upon the representations they make in the world, at least to the extent that they can monitor them through their own senses, that is, perception produced through processing in the cognitive system as outlined earlier. For example, we can monitor our own speech productions, and if we detect mistakes make a corrected utterance. However, 'slips of the tongue' may go unnoticed, and of course the mechanism by which we obtain sensory information about our own actions in the world is subject to the same processes of representation, filtering, interpreting and tidying that accompany any other perceptions, with the added complication that we have strong expectations of how our representations will be enacted in the world. To represent this, in Fig. 4.11, the percept is a limited representation of the enacted representation of the original idea, in part influenced by knowledge of what was being represented: this is the very knowledge which is absent when interpreting the public representations made by others. More permanent representations – writing, drawings, etc., may be easier to check in this way than more ephemeral actions such as gestures and speech that do not leave a permanent trace in the environment. But even here, the issues of interpretation are relevant. A person who produces an inscription that offers several readings because it can be readily understood in several ways is likely on reading back their own production (i.e. public representation) to interpret the writing according to the intended meaning and may well not detect any ambiguity.

Accessing Another's Ideas

> The stubborn fact is that conscious experience or awareness is directly accessible only to the subject having that experience, and not to an external observer. (Libet, 1996, p. 97)

Ideas, subjective experiences of products of our own thinking, cannot then be made directly available to anyone else, but rather can only be represented (e.g. described) in the public space of the external world. Reporting an idea is necessarily to produce something of a different form than an idea. The key implication here for our present purposes is that researchers only have direct knowledge of the ideas produced by their own thinking. They can only infer the ideas of others indirectly – an indirect process that is informed by the researcher's own implicit or explicit model(s) of what is involved.

This is shown in Fig. 4.12, where one individual, Jean, has an idea which he/she wishes to 'share' with another person, Lev. Jean can only do this by creating a representation of the idea in the public space they share. Lev can then form a percept of this representation, which can be used as the basis of attempting to reconstruct the idea. As a result of this, Lev produces an idea of what Jean is trying to communicate. Lev's idea is not Jean's idea, but an idea of what Jean's idea might be.

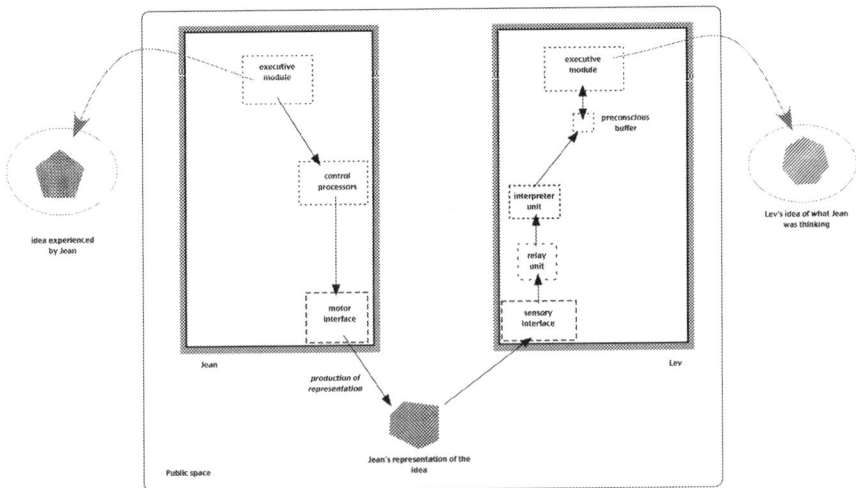

Fig. 4.12 'Sharing' ideas is an indirect process

Whether Lev's idea of Jean's idea could be considered similar enough to Jean's idea to be considered *in effect,* the 'same' idea, a shared idea, is a challenging question – how could we ever be sure? However, there is clearly an important sense in which they cannot be the same idea.

In terms of the notion of there being three worlds (Popper, 1979), then Jean's idea and Lev's idea are 'World 2' objects. However, in a sense it is misleading to talk this way, as they are not part of the *same* World 2. The public space in Fig. 4.10 represents the physical world, World 1, in which Jean and Lev exist as corporeal beings. The idea that Jean had, shown to the left of the public space occurs in Jean's subjective world (World 2J), and the idea that Lev forms, shown here to the right of the public space, occurs in his own subjective world (World 2L). There are, in effect, as many World 2s as there are creatures capable of subjective experience sharing the same physical World 1.

Research to Investigate Learners' Ideas in Science

The processes represented in Fig. 4.10 apply whenever one person interprets another's behaviour in order to understand their ideas, whether the person expressing the idea or the person attempting to understand it are students, teachers, researchers or simply two people having an everyday conversation in the workplace or shopping or at some leisure event.

However, in the case of researchers, there is likely to be at least one further step, as the researcher's purpose is not just to develop a personal understanding of the informant's ideas but also *to develop public knowledge*. (This in itself is a problematic notion, as will be explored in Chap. 10.) The researcher will therefore seek to report findings, potentially including accounts of informants' ideas. At a minimum, then, the research process will include:

- The informant expressing an idea by representing it in some form in the public space where the researcher can perceive the representation
- The researcher attempting to develop a personal understanding of what the idea being expressed was;
- and attempting to represent the researcher's own idea (of the informant's idea) – most commonly in the form of inscriptions in a research report submitted to a journal (Fig. 4.13)

If published, the 'public knowledge' is then represented in a paper, and of course readers then interpret, and develop their own ideas of, what is represented. We might say that readers form their own subjective understandings of the representation of the researcher's understanding of the original informant's idea, as mediated by the informant's public representation of that idea!

Sometimes research reports may include 'data' which more closely relates to the informant's original representation, for example, transcription of speech or handwriting, or a scanned copy of a diagram. Such data can sometimes offer close

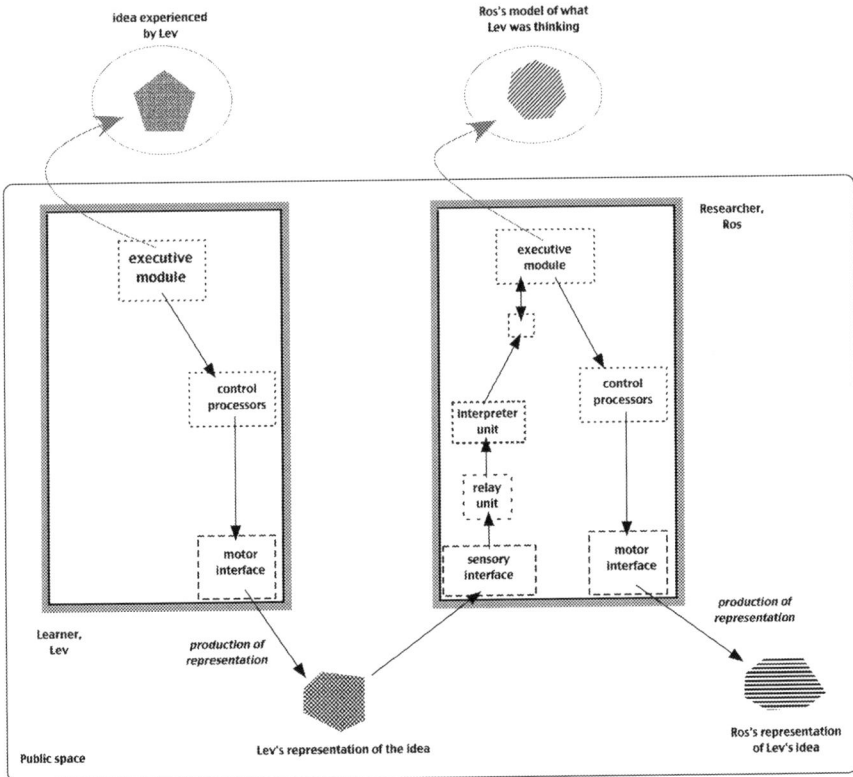

Fig. 4.13 Research reporting a learner's ideas offers accounts of second-hand interpretations of those ideas

facsimiles of the original representation although not of the informant's original idea, of course, allowing the reader to form their own alternative interpretation of the representation. However, conventions of, and constraints on, journals, and the practicalities of reading other people's research, mean that most published research reports at best present a few selected samples of the raw data upon which the researcher based their interpretations.

This process is represented in Fig. 4.14, and of course the chain becomes further extended if the original reader of the research is a teacher educator who reports the research to teachers in training, or a researcher who reports the research as part a literature review for another study or an author who discusses the research in a research review or a textbook.

Modelling Student Ideas

In this Chapter it has been argued that although we can reasonably assume that other people have ideas, much like us, we can never have direct access to anyone's ideas

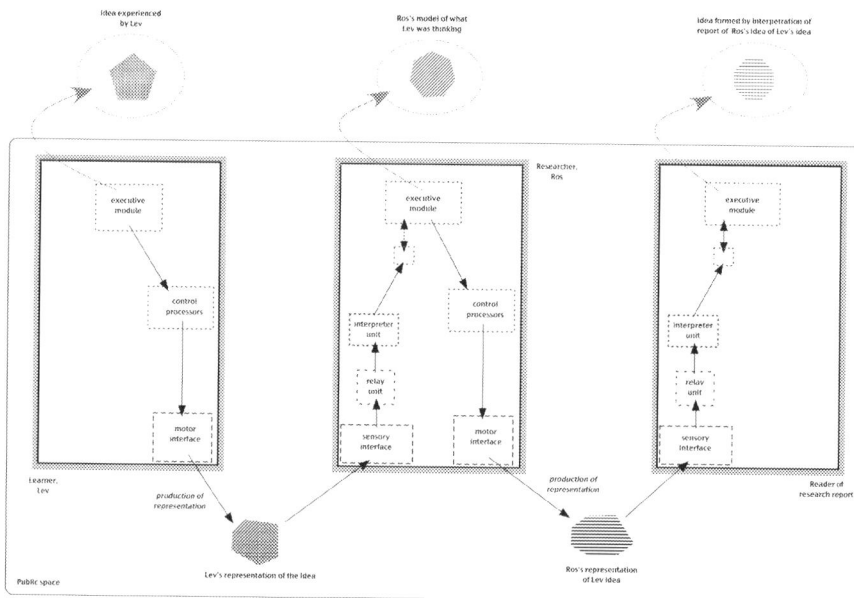

Fig. 4.14 A reader of a research report forms an idea of the learner's original idea based on a series of re-representations

but our own. It has been argued that although we can express our ideas by representing them in a public space in the external world in various forms, the representation is intrinsically very different from the idea itself and so cannot be assumed to be a perfect reflection of it. The ideas are correlates of electrical activity in the brain that can be reflected in various actions in the external world due to the body's 'motor interface'. There will be some skill involved in representing the idea through motor activity, including in speech production or writing.

Furthermore, the representation itself needs to be perceived by the person trying to understand the idea, and this involves the representation acting as a stimulus which is converted to electrical signals by the observer's 'sensory interface', then processed in the brain to be presented to consciousness as perceptions. Those perceptions may not directly lead the observer to a candidate hypothesis for what the original idea might have been, and so further conscious analysis, reflection and interpretation may be required. The process of communicating ideas is always an indirect one. This has been stressed in such detail because in our everyday lives we take such communication for granted much of the time. Arguably, human beings have evolved to communicate well enough for most everyday purposes, and in normal interactions it makes sense to largely take the processes of communication for granted rather than to analyse them in the manner undertaken in this chapter. Misunderstandings occur, but generally we feel we are able to understand what others try to tell us or at least are usually aware when we do not understand.

An argument made in this book however is that whilst research may draw upon our everyday communicative skills, it should be a technical activity where we do not take everyday processes for granted. Everyday transactions would likely be unreasonably retarded by analysing them in technical terms, as has been attempted in this chapter. In effect, in everyday life, we commonly admit the occasional flawed interpretation as the cost of quick and easy communication 'of ideas'.

Research, however, seeks to make knowledge claims that are robust and well evidenced, and a more deliberate consideration of the nature of the data and its interpretation is called for. The analysis offered here then provides an important background for thinking about the research process, as discussed in Parts III and IV of the book.

Chapter 5
The Learner's Memory

Memory is clearly a key component of learning – there would be no learning without some form of memory. Much research in science education that looks to explore students' knowledge and understanding is in effect probing aspects of the learner's memory. However, 'memory' is part of the mental register of lifeworld terms (see Chap. 2) that are generally used unproblematically in everyday communication because of the 'theory of mind' that we all acquire through normal development (see Chap. 2). That is, a person's everyday experience supports a view that I remember things and so must store those memories somewhere, and in everyday discourse I assume other people have similar subjective experiences of remembering as I do. As will be suggested in this chapter, research into memory suggests that our everyday ways of talking and thinking about memory may be inadequate for research purposes.

Whilst there have been few studies in science education which are explicitly framed as exploring memory as opposed to say conceptions, or thinking, nonetheless memory is clearly a taken-for-granted feature of a great many studies. A good deal of the research into aspects of students' ideas and thinking assumes that what is being probed is supported by some kind of stable knowledge base 'stored' in memory. Part III of this volume will consider the nature of a student's knowledge. It is useful, as preparation for that, to establish some notions about the nature of memory.

Memory 'Contents'

A number of terms are commonly used when discussing the types of memories people may have. The term *declarative* memory is as an overarching term for 'consciously accessible memories of fact-based information' (Walker & Stickgold, 2004, p. 121), whereas non-declarative memory includes both procedural memory (e.g. remembering 'how' to ride a bicycle or tie shoe laces) and so-called implicit learning that takes place without conscious awareness. (Learning is considered in Part IV of the book.)

K.S. Taber, *Modelling Learners and Learning in Science Education: Developing Representations of Concepts, Conceptual Structure and Conceptual Change to Inform Teaching and Research*, DOI 10.1007/978-94-007-7648-7_5,

Declarative memories are considered to fall into different classes. In particular, *episodic* memory relates to specific events, whereas *semantic* memory refers to memory for general information 'not tied to a specific event' (Walker & Stickgold, 2004, p. 122), that is, abstracted from the specific context of learning. Remembering one's graduation ceremony would be based on episodic memory, whilst making sense of the memory by recalling that graduation is a congregation for the award of degrees would be based on semantic memory.

Memory as a Source of Our Ideas

The previous chapter explored what we mean by ideas and therefore the nature of the focus of research that claims to explore students' ideas. It was suggested in the previous chapter that we might consider sensory information and memory as two sources of our ideas. To a first approximation, but one that we will revisit below, we may consider these two sources as distinct: sensory information (ignoring for the moment internal monitoring and regulation of the body) derives from the external world, whereas memory is an internal resource. Figure 5.1 represents this distinction using the kind of simple systems-level model used earlier in the book.

This distinction is useful, but a note of caution is important. In the previous chapter, it was described how perception involves interpretation. The apparatus that interprets sensory information is in part 'programmed' by genetics – which itself can be considered a kind of 'memory' representing the patterns of interpretation that have allowed earlier generations to operate in their environments – but partly tuned by the individual's experiences in the world. Therefore, 'memory' is involved in the subliminal processing that precedes perception.

The Physical, Functional and Mental Aspects of Memory

In Chap. 3, a three-level perspective on cognition was presented, suggesting that we can discuss aspects of learning, thinking, etc. at three different levels: the mental level of what we subjectively experience and how we commonly describe this experience (in terms of 'thinking', 'remembering', etc.); the functional systems level in terms of what the cognitive system does in processing information for us; and the physical level where cognition is considered (mechanistically) to be the outcome of processes in the nervous system that can be studied in terms of physiology and explained in terms of cells, synapses, neurotransmitters, etc.

It was suggested that although we assume a biological basis for the mind, we can often make more progress in understanding cognition by mapping subjective experience and everyday notions onto models at the functional systems level (see Table 3.4). At this level we consider the functions of different brain system components, rather than the specifics of location and physical structure.

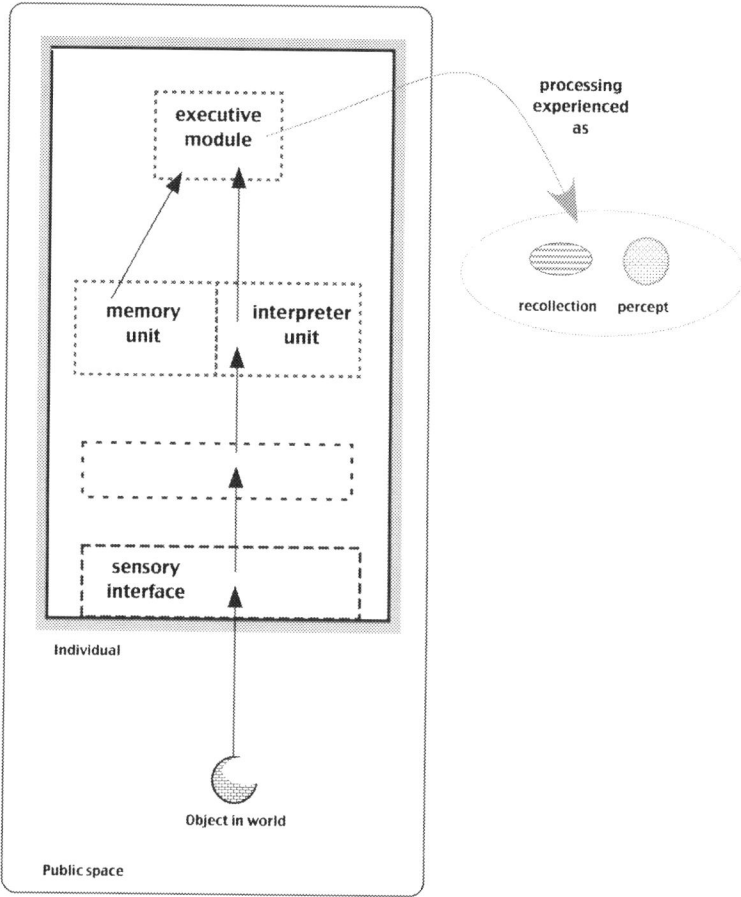

Fig. 5.1 Memory and perception as two sources of ideas

However, our assumption of a physical basis to mental life means that although we can develop models at the systems level, these should ultimately be consistent with what research into the brain has uncovered. A problem here though is that current knowledge in cognitive neuroscience does not yet enable us to specify the constraints that should apply to our functional levels (J. Driver, Haggard, & Shallice, 2007). Some of the discussion in this chapter that refers to brain function is therefore still quite provisional. A more secure basis for developing functional models is the evidence from studies of memory that can provide information on how memory functions and so set out the features a functional, systems level, model must explain.

What does seem clear, and is important for researchers making assumptions about the nature of the knowledge 'stored' in the learner's memory, is that everyday folk notions about memory may seriously misrepresent it. In everyday life we consider memory to be not only a major part of subjective experience but to be a kind of *store* of accounts of previous experience that we put away for safekeeping for future reference. Yet, studies in psychology have long suggested that memory does not (usually) work by storing high-fidelity copies of experience. Moreover, research into the brain, as discussed below, brings into question the notion of there even being a *discrete* memory store located 'somewhere' in the brain.

Our functional models, in terms of systems of components which process information in various ways, do not depend upon discrete physical (anatomically distinct) components being identified in the brain: functions can be distributed spatially without undermining the usefulness of labelling a box on a systems diagram. So, for example, in the figures in previous chapters, I have for the sake of simplicity represented distributed functions like the body's sensory and motor interfaces as each being a unit or module in the system, even though in both cases the interfaces are actually comprised of a multitude of different 'devices' spread around the body at the peripheral ends of sensory and motor nerves, respectively. Rather, the sensory interface, for example, can be considered to be a set of components with one basic type of function – representing stimuli in electrical signals. However, in the case of memory, the existence of a *functionally distinct* unit in the brain has itself been questioned, and this will influence the way the role of memory in the learner's cognition is best modelled at the systems level.

The Nature of the Learner's Memory Store

We are generally not aware of most of the previous experiences that we can sometimes access. So in that sense, although remembering is part of conscious thinking, the memory (the 'record' or 'store') is not. Memory traces can be considered at the systems level as resources that can be accessed by the executive processor that gives rise to consciousness.

At the systems level then, the cognitive system includes a component, in effect the association cortex, that is able to in some sense represent previous experience in a form that allows later access to those 'records' of prior experiences. We represent current experience in memory so that we can in some sense access (remember) it later. As subjective experience is assumed to arise from processing at a certain level in the cognitive system, this implies that in some sense the output of processing is 'stored', and can later be accessed, leading to a subjective experience (e.g. something remembered) which is to at least some degree similar to the original experience (e.g. a perception).

Of course, we have no independent way to know if our experiences of remembering previous experiences are true to the original experience – and so we tend to *assume* that when we experience particularly vivid and apparently realistic 'remembrances of things past', then these experiences are true to the original experience being evoked. However, it is also a part of everyday experience that it is not unusual for two people who think they clearly remember the same past event to actually produce contrary reports which might lead us to suspect that we cannot always rely upon what appear to be clear memories of the past. A popular musical song, 'I remember it well' by Lerner and Loewe, draws upon such an experience for comic effect.

As the physical basis of the processing that correlates with thinking at the mental level is considered to be electrical activity, then this would suggest that *either* the store must be some kind of maintenance of that particular pattern of electrical activity associated with that particular conscious experience or a representation of it in a different form. It appears that patterns of electrical activity are not themselves maintained in the brain permanently, but rather the activity level of different neural circuits shifts over time: a pattern of activity relating to an experience will die away, but may become active again at some later time.

Current thinking (at the physical, brain level) then is that the store does not consist of the original patterns of electrical activity, but rather modification to circuits of neurons which when activated can reproduce something like the original pattern. In other words (at the functional, system level) memory storage involves another process of coding of information in a new form (e.g. from electrical signals to changes in synapse configurations), and memory retrieval involves reading the original information back, decoding it (activation of the network of neurons).

So it would seem that when we remember an object or event we previously experienced, we are decoding a representation of the processing we experienced as perception, that was itself (see Chap. 4) the output from a different part of the cognitive system as a result of preconscious processing of a signal from the sensory interface, where an original stimulus had been coded into an electrical pattern.

Long-Term Memory

The system component that is considered to be responsible for this ability to access and apparently relive previous experiences is usually known as *long-term* memory (LTM), as information stored in LTM is potentially accessible for decades. Other components of the cognitive system are considered to be able to store information in the shorter term, for example, to support the 'high-level' processing (when we are conscious of thinking about something or working on a problem).

LTM includes the means to make representations of different kinds: of different modes (visual, abstract, etc.) and degrees of integration and complexity. It has been noted that 'long-term memory is replete with schemas, schemata, stories (experiences), procedures, behavioral sequences, patterns, and many other structures' (Jonassen, 2009, p. 18).

At one time LTM was considered to be complemented by another functional unit, widely known as short-term memory (STM), that was thought to be a discrete component of the system so that experiences were said to be stored in STM until they can be transferred into LTM. This idea is no longer widely accepted, and it has been argued, rather, that 'the storage of short-term memories is inextricable from the reactivation of the long-term memories that, by context or similarity, any new experience evokes' (Fuster, 1995, p. 4).

There are components of the cognitive system that store information over short periods, but with different functions to that previously assigned to STM. So what is usually called working memory (WM), and seen as an aspect of executive function, is widely accepted as an important component of the cognitive system (see below). WM relates to what is consciously being considered, rather than an intermediate memory store as STM had been understood. The buffers in which perceptual information is stored (see the previous chapter, including Fig. 4.8) can also be considered a form of short-term memory, although again this is not how STM was understood. These buffers only operate over *very* short time periods to buffer information during active processing, not – as STM was understood – over durations of minutes and hours as a kind of queue of material ready to be stored in LTM.

LTM is considered to be *in effect* of limitless capacity. Clearly if LTM is made up at the physical level of neural 'circuits', then its capacity must actually be finite, but it is generally thought that the capacity of LTM is sufficient to support its use throughout a human lifetime without becoming a limiting factor. Age may bring deficiencies in memory function, but these arise from other characteristics of the system, not a limit in capacity.

It is widely suggested that once a representation is formed in LTM, through the formation of a memory trace, it is not actively erased, although the trace might get degraded. Certainly memories that are not activated for long periods may become very difficult to access, but the trace itself may be present, even if not readily located and activated. This is suggested by experiments during surgery on conscious patients, which have shown that electrical stimulation of parts of the cortex can trigger vivid recollections of experiences that the patient claims not to have thought about for many years. As the brain does not have pain receptors, brain surgery does not require general unaesthetic. Further, as each brain has a somewhat idiosyncratic mapping of functions onto the cortex, surgeons may actually use patient feedback as a guide during surgery. Memory circuits have a threshold of activation, and in normal conditions some may receive insufficient stimulation for activation – opening the skull and directly applying an electrical potential with an electrode presumably provides a sufficient magnitude stimulus to activate an otherwise long-dormant circuit!

The key role of accessing representations during recall was, for example, noted by Wong in a study looking at how the use of self-generated analogies might support students' (in this case, student teachers) developing scientific explanations. Wong noted that

> When participants experienced difficulty explaining the phenomena, the problem was often one of access rather than availability of knowledge. For example, although two participants had learned a great deal about air pressure during their undergraduate and graduate training in the physical sciences, neither was able to access or apply appropriate elements of this knowledge in their initial explanations. For both of them, analogies triggered relevant knowledge during the task. (Wong, 1993, p. 1267)

The Cortical Basis of Memory

Perception and remembering are different functions carried out in the cognitive systems and so were shown in Fig. 5.1 as involving different modules. However, memory appears to be widely distributed in the cortex, and the basis of memory when we recall something may not be structurally distinct from the brain circuits that interpret experience to produce perceptions for us:

> Memory is a functional property, among others, of each and all areas of the cerebral cortex, and thus of all cortical systems. This cardinal cognitive function is inherent in the fabric of the entire cortex and cannot be ascribed exclusively to any of its parts. Furthermore, as the cortex engages in representing and acting on the world, memory in one form or another is an integral part of all its operations. (Fuster, 1995, p. 1)

Fuster argues that memory is 'global' and 'nonlocalizable' within the cerebral cortex, although individual memories can be 'more or less widely distributed over the cortical surface' (p. 1). According to Fuster, memory

> is made of myriad idiosyncratic associations between that common fund of specific sensory and motor memory, which already lies in primary areas at birth, and the experience that has accrued thereafter in areas of so-called cortex of association, which means practically anywhere else in the neocortex. (Fuster, 1995, p. 1)

Fuster offers the following basic notions about memory (p. 2):

- A memory is in effect a network of neocortical neurons and the connections linking them together.
- A memory is formed by the concurrent activation of ensembles of neutrons (which represent aspects of the internal and external environment and motor action).
- The networks are modifiable by further experience.
- Individual neuronal ensembles may be part of multiple networks and so a range of representations (in each of which the ensemble is part of a different network).

Furthermore, according to Fuster, 'the cortical substrate of memory, with its many potentially infinite representational networks, *is very nearly identical* to the connective cortical substrate for information processing, in perception as in action' (p. 2, *emphasis added*). This commonality between the substrate for 'storing' memories and for processing information (see Fig. 5.2) is potentially significant for how we think about memory, that is, although it may be helpful to think of memory as functionally different to information processing, this could be misleading.

So in this respect our cognitive systems are quite different from an electronic computer where information that is stored and accessed in a discrete memory unit such as a hard drive or a USB pen drive is only changed when it is deliberately deleted or overwritten. The storage devices in our computer are just storage devices, and information has to be copied from them into the computer's working memory for processing. In humans, however, a good deal of preconscious processing occurs in the 'store' itself, and potentially modifies what is stored, so memories accessed by consciousness cannot be considered to have been securely stored since they were initially laid down.

From a biological and, in particular, evolutionary perspective, the notion that the same physical components may develop several functions will not seem surprising. The evolutionary process seems to often involve existing structures being recruited

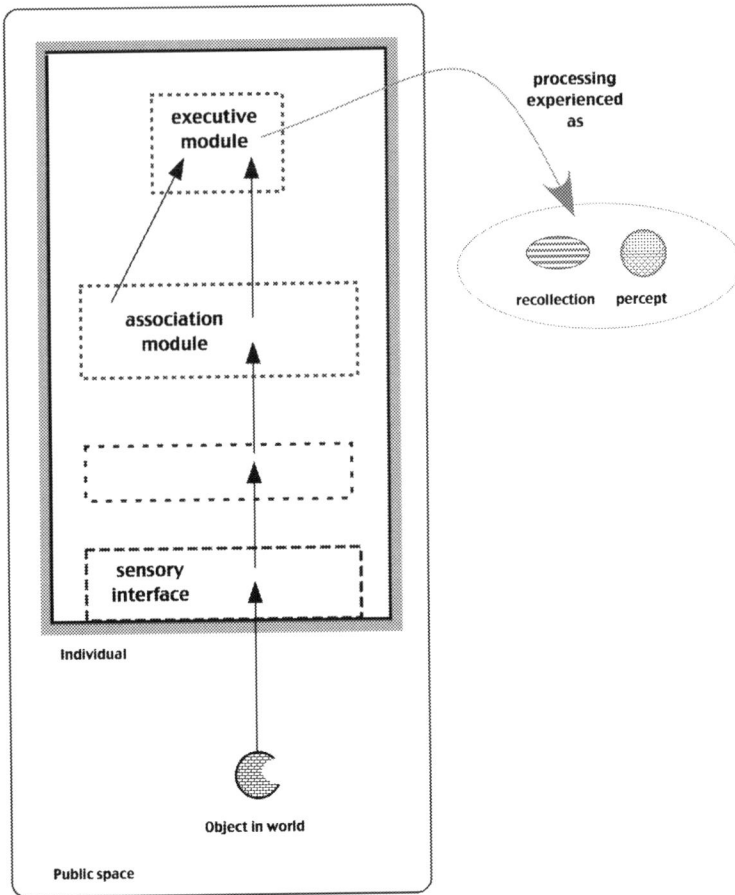

Fig. 5.2 Considering memory and perception to be different functions of the same brain structures

for new roles. However, this interpretation of the physiology may be quite significant for learning and remembering. Arguably:

- All of our perceptions are based upon the activation of combinations of existing neural networks, which in effect reflect memories of previous experience.
- The activation of neural networks during perception modifies neural connections and leaves those networks (which are in effect components of our memories) changed.

The former point is highly relevant to models of learning and in particular the basis of constructivist approaches to teaching and learning (see Chap. 15). If everything we perceive is based upon the activation of combinations of previous memories that have been 'judged' (in the cognitive system, through the activation of particular networks in the cortex) to give the best match to current sensory inputs, then there is a strong biological basis for constructivist pedagogy. Constructivist ideas about teaching emphasise the importance of working from the learner's current ways of understanding and thinking about topics (Taber, 2009b).

Similarly, if our memories are being modified all the time, then they cannot be considered to be accurate records of specific experiences, but rather should be seen more as some kind of democratic averaging of our interpretation of past experiences. This is an extreme suggestion of course, and it may seem to be over-stating the case. However, in principle, *we cannot assume memories are accurate representations of distinct, specific experiences*. This is something that will inform the discussion of students' recollections later in this chapter.

It would also seem that if, as Fuster reports, information processing uses the same neural networks that act as the basis for memory, then studies into student thinking cannot in principle be separated from studies into memory. That is, the idea that there is a 'content free' processing apparatus into which the researcher can feed input to explore thinking is false. This point will be important when we consider the nature of research into student thinking (see Chap. 7).

Active Memory and Focus of Representation

The networks in the association areas that provide the basis for remembering may be active without us being aware of memories. The pattern of 'active memory' in various part of the cortex at any one time has been described as a form of parallel processing, and most of this activity occurs without conscious experience (Fuster, 1995). In a sense then, it seems memory may be active without our awareness. This is something that would not surprise the adherent's of Freud's psychological theories. In contrast, a limited portion of this active memory that has been described as the 'focus of representation' and is considered to be activated by serial processing is linked to conscious experience (Fuster, 1995, p. 5). This is represented in Fig. 5.3.

Figure 5.3 offers two possibilities (metaphors) for how this might work. The left-hand figure is intended to imply that the 'focus of representation' involves transferring the representation from the active network and re-representing it in the executive module (e.g. within working memory, see below). However, another possibility is that the executive acts more like a metaphorical 'spotlight' highlighting the focus of representation, at which point it becomes conscious.

Memory Is Reconstructive

> One of the least controversial – but most important – observations is that memory is not perfect. Instead, memory is prone to various kinds of errors, illusions and distortions. (Schacter & Addis, 2007, p. 27)

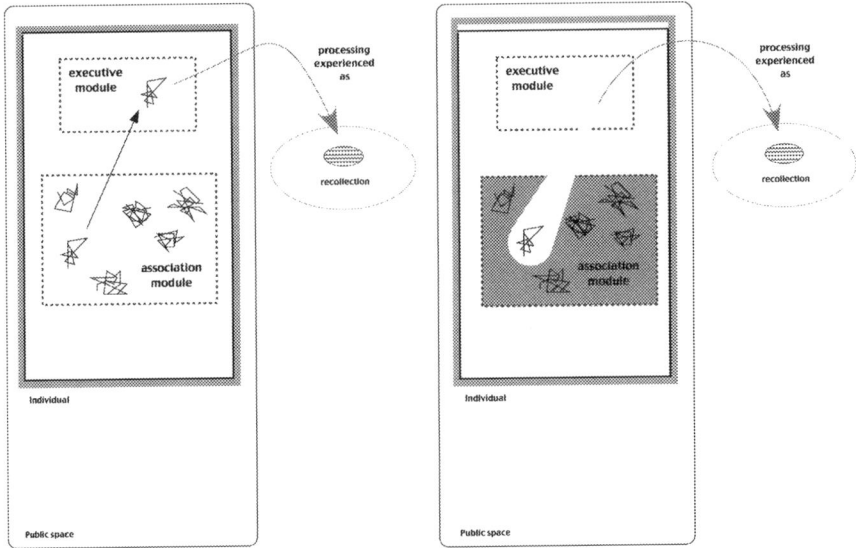

Fig. 5.3 At any one time a range of networks representing memory traces will be active, but only one will be the 'focus of representation'

An important aspect of memory function, consistent with this way of thinking about memory within the cognitive system, is that remembering is a constructive process (Parkin, 1987). Certainly in the case of what is called episodic memory when we recall an 'episode' of previous experience, research suggests that what we recall may be considerably different from what is objectively recorded to have been the case – although we are not aware of these distortions – as what is presented to consciousness usually seems a complete account. In the everyday discourse of the lifeworld, there is a common distinction between remembering something or not remembering (forgetting), but it is usually taken for granted that what we clearly remember will be accurate. On occasions where what we remember clearly appears to be contradicted by strong external evidence, we are surprised and sometimes disturbed as what we feel we know based on our strong recollections is a key part of our foundations for understanding and acting in the world. Despite the common folk notion of memory as fallible but largely reliable, research suggests that 'memory is not a literal reproduction of the past, but rather is a constructive process in which bits and pieces of information from various sources are pulled together' (Schacter & Addis, 2007, p. 27).

Extreme cases that are found in clinical examination of patients with certain medical conditions are referred to as *confabulation*, that is, 'fabricating an answer to cover up a memory deficit' (Parkin, 1987, p. 114). Patients apparently give reports that they experience as genuinely remembered, although actually incorrect. Kopelman (1987) reports one patient insistent that she needed to get home as it was part of her routine to cook for her mother – although the mother had actually been dead many years. As in this example, confabulation seems to be primarily a

Table 5.1 Examples of confabulation reported by Kopelman (1987)

Source account	Examples of 'recalled' detail reported
Anna Thompson of South Bristol, employed as a cleaner in an office building, reported at the Town Hall police station that she had been held up on the High Street the night before and robbed of 15 pounds. She had four little children, the rent was due and they had not eaten for 2 days. The officers, touched by the woman's story, made up a purse for her	She was a cleaner at a hospital who was stopped by police on the High Street* Anna Thompson of Sussex* Aged 44* The police made up a sum of 50 pounds* Thieves took her money near a railway station She has a little boy, aged 2

phenomenon of episodic memory, rather than semantic memory (Barba, 1993). Clinical examples of confabulation tend to be considered as either 'momentary', that is, 'sensible but untrue' accounts based on genuine memories inappropriately accessed (reporting genuine but distant experiences as if recent or current) or 'fantastic', involving reports that 'are often markedly bizarre and bear little, if any, relation to real events' (Burgess, 1996, p. 360).

Although confabulation is particularly associated with patents with memory deficits, similar (non-fantastic) confabulation seems to be common in people with normal memory function. Kopelman reports a study comparing a group of patients diagnosed with particular medical conditions with a healthy comparison group. As part of the procedure, the participants were read a short story and asked for immediate and delayed (45 minutes later) recall. The participants in the healthy group were also asked for a further report a week later.

The examples reported in Table 5.1 are from the 'normal' healthy participants, and those marked by an asterisk (*) were from the *immediate* recall condition. Kopelman found that details confabulated in an early report tended to be repeated during later reports. It seems that the non-fantastic forms of confabulation known to be common in certain memory-impaired conditions may actually reflect a characteristic of normal memory. To clarify, it seems that *normal people when presented with an account, and asked for almost immediate recall, will not only (as we might expect) miss some details, but will actually 'remember' additional details that were never part of the account, and when they do those embellishments will then often be recalled on later occasions as being part of the original information.* This account might well ring true to teachers, where common professional experience is that what pupils remember being taught often bears only limited similarity to what was actually presented by the teacher.

One interpretation of why memory is so fallible is that when the memory trace is activated, the pattern of neural activity usually under-specifies the original experience that is being remembered. There is here, then, a kind of 'filling-in' (see Chap. 4) to construct a feasible account from the fragmentary information available. An alternative, if potentially complementary, interpretation offered by Schacter and Addis (2007) sees the basis of this infidelity as adaptive. They suggest that the apparatus used for remembering is also that used for imagining future events and so has to be flexible rather than just offering a replica of previous experience. Presumably, a system that is able to use past experience to help us imagine what seems most likely to happen in the future automatically generates recollections of past events on the basis of what it seems most likely will have happened – and so tends to produce recollections higher on credibility than on accuracy.

If this is accepted as part of normal memory function, then it suggests that human memory is indeed quite different from the folk model account of some kind of storage device, where we put things in and later take 'them' out. That way of thinking implies fidelity as the memories are talked about as having the ontological status of objects that have permanence and are just put away for safekeeping. The research reviewed above offers a very different notion of a memory as a subjective experience created anew from the activation of the current state of a continuously evolving substrate. In terms of an analogy, memory is less like opening a computer file to recover information previously saved, than accessing a shared Wiki page that is constantly being edited and updated in the background.

It is important to note both that these processes are still not well understood, and that there may be more complexity than any simple description such as that given above would suggest. For example, although the processes described here as the basis for memory formation – interpretation of experience through, and then modification of, existing memory circuits – seem to imply that usually our memories will tend to merge and blur into each other, there are (e.g. traumatic) occasions when very vivid memories will form, as high levels of adrenalin seem to support the formation of especially strong memory of events, that is, those that may colloquially be said to be 'burnt into memory' (Elbert & Schauer, 2002).

The key point here, however, is that in remembering, as in perception, the information reaching the executive processing apparatus providing conscious experience should not simply be considered to be some kind of pure account, but rather an interpretation or impression, and – as with perception – we are not aware of the 'joins' and so are unable to distinguish which aspects of the recollection are drawn directly from the particular experience we seek to remember and which components have in effect been preconsciously 'filled-in' to complete the account.

This effect has been studied, as it has considerable significance, for example, in interpreting eyewitness accounts in court cases. A witness can honestly, that is, to the best of their ability, report remembering aspects of a crime that are actually not based on their original experience of the event. It is not unusual for different eyewitnesses to give 'factual' accounts that are inconsistent and contradictory even when those involved are asked to only report what they clearly remember.

False Memories

It has also been suggested that during periods of intense questioning, such as in police interviews, the original memory traces may be modified in such a way that later (e.g. in court) the individual 'remembers' features that they had no knowledge of during earlier questioning. This seems consistent with the account of memory discussed here: questioning leads to activation of networks associated with the earlier experience and modifies them by associating any new ideas or information presented. To caricature this, if the witness is asked enough times whether they saw a gun during an incident, so that the notion of a gun becomes strongly associated with their memory of that incident, then months later when they appear in court, their recollection of the original incident could now include a gun that they earlier denied having seen.

So 'exposure to new information prior to recall can result in incorporation of the new data into the originally perceived events' (Holdsworth, 1998, p. 115), and this has led to claims that many cases of childhood abuse which have only been 'recovered' many years later in therapy are actually the result of suggestion and not recollections of actual events. This idea of there being a 'false memory syndrome' associated with psychotherapy is subject to some controversy, but there is little doubt that false memories are in themselves common.

Arguably, these findings could be relevant to approaches to assessment in science education (and more widely). Sitting formal assessments may be just the sort of setting where there is a premium on giving complete answers and where students may commonly be told it is better to 'write something' than leave an answer. This might be a strong context for encouraging students to produce full answers even when memory does not support this. Tests can act as learning experiences because students use their available knowledge to answer novel questions and may develop new insights – but they also seem ideal contexts for presenting false memories, which once generated may be more readily recollected in future. This might suggest a hypothesis for empirical investigation: that the use of regular summative assessments as part of teaching is more likely to encourage the development and establishment of new alternative conceptions than formative approaches to assessment that lead to immediate feedback – that is, teacher questioning, followed by examination and evaluation of responses during the same teaching episode.

False Memories in Science Education Research: An Example from the Literature

Studies in science education that involve asking students about their understanding of scientific concepts either implicitly or explicitly call upon their memories. If the focus of a study is primarily about current understanding, then the reconstructive nature of the memory process is not problematic – although it is of course a key factor in determining the nature of that understanding. However, research that asks questions about earlier learning experiences needs to be informed by an understanding that student reports may be true *to their recollections* without necessarily being accurate accounts of the original experiences.

Fig. 5.4 Unipolar model of electric current: students may understand a circuit as allowing something to travel from one end of a cell

Fig. 5.5 Clashing currents model: students may understand a circuit as involving clashing currents originating from the two poles of the cell or battery

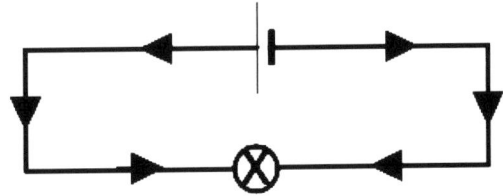

Anyone who has spent much time talking to students about their learning in science is likely to have come across various reports that at face value suggest teachers having taught a wide range of incorrect scientific ideas. In some case, the accounts may well be accurate, as research shows that sometimes teachers hold similar alternative conceptions to students, and so may well be a source of alternative conceptions acquired by students (Taber & Tan, 2011). However, it is likely that often students are either recalling teaching that was misconstrued at the time or constructing a recollection that has more coherence for them than the original teaching.

A classic example of this second case, memory being a distorted version of what was experienced at the time, is reported in the literature by Gauld (1986, 1989). Gauld describes teaching designed to shift pupils' thinking about electrical circuits. It is very common for students' ideas about what is going on in electrical circuits to be inconsistent with the scientific models represented in the curriculum. A particular difficulty, when considering series circuits, is to appreciate that the current is the same at all points in the circuit.

Research suggests that it is quite common for students' intuitions about electric circuits to fit one of a number of patterns. Early ideas might relate to a 'unipolar' model (that something passes along one wire from one pole of the cell to light a bulb) or a 'bipolar' model where there are 'clashing currents' – that current has to come from both poles of the cell (perhaps 'positive current' and 'negative current') and somehow interacts when it meets at the bulb (see Figs. 5.4 and 5.5).

However, the scientific model presented as target knowledge in the school curriculum has current flow in one direction all around a simple series circuit. Moreover, the current is constant around the circuit. In the curricular model, current is the flow of charge, and the amount of charge flowing in a circuit depends upon overall characteristics of the system: that is, the electrical potential difference (p.d.) across the circuit and the total resistance. The charge flowing in the electric field set up by the p.d. transfers energy around the system.

Fig. 5.6 Model of
diminishing current flow
in a series circuit

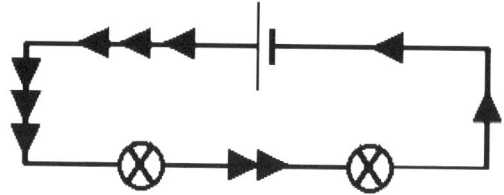

Fig. 5.7 Model of constant
current flow around a series
circuit

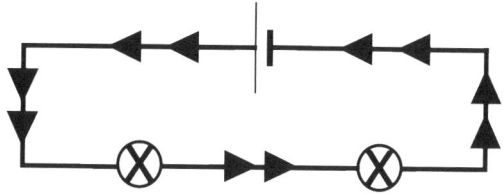

So in a series circuit with several lamps, the resistance of the individual lamps will collectively determine the overall resistance of the circuit, which will determine the current flowing when there is a particular p.d. across the circuit. That current will be the same all around the circuit, although if the lamps have different resistances (and so power ratings), the energy transferred in each lamp will be different.

For many students, this is a difficult model to understand, because they may not have a clear distinction between the charge flow (current) and the energy being dissipated in the circuit. Moreover, rather than thinking about a circuit 'globally' as an interacting system, students commonly think about the circuit sequentially (i.e. current flows from the cell, and first it reaches this lamp where it causes the lamp to glow, and then…). Taken together, this leads to a common notion that current diminishes around a circuit, with some being used up at each load (see Fig. 5.6). This is in contrast with the model presented in school science, where – as in Fig. 5.7 – the same amount of current flows on both sides of a load.

It should be noticed that the discussion here illustrates a number of features that reflect issues relating to way we discuss learners' ideas as highlighted in this book. In particular, the figures presented are *my* representations of common ways that students think about this topic, according to the literature. So they are based on my recollections of my reading of the accounts of other researchers, reporting their own understandings and interpretations of data collected from students – see Chap. 4. One assumption here is that the thinking of particular students can be similar enough to some of their peers to allow classification into a small number of models of 'common' ways of thinking/understanding. This is something that will be considered in the next chapter.

The work Gauld discusses concerns teaching to shift student thinking away from the models that students commonly intuit, towards the model that is set out as target knowledge in the curriculum. Gauld elicited students' models before and after a 'critical' lesson and then followed up with interviews about 3 months after the end of the topic. A critical lesson is meant to be in some sense analogous to a critical experiment that (in principle) must support a hypothesis for a theory to be accepted: although the notion that there really are such crucial experiments has been challenged (Lakatos, 1970). The basis of this critical lesson in teaching about electrical circuits was a demonstration that ammeter readings taken 'before' and 'after' a component such as a lamp were the same, so that no current was being 'used' or 'lost' in the component. In order to focus student attention to the significance of the demonstration, the predict-observe-explain sequence (White & Gunstone, 1992) was used:

> The role of the ammeter as a current measuring device was discussed and pupils were asked to predict, on the basis of their preferred model, the relationship between readings on ammeters placed before and after the globe. When the meters were connected (during the 'critical' lesson) students discussed ways they could make sense of the meter readings, especially when these conflicted with expectations based on their original ideas. One pupil introduced the notion of a 'carrier' model in which the 'material' stored in the battery and 'consumed' in the globe does not flow by itself through the wires but is transported by a second material substance which acts as a carrier. (Gauld, 1986, p. 50)

So, in other words, for those students who expected the 'electricity' to be used up in a lamp, the finding that the current was the same after passing through a lamp should be surprising and so needed explaining. The student's 'carrier' model, described by Gauld, solves this by positing a distinction along the lines of the curriculum model: that understanding the circuit requires thinking in terms of a 'carrier' (cf. current – charges flowing around the circuit) which is conserved, but which carries something else (cf. energy) which is 'consumed' in circuit components.

Gauld's work was informed by a previous study (by Cosgrove and Osborne) which had found that out of a class of 15 students, only one had demonstrated the conserved current model – that is, was classified as using the conserved current model – before a critical lesson, but that this had risen to all but two of the class afterwards. The lesson seemed to have been very successful in persuading students of the merits of the curriculum model. However, 1 year later, only just over half of those students (seven) were considered to retain the conserved current model (Gauld, 1989). Cosgrove later reported how:

> In classroom studies these models were elucidated and challenged, students and teachers devised a test to discriminate amongst these models, using two ammeters, locating one on each side of the light. On noting that the current shown was the same on each meter, there were different reactions. Most students reported that this outcome was strange…For a time, others appeared to agree with the idea indicated by the critical test, but they then reverted to previously held ideas…. Some students took up a different idea (changing from [diminishing current] to [clashing currents], for example), and others acknowledged the likely value of the [conserved current] idea to scientists *while preferring their own idea*. (Cosgrove, 1995, p. 296, emphasis added)

Gauld found that among the class of 14-year-old boys in his study, 4 of 29 held a conserved current model before the critical lesson rising to 25 afterwards. Then, about 3 months after the class had completed the teaching sequence, Gauld interviewed 14 of the students from the class. Only one of these was considered to demonstrate the curriculum model, although ten of the pupils appeared to have adopted notions of a carrier, albeit in conjunction with other models. Indeed half of the interviewees (seven) were considered to present a version of the unipolar model with a carrier, although none of the students had been considered to hold this type of model before the critical lesson.

So although the critical lesson, with its demonstration of the conservation of charge, appeared effective at shifting student thinking towards the curriculum model (at least as judged by interpreting students' written responses) at the time, over the following months many of the students apparently came to think in terms of a different model for understanding circuits – nearly always one inconsistent with the critical demonstration of current conservation.

Cosgrove and Osborne had warned that

> Even in situations where it is possible to provide a critical experimental test which invalidates all but one of the views proposed, some learners will not necessarily change their views at this stage. They still find their own view satisfactory because it links more coherently to other ideas they hold. (Cosgrove & Osborne, 1985, p. 107)

Gauld found that one feature of students' explanations for their preferred models was their recollections of the empirical evidence from their lessons, including their memories of the critical lesson. However, those recollections were not necessarily accurate. So, for example, 'pupil P4' had considered the conserved model, but then dismissed it because it had 'proven wrong by the meters' (Gauld, 1986, p. 51):

> "For a number of students the coherence of their arguments depended on the fact that their *memories of* meter readings (or of the relative brightnesses of two lamps in series) were not correct. The reason why pupil P4 could appeal to the meter readings to support [the diminishing current model] was that he had *an apparently clear memory* that the second meter gave a larger reading than the first as would be expected if the lamp had consumed a certain amount of electricity. On the other hand P14 rejected [the diminishing current model] because his memory was that the meter readings were equal. In many cases consistency existed because '*memories*' *were apparently reconstructed from implications of the adopted model*. (Gauld, 1986, p. 52, emphasis added)

The relationship between recalled meter readings and preferred models was not straightforward, so that Gauld reports how one pupil 'used a memory of equal meter readings to support [the unipolar carrier model] because he believed that meters measured the flow of carriers' whereas another pupil 'could justify the same model by appealing to a memory of unequal meter readings since he believed that the meters measured the amount of load carried' (Gauld, 1986, p. 53). Gauld concluded that 'while the crucial importance of the empirical evidence was implicitly acknowledged by all, in a number of cases the coherence of the point of view was achieved because "memories" of empirical evidence were apparently reconstructed to be consistent with the model adopted' (Gauld, 1986, p. 53).

Research which follows students' thinking over extended periods of time (months, years) can offer insight into the way that memory operates in learning science. This has not been a common focus of research, which is unfortunate because Gauld's study suggests that this is a topic very worthy of researchers' attention. It is perhaps unsurprising that students persuaded of the value of a novel scientific way of thinking about a phenomenon, in the social context of the classroom where teacher authority and peer pressure may be significant, may often later revert to earlier ideas when the evidence and arguments presented in class seem less persuasive.

It would perhaps not be surprising if students could not recall a demonstration that seemed crucial to change minds at the time. However, Gauld's work suggests that it is possible for students to perceive and understand classroom demonstrations as strong evidence for the scientific model during a lesson, yet, later, shift away from that way of thinking and come to (falsely) *remember* the outcome of a critical demonstration as supporting their own alternative way of understanding. If this effect were found to be common across science learning contexts, then this would seem to be potentially important for designing teaching schemes in science if we hope for science learning that survives well beyond the end-of-topic test. Perhaps critical events only remain critical with suitable regular teacher reinforcement: the immediacy of a surprising outcome motivates seeking an alternative understanding now but soon becomes part of our background experience, and often modified to better fit our prior thinking.

Forgetting

Forgetting is another of those terms that has a clear enough meaning in everyday life, without being understood in a very precise way. In the lifeworld, we talk about having forgotten something as though this is an unproblematic phenomenon – perhaps as if some of the 'objects' placed in our mental storeroom have either been removed or have simply got lost among all the clutter.

Classic experiments in psychology demonstrated that when memory is tested, in general the amount of material that can be recalled diminishes over time, and this finding has been widely reproduced: 'memories seem to fade with the passage of time. Many experiments have studied memory loss as a function of time... Initially, forgetting is rapid, but memories continue to worsen nearly forever' (Anderson, 1995, p. 227). However, despite this familiar effect, hypermnesia, that is, 'improvements in net recall levels associated with increasing retention intervals' (Payne, 1987, p. 9), has also been reliably demonstrated under test conditions. It should be noted that much clinical testing of memory has used target material which has no particular personal relevance to the 'subjects' being tested, and such learning may not always be *meaningful* in Ausubel's (2000) sense – although cynics might argue that such testing conditions may be quite relevant to the way some pupils experience school learning.

In everyday discourse, to forget is to fail to remember something at an appropriate time, but often also to suggest that something previously represented in memory is no longer there. Forgetting to go to the post office on the way home from work, when one planned to do so, although it was not part of a usual routine, would normally be considered as an absent-minded aberration; whereas forgetting that there was a post office in a town one knew well would be unusual and might be considered to be indicative of some more serious organic memory problem. No longer being able to recall a poem, one learnt verbatim decades before at school provides an intermediate case, although whether 'cannot remember' should be equated to 'have forgotten' is less clear-cut: if remembering the poem is important, then it is likely that although it cannot be immediately remembered, given enough time elements of the composition will begin to be accessed. Our experience of remembering and forgetting seems consistent with the models of perception and memory discussed above: memories are not actively 'erased', but memory traces that have not been called upon for a long time may have a higher threshold for activation.

Significance of Forgetting for Science Education

Certainly it is a commonplace of science teaching or indeed any teaching that when students are tested, they may often fail to produce the hoped-for answers, even when the same students had apparently been able to demonstrate they knew the correct answers previously in class. As Gauld suggests (see above), it may be quite difficult to elicit evidence of 'learning' that had been demonstrated soon after the event some time later.

This is a particular issue for science education, because often the scientific models being presented in class are not the only ways of thinking about an issue for the students. Where students come to class with strong ways of thinking about a phenomenon, then science education is intended not only to offer an alternative way of thinking, but one which will be selected and applied in appropriate contexts. New science learning may well be in competition with existing well-established alternatives.

A Case of Learning and Forgetting in Science Education

This was certainly the case with one student I worked with over a period of time. Tajinder (an assumed name used in the study) was studying A levels (university entrance level qualifications) in an English college of further education, and one of his chosen subjects was chemistry. He was one of the students who agreed to participate in a project to explore students' developing understanding of chemical bonding (a key concept area in chemistry) during the A level course. Tajinder was very generous with his time, as well as being highly motivated to do well on his course as he hoped to read a medically related subject at university. He was inter- viewed in depth (many of the interviews exceeding an hour) over 20 times over the two academic years of his course, providing an extremely rich data set.

This allowed the researcher (i.e. me) to develop quite detailed ideas about both how Tajinder thought about core aspects of chemistry at the start of his course and how these shifted during his college course. This work is reported in some detail elsewhere (Taber, 2000b, 2001b), and here I will just briefly outline key features of the case.

At the start of his college course, Tajinder commonly explained key aspects of chemistry (bonding, reactions) in terms of an explanatory principle that was based upon the notion of atoms achieving octets or full shells of electrons as the driving force for bond formation or chemical change. From a scientific perspective, this is an alternative conception, although it is related to the generally valid heuristic that stable chemical structures normally exhibit such electronic configurations. As this 'octet rule' does widely apply, the reactants in most chemical reactions can already be understood as 'obeying' the rule, so it has little value as an explanation for *why reactions occur*. Although this can be considered an alternative conception, it is one that seems to be very widely acquired among chemistry students (Taber, 1998a, 2013d), perhaps in part because it fills an 'explanatory vacuum', as no scientifically based explanation for why reactions occur tends to be taught in introductory chemistry lessons.

During his 2-year college course, Tajinder met more scientifically acceptable models and developed two further explanatory principles to explain chemical phenomena: one based on the idea that systems tend to shift to minimise energy and one based on how chemical processes can be considered to be due to Coulombic (electrical) interactions between quanticles (molecules, ions, electrons, etc.). During much of Tajinder's course, he would draw upon this repertoire of three explanatory principles, sometimes quite flexibly, and he did not seem to see it as problematic to use these ideas as a kind of conceptual toolkit for building explanations. That is, he operated with manifold conceptions of why chemical processes occurred (Taber, 2000b). Tajinder's minimum energy and Coulombic forces explanatory principles are closely related from a physical perspective: that is, they can be seen as being alternative descriptions of the same processes but in terms of either energy or forces. However, Tajinder was not studying physics, and for him these were two alternative explanatory principles, not different ways of describing the same underlying principle.

By profiling Tajinder's use of these three principles over time, it was possible to see shifts in the use of his three explanatory principles. In particular, he gradually came to use the Coulombic principle as the preferred basis for explanations, and his use of the octet rule principle became less frequent. He had entered the course attempting to explain most bonding-related phenomena in terms of atoms seeking octets/full shells, and he continued to use these ideas throughout his course. However, in the context of the topics and problems considered at A level, this scientifically invalid approach is often unhelpful, whereas much can be successfully explained in terms of electrical interactions between quanticles. The shifts were understood as reflecting the broader conceptual ecology in which Tajinder thought about chemical issues as his knowledge of the subject and experience of trying to make sense of new learning grew (Taber, 2001b).

Some time later, almost 4 years (46 months) after successfully completing his course, Tajinder was interviewed again. He was in the fourth year of a medically related degree course, which although it required chemistry as a prerequisite had, at least in Tajinder's view, given him limited reason to engage with the chemistry he had studied at college. Again, this has been reported in some detail elsewhere (Taber, 2003), but the key findings were:

- Although Tajinder reported not having engaged in thinking about the target concepts to any extent since completing his chemistry course almost 4 years earlier, he was immediately able to answer many questions, apparently drawing upon his previous learning of the subject.
- However, if Tajinder's changing pattern of thinking during his earlier college course is considered as demonstrating progression in learning – that is, he had expanded his repertoire of concepts, he had refined his use of explanatory principles, etc. (Taber, 2001b) – then his responses in the deferred interview suggested a regression in his thinking: that is, his responses in the delayed interview more closely reflected his explanations in the interviews at the start of his college course, rather than those near its successful conclusion.
- There was also evidence that despite this sense of having 'slipped back' to earlier, less advanced ways of thinking about chemistry, Tajinder's thinking had also undergone a new shift since the previous interviews undertaken prior to final college course examinations.

An important caveat in this study is that the single follow-up interview, undertaken several years after regular engagement with the topic, cannot be considered to demonstrate the full potential of Tajinder's response repertoire at that later time. That is, had it been possible to carry out a sequence of interviews with short delays – perhaps three or four interviews over the course of 10–14 days – then there may well have been further quite rapid shifts in patterns of responses. This possibility follows from (as suggested by the literature reviewed above) seeing what we remember in any specific situation as less about what is represented in the brain than about what memories can be accessed by activating those neural circuits modified by the original learning. The experience of the initial interview might have itself shifted the threshold for subsequently activating other long-dormant representations. This would be a useful direction for future research – assuming the availability study participants prepared to subject themselves to extended sequences of interviews on topics they have long ceased to have needed to study.

Regression in Learning

Given this important caveat, it may not be helpful to suggest that Tajinder had 'forgotten' a lot of his chemistry: rather that a good deal of learning that had previously been demonstrated in research interviews was *not brought to mind* in the deferred interview. His *performance* in terms of offering explanations relating to curriculum models was inferior to what had been demonstrated when he was last actively studying the subject. In practical terms it is less significant *whether or not* someone still retains representations of earlier experience than whether they can activate such representations in particular contexts and so access the memories.

Forgetting is best understood as a context-dependent phenomenon (Parkin, 1987) – in particular contexts we do not access certain memories.

The report of this case here is also subject to all of the considerations I raised earlier in the book. I report here key features of *my* mental models of Tajinder's understanding and thinking at various points in our study. These models are based upon the representations he made in the public domain (mostly talk, some drawing) which were then interpreted through my existing expectations and ways of making sense of the world. Moreover, a very rich data set was subject to being summarised in a suitable form for reporting in the literature, leading to the omission of a great deal of detail and nuance.

Given this, it is possible to consider the most salient patterns in terms of the three main explanatory principles identified in the original sequence of interviews. At the outset of his college course, Tajinder called extensively upon his octet rule explanatory principle to explain aspects of chemistry he was asked about in interviews. By the end of his course, he had a repertoire of three, essentially distinct, explanatory principles, and of these the Coulombic forces principle tended to be applied most often. This was understood as a shift from relying upon an alternative conceptual framework to increasingly offering explanations better matched to the target knowledge presented in the curriculum.

However, in the deferred interview Tajinder used the inappropriate octet rule principle extensively, whilst calling upon ideas about the electrical forces between quanticles much less. This therefore represents a regression in performance to something closer to his level of understanding on entering the college course. This was reflected in the way Tajinder discussed different categories of chemical bond. Not only is the common octet alternative conceptual framework inconsistent with scientific models (Taber, 1998a), it also restricts explanations of bonding to two main categories (covalent and ionic), with some further categories (polar and metallic) understood as deriving from covalent and/or ionic bonds. The evidence from the deferred interview was that

> other types of bond he had learnt about during his college course and discussed in some detail in interviews had been 'forgotten', i.e. if traces of this knowledge were held in memory then those traces were not activated, and *the memories were not accessed*, during the interview. (Taber, 2003, p. 270)

In this case, it seems that the ideas Tajinder had about the nature of chemical interactions (bonding, reaction etc) when he entered his college course from school study were well represented in his LTM in ways that made them readily accessed. Subsequent study slowly persuaded him of the greater explanatory value of new ideas he developed in response to college teaching, so that the changes in his LTM led to his more frequently applying those ideas – and particularly explanations in terms of electrical charges.

Yet presumably further changes took place in LTM over the following years such that later it was the representations of his *earlier ideas* that tended to be activated and drawn upon in the later interviews. If we think in terms of the internal representation of chemical knowledge, then it would seem that Tajinder's conceptual structure developed during his chemistry course to become more consistent with target knowledge in the curriculum, but then slipped back to become closer to its earlier state.

This is a single study, and unfortunately there is no way to know whether subsequent learning – either on his university courses or due to other life experiences – would have given reason to activate and reinforce his longer established patterns of thinking, despite his own recollection that this was not an area of study he had considered much in the interim. However, given the common experience that learners' performances in demonstrating understanding and knowledge of taught scientific models do often regress after completing a course (module, topic) of study, further research into the nature of 'forgetting' science learning is indicated.

A Tendency to Integration

Before leaving Tajinder's case it is also important to note that his deferred interview did not only suggest a regression in thinking but also a development which in principle might seem more promising. Although Tajinder used his Coulombic force principle to a limited extent in the deferred interview, he did use his minimum energy to a noticeable extent, alongside the octet rule principle. The minimal energy principle was more aligned with scientific thinking than the octet rule principle, although it was necessarily used in a rather tautological way: that as changes occur to minimise the energy of a system, then if a change has occurred, it was because the system has evolved to a lower energy state.

Of particular interest though is that whereas during his college course, Tajinder perceived his three explanatory principles as distinct complementary narratives that could be selected from to build explanations in chemistry; in the delayed interview Tajinder tended to link his octet rule principle with his minimum energy principle: these were now less alternative explanations than different facets of the same explanatory scheme – changes occurred so that atoms could obtain full electron shells, *because* this would lead to a lower energy state.

Presumably, during the intervening years, Tajinder's representation of his thinking about chemical principles had been modified so that on activating these representations there was now a strong linkage between what had been two discrete models of explanation. This is particularly interesting, as during his course he had never demonstrated that he had in a similar way linked the minimum energy principle with the Coulombic forces principle, despite *that* being more scientifically appropriate.

Again, it is important not to seek to generalise from a single case, and more detailed longitudinal studies of individual learners would be valuable here. However, what is suggested is that over a period of time when representations are not being actively drawn upon in thinking, there is modification of those representations, and in this case part of that was towards a greater integration of concepts (see Fig. 5.8).

However, this finding is consistent with some ideas about the cognitive system which suggest that a key aspect of the way the system is organised is that it

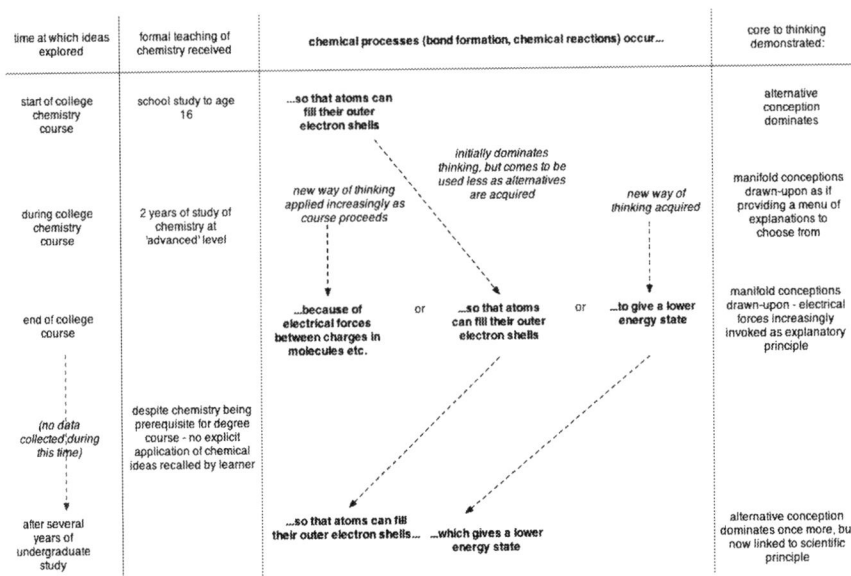

time at which ideas explored	formal teaching of chemistry received	chemical processes (bond formation, chemical reactions) occur...	core to thinking demonstrated:
start of college chemistry course	school study to age 16	...so that atoms can fill their outer electron shells	alternative conception dominates
during college chemistry course	2 years of study of chemistry at 'advanced' level	new way of thinking applied increasingly as course proceeds — initially dominates thinking, but comes to be used less as alternatives are acquired — new way of thinking acquired	manifold conceptions drawn-upon as if providing a menu of explanations to choose from
end of college course		...because of electrical forces between charges in molecules etc. or ...so that atoms can fill their outer electron shells or ...to give a lower energy state	manifold conceptions drawn-upon - electrical forces increasingly invoked as explanatory principle
(no data collected during this time)	despite chemistry being prerequisite for degree course - no explicit application of chemical ideas recalled by learner		
after several years of undergraduate study		...so that atoms can fill their outer electron shells... ...which gives a lower energy state	alternative conception dominates once more, but now linked to scientific principle

Fig. 5.8 Research from a case study suggests that an initially dominant alternative conception applied with decreasing frequency during a chemistry course later came to dominate thinking again (but with a potentially important modification) several years after completing formal study of chemistry

inherently works towards greater coherence and integration: that is, that over time the brain tends to modify representations to support more generalised and self-consistent thinking. A widely accepted notion here is that of memory consolidation, and some commentators suggest there is also a distinct additional phenomenon called memory enhancement.

Memory Consolidation

The formation of memories does not seem to be a single event, but an ongoing process, such that memories can become more robust over time (Parkin, 1993; Wiltgen, Brown, Talton, & Silva, 2004). Studies suggest that 'new memories are initially vulnerable but are gradually strengthened over time' (Wiltgen et al., 2004, p. 101). The phenomenon of memory consolidation has been described as 'the progressive stabilization of items in long-term memory' (Dudai & Eisenberg, 2004), so that an item 'becomes increasingly resistant to interference from competing or disrupting factors in the absence of further practice, through the simple passage of time' (Walker & Stickgold, 2004, p. 122). It is widely thought that this involves more than one stage, with a quick initial step, 'followed by subsequent, "elaborative" consolidation in which the new memory trace becomes more fully integrated with pre-existing memories' (Parkin, 1993, p. 24).

Consolidation has also been described as 'the transformation of memory from short-term memory to long-term memory' (Bourtchouladze, 2002, p. 23). Yet, as we have seen above, it is now considered above that experience gives rise to memory by modulating *existing* neural networks, which suggests that models of memory based upon a *discrete* short-term store, where memories are kept prior to being transferred to a long-term store, are not feasible. At first sight, this appears inconsistent. However, Fuster comments that

> The classic distinction between short-term and long-term memory is operationally valid but, from the point of view of cortical topography questionable. These two forms of memory share much of the same cortical substrate and simply reflect different states of that substrate. (Fuster, 1995, p. 4)

In other words, memory seems to operate *as though* memories are initially stored in a temporary store before being transferred to a more permanent store, but the distinction is not actually about where the memory is located in the brain, but about the way in which it is connected into the system. Research suggests that consolidation is not about 'relocating' a memory, but providing a permanent binding of its components. So-called 'declarative' memories (see above) are thought to have 'temporary dependence on structures in the medial temporal lobe' (MLT, in particular, the hippocampus) but over time can be considered to be 'stored in neocortical circuits without a significant MTL contribution'. So it seems that at the physical level of description, a specific area of the brain, the MTL takes on a *transient* role in linking together the spatially discrete parts of a memory trace in different locations in the cortex. The MTL is able to facilitate such binding relatively immediately, but has a limited capacity. Activation of one part of such a network, through its connections through the MTL, activates the other components which build up *direct* linkages – so that over time the MTL's role becomes redundant (Alvarez & Squire, 1994; Wiltgen et al., 2004). The development of direct linkage within the neocortex is much slower than the immediate links provided through the MTL, but not limited by the same capacity issues – so the short-term and longer-term mechanism for binding a memory are complementary.

In this model, consolidation occurs by the concurrent activation of the physically separated regions of neocortex coordinated by the MLT, leading to the strengthening of the direct linkages (Alvarez & Squire, 1994). It was suggested earlier in the chapter that there is always a wide range of cortical activity, most of which is not consciously attended to (see Fig. 5.3), and consolidating processes are one reason for such activity. It has been suggested that each time a person accesses a memory, this activity itself should be considered as a cognitive event which will leave a trace in the system, in effect facilitating more ready subsequent access to the memory (Parkin, 1987).

Memory Enhancement?

It has long been suggested that one possible role for sleep is linked to the operation of memory: for example, that perhaps there is some important 'housekeeping' that occurs during sleep to maintain proper functioning of memory. When the model of a discrete STM was popular, it seemed viable that transfer between STM and LTM occurred during sleep, but this idea has been replaced by the notion of consolidation discussed above. Although it *could* be conjectured that consolidation primarily occurs during sleep, according to Vertes (2004, p. 145), 'there is simply not enough evidence, or evidence of sufficient weight, to maintain that one of the functions of sleep is memory consolidation'.

However, it has been suggested that as well as stabilisation of memory during consolidation, there may be additional 'mechanistically distinct' processes of memory *enhancement* (Walker & Stickgold, 2004, p. 122), said to 'occur primarily, if not exclusively, during sleep' and responsible for 'restoring previously lost memories' and 'producing additional learning' but 'without the need for further practice'. Perhaps advice to those studying to get a good night's sleep has some basis in the physiology of memory.

Working Memory

Earlier in the chapter the notion of long-term memory (LTM) was discussed. LTM is the name given to the process by which we can access accounts of previous experience, whether that be episodic memory of our personal biography or declarative learning of various material: facts, procedures, abstract ideas and so forth. It was suggested that LTM is not so much a separate storage facility within the cognitive system, but more a core function of the cortex in general, and that it should probably be understood less as a means of making records of our experiences than the way that the cognitive apparatus that has evolved to support decision-making that is dynamic – constantly being modified by experience and our interpretation of that experience. That is, the adaptive value of memory in the brain is not that it can offer an accurate record, and indeed we have seen that this is not always so, but rather that *it provides an iterative and constantly updated interpretation of experience that can support future action*. Clearly, for memory to be adaptive in supporting survival in the environment, it must in some sense offer an effective basis for decision-making: but that need not mean high-fidelity recollections of specific past experiences – rather there seems to be a constant process of seeking integration and coherence that may lead, perhaps simply as a side effect, to objectively inaccurate memories that seem subjectively to be true records of our past.

In other words, just as it was suggested in Chap. 3 that consciousness might be best considered a side effect of the development of more complex and sophisticated cognitive processing during evolution, it may also be possible that conscious memories may best be considered as incidental side effects of the evolution of a plastic cognitive system able to modify itself in response to feedback from previous decision-making. If that were the case, then we might do better to treat our explicit memories as little more than mythical accounts of our past: that is, like a myth, a memory may represent some important truth, but packaged into a poetic or narrative form that is most suitable for informing future behaviour rather offering a literal account of the past.

The second component of memory which is generally accepted to be a distinct part of the cognitive system is very different from LTM and is commonly known as working memory (WM): this is 'a system for the temporary holding and manipulation of information during the performance of a range of cognitive tasks such as comprehension, learning, and reasoning' (Baddeley, 1986, p. 34). WM is considered to be a 'fluid' component of the cognitive system, with its contents readily updated compared to the 'crystallised' components 'capable of accumulating long-term knowledge' (Baddeley, 2000, p. 421), that is, LTM.

Whereas LTM has effectively infinite capacity, and representations in LTM are retained, if subject to modification, indefinitely, WM has a very limited capacity and is much more transient – being about what is currently the focus of attention: 'working memory refers to the temporary retention of information that was just experienced or just retrieved from long-term memory but no longer exists in the external environment' (D'Esposito, 2007, p. 7).

As with LTM, there is some disagreement over whether WM should be considered to be discrete and localised at the physical level (Andrés, 2003), or – akin to thinking about LTM – '…neither a unitary nor a dedicated system' but rather 'an emergent property of the functional interactions between the [prefrontal cortex] and the rest of the brain' (D'Esposito, 2007, p. 7). However, WM has become a widely accepted construct for describing part of a person's cognitive functioning. It is generally considered as a distinct *functional* component of the cognitive system: 'working memory continues to provide a highly productive general theory [given that] human thought processes are underpinned by an integrated system for temporarily storing and manipulating information' (Baddeley, 2003, p. 837).

Executive Function of WM

As with LTM, uncertainty at the physical (anatomical) level about exactly how WM could be located and embodied does not undermine the functional arguments for considering it a component of the cognitive system. When considering cognition as a system, WM is an important system component. Quite a lot is known about how WM operates. Baddeley describes the assumed function of WM as 'a limited capacity system, which temporarily maintains and stores information, supports human thought processes by providing an interface between perception, long-term memory and action' (Baddeley, 2003, p. 829).

WM is able to access various types of information (Jonassen, 2009), which can be considered and reflected upon: 'these internal representations…can be subjected to various operations that manipulate the information in such a way that makes it useful for goal directed behaviour' (D'Esposito, 2007, p. 7).

Models of WM

WM is considered to be a transient type of memory, concerned with what can be 'held in mind' rather than what is represented permanently in neural circuits – although it draws upon information represented in that way. To support the executive functions of WM, it appears to have its own system of buffers that allow certain types of information to be held available for processing.

Alan Baddeley (1986, 1990, 2000, 2003) has developed a model, originally proposed with Hitch (Hitch & Baddeley, 1976), which has been widely adopted, in which WM has a central executive processor, supported by 'slave' systems which retain information 'through active maintenance or rehearsal strategies' (D'Esposito, 2007, p. 7): Baddeley's (2003, p. 837) 'multi-component model'. This is represented in Fig. 5.9.

Two particular types of buffer or slave systems are well-established features of the model. One holds images and is called the *visuo-spatial scratch pad*; the other rehearses a limited about of auditory information and is known as the *articulatory* or *phonological loop* (Parkin, 1993). More recently Baddeley has argued that it is useful to acknowledge a third buffer as part of WM, the *episodic buffer*.

The Visuo-spatial Sketchpad: The visuo-spatial sketchpad allows the person to hold in mind, and manipulate, imagistic representations. This supports a form of nonverbal intelligence. Baddeley (2003) notes its importance in engineering and architecture and gives the example of the type of thought processes that Einstein reported as crucial to his theorising.

The Phonological Loop: The phonological loop acts as store, which holds memory traces from a short period: a matter of a few seconds. However, it is supported by an 'an articulatory rehearsal process that is analogous to subvocal speech' (2003, p. 830). In effect the data in the store is 'played', and the output fed back into the store to refresh it. This store allows us to rehearse small amounts of information, such as a telephone number we have just heard.

The Episodic Buffer: Baddeley (2000, p. 421) proposed the additional component of the episodic buffer to fill an apparent gap in the WM model. The episodic buffer is seen as a 'modelling space', but one that is able to handle different types of information ('by using a common multi-dimensional code'), and so 'serves as an interface between a range of systems'. In the model (see Fig. 5.9), the buffer is shown as a separate system component, but Baddeley comments that this could alternatively be seen as the 'the storage component of the executive', providing it with 'a global work-space that is accessed by conscious awareness' (Baddeley, 2003, p. 836). The buffer is considered to have an 'integrative' role and to have capacity in terms of the 'number of multi-dimensional chunks' it can handle (Baddeley, Hitch, & Allen, 2009).

The Central Executive: Baddeley acknowledges that although the central executive is the key component of the model, it has been subject to less empirical investigation than the visuo-spatial sketch pad or the phonological loop (Baddeley, 1986) and indeed was originally 'simply treated as a pool of general processing capacity, to

Fig. 5.9 A model of working memory comprising a central executive processor plus small capacity buffers that can store different forms of information (After Baddeley, 2003)

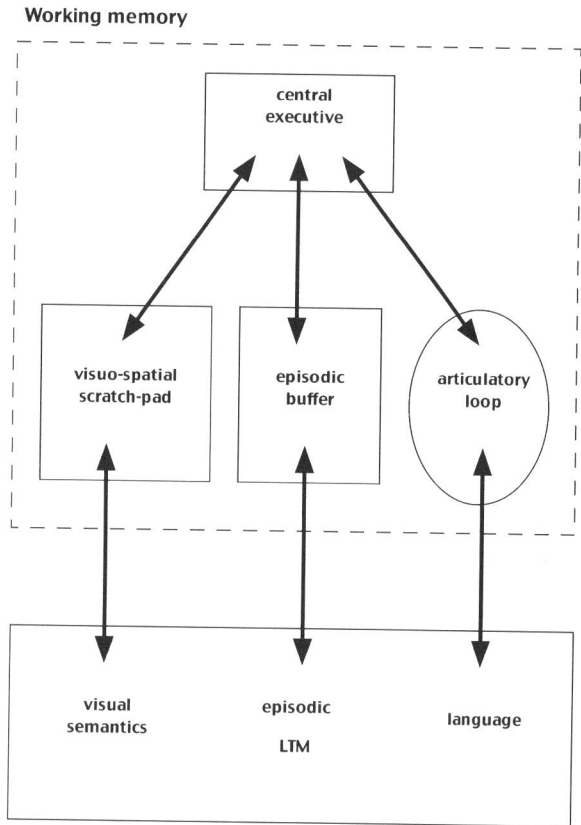

Working memory

central
executive

visuo-spatial
scratch-pad

episodic
buffer

articulatory
loop

visual
semantics

episodic

LTM

language

which all the complex issues that did not seem to be directly or specifically related to the two sub-systems were assigned' (Baddeley, 2003, p. 835).

The precise operation of the central executive has come into focus a little more with the proposal of the episodic buffer as a system component (Baddeley, 2000). The central executive is considered to control the episodic buffer and be responsible for binding information to support the development of episodic memory. That is, an episode recalled from memory and based on representations in LTM may involve images, semantic content, temporal sequencing, etc.: the connecting together of these different sorts of information, so that they can be associated in LTM, so to later be accessed consciously as an episode of experience, depends upon the central executive acting upon information temporarily held in the episodic buffer.

The central executive is conjectured to be able to retrieve information from the LTM so that it becomes consciously available (it 'attentionally' controls the episodic buffer, cf. directing the 'spotlight' in Fig. 5.3) and to process this information (see Chap. 7):

The executive can, furthermore, influence the content of the store by attending to a given source of information, whether perceptual, from other components of working memory, or from LTM. As such, the buffer provides not only a mechanism for modelling the environment, but also for creating new cognitive representations, which in turn might facilitate problem solving. (Baddeley, 2000, p. 421)

The Role of WM in the Cognitive System

Baddeley's WM can be seen as providing the executive processor, considered as a core part of the cognitive system (see Chap. 3). It interacts with LTM in the manner envisaged when discussing the nature of students' ideas (Chap. 4), with LTM having the dual role of interpreting and representing previous experience. The buffers that act as part of WM seem to be positioned in the system at the level of the preconscious buffers, that is, between LTM and the executive module, which temporarily store perceptual information that may be attended to by the executive. A synthetic representation is shown in Fig. 5.10, where the executive is shown as calling from buffers information from current sensory input and from past experiences no longer current, but represented in LTM.

WM Capacity

WM is considered to be subject to severe restrictions on its capacity to process information, as if it has a very limited number of 'slots' for data. Although there is variation between individuals, and some apparent expansion early in life which may reflect access issues rather than actual capacity, a common value that is quoted as typical is 7 ± 2: that is, that most people can 'keep in mind' from 5 to 9 distinct quanta of information at a time. This number derives from pioneering work by Miller (1968). Memory 'span' is commonly determined by tasks such as the 'digit span technique', which explores the length of a string of random digits that can be reliably reported back in the correct sequence, immediately after a (verbal or visual) presentation, before the subject begins to make mistakes. The largest length of string that a person can reliably report is their digit span. It has been found that for 'normal' adults (those without some kind of mental deficit/impairment), this is usually in the range 'seven, plus or minus two' (Parkin, 1993).

More recent research (Cowan, Chen, & Rouder, 2004; Mathy & Feldman, 2012), whilst supporting Miller's general principle, suggests that Miller's magic number may actually overestimate the number of available 'slots' in working memory, because of our natural tendency to spot patterns that allow us to 'chunk' information. It seems that the real capacity of working memory may actually be more commonly around 4 rather than 7, because even when 'arbitrary' information is used to test memory span, people tend to be able to impose some order on the material to be remembered and so 'chunk' it to some extent. That is, because part of the 'arbitrary' stimulus is often recognised as similar to something in LTM, it is processed more

Fig. 5.10 Working memory as part of the cognitive system

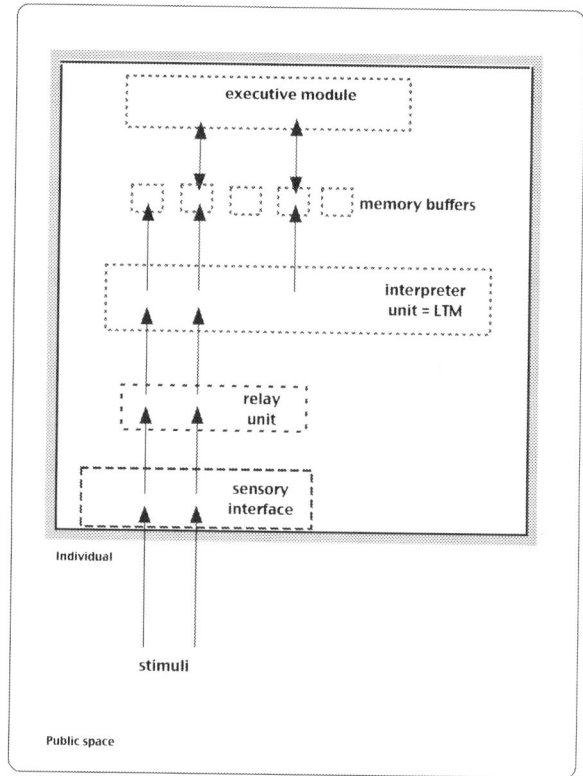

efficiently. (Chunking is discussed further later in the chapter.) It is suggested that the actual capacity may be better described as 4 ± 1 (more in keeping with what is often measured in young children), but appears to be 7 ± 2 when measured in most adults due to automatic strategies used to chunk information.

The model assumes that in such test situations the subject repeats the list of digits over in their mind (using the articulatory loop) and fails when the number of items in the list exceeds the capacity of that articulatory system (Baddeley, 2003). A useful, if crude, analogy would be a loop of magnetic recording tape used as a note-taker, but only having a few seconds of playing time before any new input erases the recording of information already made. Another analogy would be the 'roller' chalkboard that used to be commonly used in schools and which was a vertical loop of writing surface such that the teacher periodically moved the board 'up' to reveal new board from below: once the board had completed one roll, the teacher needed to erase previous inscriptions to make space for further notes.

The visuo-spatial scratch pad also has limited capacity, which according to Baddeley is typically only three or four objects. Baddeley (2003, p. 838) suggests that this is normally sufficient as our visual worlds tend to be largely stable, although this can lead to the phenomenon of 'change blindness, whereby objects in scenes can change colour, move or disappear without people noticing'. Many readers will have seen the amazing video set up to demonstrate this effect and reported by Simons and Chabris (1999). The video shows a small group of people, wearing white or black shirts (three 'players' in each team), practising basketball. In the recording they move about (the two teams intermingling), and the members of each team pass a ball between them. Importantly the observer is given a specific task to attend to, for example, silently counting the number of passes completed by players in white shirts. However, the video includes an unexpected, and incongruent, event that observers are not forewarned about. At the end of the test, the observers are asked to report on the explicit task, but also asked about what else they noticed. Simons and Chabris used several variants of the task, but essentially they found that about half of the participants attending to the set task failed to notice the 'unexpected' event.

When watching the footage without priming on what to attend to, this can seem quite incredible. In one version of the task as the players mill about (in a confined area), a person in a gorilla suit slowly walks into scene among the players, pauses at the centre of the scene, turns to the camera, gestures by beating its chest, turns, then slowly walks off the other side of the scene, again threading between players. The 'gorilla' is clearly visible in shot for almost 10 s. Half of the observers carefully watching the film failed to notice this!

As with the filling-in phenomena (see Chap. 4), this demonstrates the extent to which our mental models of the world may be partial whilst seeming to us quite complete. The significance of this effect for teaching and learning is clear – especially perhaps when students are expected to notice particular effects in practical work (cf. R. Driver, 1983). Clearly one key feature of teaching is to help learners focus their attention on what is salient according to the curriculum.

Visual Memory and Eidetic Imagery

Although it is usually considered that the visuo-spatial scratch pad has very limited capacity, there is a phenomenon known as eidetic imagery and linked to the common notion of 'photographic memory' where people appear capable of holding much more visual information in mind. This eidetic imagery is said to be 'a rare phenomenon in which the individual seems able to form vivid and detailed images which are experienced as if they were actual percepts' (Parkin, 1987, p. 53). It is thought that this type of visual memory may be more common among children, usually being lost by adulthood.

Educational Significance of WM Capacity

It is notable that whereas for practical purposes LTM can be considered to have an infinite capacity and certainly retains the ability to be modified by experience throughout life; WM is, at any one time, only able to handle a relatively tiny proportion of the potential information represented in LTM. This limited capacity can be considered a severe restriction on cognition and to act as the limiting factor in many learning situations.

For example, tasks set in class may be too difficult for students because their apparent complexity exceeds the processing capacity of the students' WM – there is just 'too much to hold in mind at once' to complete the task. This links with key education ideas such as the importance of metacognition which can allow people to develop strategies based on sequences of steps, each of which is individually manageable within WM, and the notion of 'scaffolding'. Scaffolding is a teaching technique which allows the novice learner to initially rely on others to provide the strategy as they master individual steps of a procedure: support which may later become redundant due to the possibility of 'chunking' as discussed below.

Although this severe limitation on WM seems to be a disadvantage, it has been argued that it could be adaptive. It is interesting in this context that whilst having a 'photographic' memory would seem likely to be a very useful attribute, as noted above, eidetic memory, 'as the memory capability to retain an accurate, detailed image of a complex scene or pattern' (Inoue & Matsuzawa, 2007, p. R1005), seems to generally be lost during human development. Moreover, there is some suggestion that humans have a more limited visual working memory capacity than chimpanzees, so that 'young chimpanzees have an extraordinary working memory capability for numerical recollection better than that of human adults' (Inoue & Matsuzawa, 2007, p. R1005).

Sweller (2007) has suggested that the capacity of WM relates to the optimum solution in a balance between the need for an organism to respond to novel information and yet to maintain a stable basis for action deriving from previous experiences. That is, in an environment that is generally stable over the medium term, it does not make sense to keep fundamentally changing one's models of that environment, although there does have to be some flexibility to learn from new experience. Sweller's hypothesis is that the limited capacity of WM is adaptive because it protects us from having models of the world that are *too* labile. This is possible because of the way processing capacity is understood: that is, what exactly the quanta of information are, that is, what exactly *it is* that cannot exceed 7 ± 2. This relates to a process known as 'chunking'.

Chunking

Although processing capacity in human cognition is limited to working with around 7 units of information or perhaps slightly less, these units are not themselves of a fixed size or complexity: so a single unit of information can in some circumstance be rather complex. For example, people are able to recall more words when asked to recall sentences than when asked to recall simple lists of words (Baddeley et al., 2009) – and this is considered to reflect the ability to treat several words as a single unit of data because of the ability to perceive meaning in the string of words

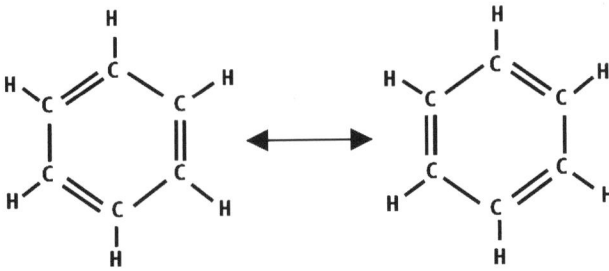

Fig. 5.11 Perceived complexity of information has a subjective component

(i.e. the sentence). This phenomenon is called 'chunking' as in effect the cognitive system is able to 'chunk' together different elements into a whole, which can then be handled as a single unit.

Chunking is not a means of bypassing processing capacity limits in an ad hoc fashion, or else it would be meaningless to quote a limit, and there would be no issue of WM capacity. Rather, information which is closely associated, that is, which *has become* closely associated by previous cognitive activity, can be treated as a single quantum of information for the purposes of WM. Baddeley comments that 'chunking results in an immediate memory span for sentences of about 15 words, compared to five or six unrelated words' (Baddeley, 2003, pp. 835–836). This refers however to a speaker of the language concerned (not someone asked to recall a sentence in an unfamiliar foreign language), who therefore interprets the sentence as something more than a string of meaningless words. That is, because of the representations of previous experience in long-term memory, which acts to interpret new sensory information, the auditory information sensed is perceived in meaningful ways in the processing that occurs before it reaches WM.

For a visual example, consider Fig. 5.11:

It makes little sense to ask 'how complex' the image in Fig. 5.11 is, in terms of whether it would overload a person's WM, without considering the particular context of that individual's LTM. Someone with a reasonable background in chemistry will perceive the image in a very meaningful way as representing the resonance between the two most important canonical forms of benzene, the Kekulé structures. Someone perceiving the information in this way is interpreting it in terms of previous experience (i.e. learning), and it is being associated with (and activating) existing neural circuits that already represent this idea. Such a person can readily process the image, for example, if an unkind lecturer asks them to inspect the figure near the start of a presentation with the warning that questions will be asked later.

When some time later in the session, the lecturer requests course members to sketch the image shown earlier, someone who perceived it as having this meaning will be able to reconstruct the image from having associated the image with the previously represented idea. Such a person is likely to be able to sketch a good approximation of the image. Where there are discrepancies, these are likely to be chemically insignificant – perhaps reversing the position of the two hexagons, perhaps using an equal sign in place of the double-headed arrow. Other course members,

Fig. 5.12 A pattern that
could exceed WM capacity

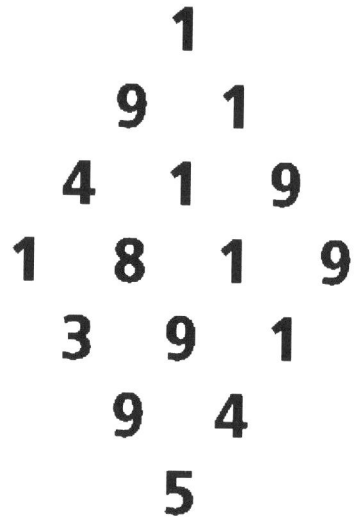

$$1$$
$$9 \quad 1$$
$$4 \quad 1 \quad 9$$
$$1 \quad 8 \quad 1 \quad 9$$
$$3 \quad 9 \quad 1$$
$$9 \quad 4$$
$$5$$

who perhaps have limited background in chemistry, will often struggle to reproduce a close approximation to the original image – perhaps offering something that looks superficially like a chemical structure, without being feasible in chemical terms.

I have used this example in a presentation on teaching and learning to groups of graduates at Cambridge University who are taking on teaching roles, for example, doctoral students taking on supervision roles in labs or working with undergraduates. The level of success of the course members in reproducing the image from memory varies from a good facsimile to abandoned attempts involving a few apparently random lines and letters. It is clearly not fair to consider this reflects upon the differential memory of the different graduates in any absolute sense: but only *in the particular context of this image*. Graduates specialising in medieval literature or law or cultural anthropology do not have the same cognitive resources to interpret the image as natural science graduates who have studied organic chemistry at a high level. On a different choice of task, levels of performance might be very different within such a multidisciplinary group.

Readers might wish to consider the following questions in terms of the image presented in Fig. 5.12.

- Would this exceed your working memory capacity?
- Is it possible to look at this image a few seconds, and then look away, but reproduce the image accurately?

I suspect that (without spending time studying it, and 'committing it to memory') most readers would not be able to reproduce the image accurately unless they are able to recognise some kind of familiar pattern beyond an apparently random array of numbers.

This particular pattern does seem to offer some intriguing features: the diamond shape shows some symmetry in how the digits are arranged into seven different numbers (1, 91, 419, etc.). Moreover, some of the numbers seem to repeat in particular ways, which might aid memory if those patterns are meaningful for individual readers. Whether this provides enough order to allow individual readers to recognise patterns associated with existing representations in memory that can organise perception of the image within WM capacity (i.e. 5–9 units for most people) is an open question. The author's hypothesis is that attempting to reproduce the pattern without errors (and based on what can be held in WM) is likely to prove difficult for many readers.

However, if the pattern of digits in Fig. 5.12 is construed differently – for example, as initially a set of four four-digit numbers that have then been reorganised into a diamond pattern – then the pattern may 'fit' within WM capacity if those numbers can be associated with existing representations in LTM. That should be possible for most readers as the four-digit numbers are not arbitrary and form a sequence that has existing associations for many people. The sequence of digits presented in Fig. 16.1 is likely to be familiar enough for most readers to easily succeed at the task. The pattern in Fig. 5.12 is inherently no more complex than that in Fig. 16.1, although it may be less easy to perceive it in meaningful terms.

This would seem to be an issue about perception, rather than memory. The same sequence of 16 digits becomes easier to remember if it is perceived as four meaningful strings rather than as five meaningless ones. However, it is important to recognise how the cognitive *system* works *as a system*: the effective functioning of limited capacity WM is supported by the interpretive processes of perception. These ideas link with the work of the Gestalt psychologists, who explored how perception organises the sensory field (Koffka, 1967), so that, for example, there is often the identification of a 'figure' that stands out from the background or 'field' (see discussion of perception in Chap. 4).

Memory Techniques

An extreme example of 'chunking' in memory was reported by Ericsson, Chase, and Faloon (1980, p. 1181), who discussed the case of a student who they described as having 'average memory abilities and average intelligence for a college student', but who over a period of 20 months of regular testing in the laboratory managed to demonstrate an increased performance in digit span from 7 digits to 'almost 80 digits'. Despite this extraordinary feat, when partway through the training, the undergraduate was tested on consonants, rather than digits; his memory span was found to be six letters. Ericsson and colleagues report that their subject had not expanded the capacity of working memory, but rather relied upon mnemonic techniques to recall the sequences of digits (cf. 1914191819391945, see above). They concluded that 'with an appropriate mnemonic system, there is seemingly no limit to memory performance with practice' (p. 1181). The ability to increase retention of learned material through rehearsal is well established, although Parkin (1987, p. 12) warns that 'the rehearsal concept places too great an emphasis on the role of intentionality in learning. It is a perverse fact about human memory that we often remember things we would rather forget and forget things we want to remember'.

The use of mnemonic techniques is often recommended to support student learning. The author remembers the phrase 'Now Must I Go Right Round England' being recommended as a way of remembering the characteristics of living things (i.e. Nutrition, Movement, Irritability, Growth, Respiration, Reproduction, Excretion), although the alternative 'MRS GREN' now seems preferred in English schools (with 'Sensitivity' replacing 'Irritability'). Similarly 'Richard Of York Gave Battle In Vain' still seems well known as a way to remember the order of the standard colour names used to describe a rainbow. These devices are often considered useful even though MRS GREN seems a rather arbitrary mnemonic, and very few school children today have any notion of who Richard of York might have been. Similarly the term OIL RIG is taught to students to help them remember that 'oxidation is loss [of electrons, and] reduction is gain [of electrons]', despite the phrase having no direct relevance to the semantic meaning being learnt. When such phrases are used repeatedly, they do seem to be readily remembered and aid learners to access the information they are meant to code.

Whilst it is possible to 'train' memory by using a set of standard techniques (such as memorising a list by imagining a familiar journey and visualising each item on the list in a sequence of places on the journey), the fallibility of normal 'untrained' memory may sometimes be a blessing. Luria (1987) reported an extreme case of pathological memory where an individual with apparently naturally highly developed mnemonic skills who was able to recall extensive complex and meaningless information presented in a clinical setting verbatim years later was largely dysfunctional due to the burden of remembering a morass of detail of everything in his life – apparently making it very difficult to focus on what was significant at any particular moment. The tendency to fail to remember most of the detail of our experiences, just like the tendency to 'remember' tidied accounts that may lack fidelity to original events, may be better considered a result of evolution leading to cognition that is generally adaptive for humans, rather than being seen as flaws in the cognitive system.

The Mnemonic

At first sight, mnemonic techniques appear to be based on arbitrary associations, which might seem at odds with the common wisdom that meaningful learning is more robust than rote learning (Ausubel, 2000), and that so-called deeper, semantic learning leads to a 'better, more durable memory trace' (Parkin, 1987, p. 25),

> The superiority of semantic over non-semantic processing was demonstrated in numerous studies and led to the generation of a principle: namely, that the probability of remembering something is a positive function of the depth to which it was processed. (Parkin, 1987, pp. 25–26)

Yet, actually the basis of mnemonic techniques is to form an association that becomes 'meaningful' for the learner, between otherwise arbitrary information and an aspect of existing knowledge structures. Brahler and Walker (2008, p. 223) tested 'the effect of illogical word associations on the recall of Greek and Latin word roots comprising scientific terminology' through the use of a mnemonic system, 'Medical Terminology 350', that is designed 'to facilitate the recall of factual information by linking new material to an existing framework of life-long knowledge' (p. 219), and used in teaching medical terms. This system

> creates an association to a Greek or Latin scientific word part by introducing a sound-alike keyword (audionym) and related visual image that is familiar to the learner but, however, unrelated to the medical term being learned. The visual image is then altered "illogically" to link the word part to its meaning. Medical Terminology 350…postulates that the "crazier" or "more illogical" an association, the better it is to help recall, retain, and remember over a long period of time and in essence "learn" the meanings of word parts comprising medical terms. (Brahler & Walker, 2008, p. 219)

Brahler and Walker found that the 'illogical association' was 'an effective way to facilitate and improve the recall process because this tool effectively links the new material into an existing framework of knowledge and familiar associations' (p. 223).

Overview: Modelling Memory

Memory is clearly a major feature of human cognition and a major issue in considering student learning. Despite this, it seems to have had limited attention as a research topic in its own right in science education. This is unfortunate because, as the present chapter suggests, the notions of memory that we may often take for granted in everyday life, actually, are somewhat at odds with the way research suggests memory actually functions. Research into student learning in science needs to acknowledge this, so that models of student learning are consistent with how memory actually functions. In particular, such models need to acknowledge that memory is fundamentally a reconfiguring of the higher-level processing apparatus, rather than a discrete store somewhere where memories are located in some encapsulated form until they are accessed. This is important because remembering is not (as we might naively assume) a process of getting something out of a store, but of reconstructing an account from available resources.

Memory tends to be accumulated in an iterative way, rather than as records of a discrete sequence of experiences: we form memories to the extent that current experiences modify the existing substrate for thinking patterns available, and later thinking will further modify those patterns. In other words, when we form memories we are compromising both the current experience and prior memories as the cognitive system looks to interpret the current in terms of prior experience. This makes sense if memory is understood as the process of ongoing modification of our processing apparatus to better fine-tune it to our environment to support work that is largely undertaken at a preconscious level, rather than as an accurate record of specific prior events for conscious inspection and consideration.

Having said that, clearly memory can sometimes provide quite detailed recollections of specific episodes that can be shown to be largely objectively correct – and this suggests a fruitful area of work to explore to what extent memory in science learning operates in terms of episodic memory, and whether there might be very significant individual differences in this. For example, at the present time, science teachers in some educational systems spend a large proportion of lesson time allowing students to undertake hands-on laboratory activities. Part of the rationale for this is that students will remember the specific practical work when they are involved in manipulating the equipment and making the observations themselves. Research however suggests that many secondary students can only offer very vague recollections of a small number of the many specific science practicals they have undertaken during secondary science classes (Abrahams, 2011). Arguably, teachers are operating with lay ideas about how student memory operates – as if a store of records of past experiences – and research is needed to find out how to best fit teaching to the actual operating characteristics of the human cognitive system with its built-in drive for ongoing and on-line updating of its interpretive networks rather than building up sets of accounts of past experiences to refer to.

Chapter 6
The Learner's Understanding

> How can one be sure that someone else understands a concept? Perhaps one cannot. For that matter, how can one be sure that one understands a concept oneself? Again, perhaps one cannot. (Nickerson, 1985, p. 229)

One focus of much research into student thinking in science is the extent of the 'understanding' of science, and the literature includes many claims about aspects of learners' understandings in science. However, 'understanding' is part of the common mental register that I have argued here is generally taken for granted in educational research (see Chap. 2), and so it is usually assumed that this term does not need to be defined. So the authors of a recent study that claimed to have 'developed a new working model to help visualize the relationships among opinion, understanding and evaluation while learning about a socioscientific issue' (Witzig, Halverson, Siegel, & Freyermuth, 2011) did not feel the need to explain what they meant by this core focus of their study.

As suggested before, the 'theory of mind' that we are all considered to develop to make sense of human behaviour (see Chap. 2), and which underpins the language for making sense of our own mental experiences, is almost transparent when we apply it to other people. So in their research report, Witzig and colleagues claim:

> In addressing our current RQs [research questions], we have extended our current knowledge of the interactions among opinion, understanding and evaluation while identifying areas that need additional investigation. Further research in this area is needed to continue to close the gap in our understandings about how topics and evaluation criteria influence students' opinions and knowledge. We believe that our working model ... has assisted in advancing our understanding in this area and imagine that other researchers investigating source selection and evaluation of SCR [stem cell research] and other SSI [socioscientific issues] topics can contribute to this. (Witzig et al., 2011, p. 21)

Here the word understanding is twice used to describe an aspect of the authors' own making sense of the world, as well as being used to label what is inferred from research data about the mental experiences of the (in this case undergraduate) students they studied. Use of the mental register allows ready communication,

K.S. Taber, *Modelling Learners and Learning in Science Education: Developing Representations of Concepts, Conceptual Structure and Conceptual Change to Inform Teaching and Research*, DOI 10.1007/978-94-007-7648-7_6,
© Springer Science+Business Media Dordrecht 2013

but the use of a fuzzy everyday concept without any technical definition limits the precision of the claims made in a study. A consideration of what can be meant by this term offers a useful illustration of the argument made in Chap. 1 that it is important to consider the ontological and epistemological assumptions underpinning research in science education.

The Meanings of Understanding

'Understanding', like 'ideas' and 'remembering', is another of those terms that are drawn from everyday discourse and so generally assumed to be widely understood (sic) and not need defining. However, as Newton (2000, p. 15) has suggested, 'the word "understanding" commonly denotes a variety of mental processes, states and structures'.

The view taken in this volume is that it is important to specify exactly what our research is about and to ensure that authors of research reports and their readers hold sufficiently shared meanings for terms to aid effective communication. As discussed in Chap. 4, a research report involves the representation of the ideas of the researchers in a physical format, which – after being re-represented into nervous impulses by the reader's own 'sensory interface' – will be interpreted before reaching the level of the cognitive system which leads to conscious thought (see Fig. 4.12).

Understanding the Meanings of Others

However, it seems clear that in common discourse we can refer to someone understanding in two significantly distinct ways. According to one dictionary of psychology, 'understanding' is

> Apprehension of the meaning of phenomena, words, or statements; often employed loosely and indefinitely, as some sort of agency; general term, covering functions which involve apprehension of meaning. (Drever & Wallerstein, 1964, p. 306)

If we take 'apprehension of meaning' as a guide, then this suggests that understanding involves a person appreciating *what another person means* by a word or statement. However, in science education, the term is also commonly used in another sense.

Making Sense of the World

White and Gunstone (1992) explored the nature of understanding in a book setting out suitable methods (particularly for teachers) to probe students' understanding. They noted that '…understanding is an elusive quarry. Teachers and students want to secure it, but how can they do so, and how will they know when they have succeeded?'

(p. vii). White and Gunstone saw understanding as depending upon resources from previous learning: 'to understand a concept you must have in your memory some information about it' (p. 3). They also saw understanding as multifaceted and evolving:

> Our definition of a person's understanding of democracy is that it is the set of propositions, strings, images, episodes, and intellectual and motor skills that the person associates with the label 'democracy'. The richer this set, the better its separate elements are linked with each other, and the clearer each element is formulated, then the greater the understanding… understanding of a concept is not a dichotomous state, but a continuum…Everyone understands to some degree anything they know something about. It also follows that understanding is never complete; for we can always add more knowledge, another episode, say, or refine an image, or see new links between things we know already. (White & Gunstone, 1992, pp. 5–6)

White and Gunstone's definition takes account of how understanding is a complex and dynamic entity. In this sense White and Gunstone offer a good example of setting out the ontological nature of what they understand (sic) by their theme of 'understanding'. Research into learners' understanding of aspects of science will inevitably be complicated by the complexity of that research focus.

Two Perspectives on Understanding

Indeed, in terms of the ontological nature of 'understanding', there seems to be something of a difference in the meaning of 'understanding' between the Dictionary of Psychology (Drever & Wallerstein, 1964) definition and the way the term is used in the guide to probing student understanding (White & Gunstone, 1992).

'Apprehension of the meaning of phenomena, words, or statements' implies the question of *whether* the individual does understand 'the' meaning of something. This implies the existence of some standard by which 'understanding' can in principle be judged – whether this is the intended meaning of a specific author or what is taken as a consensus meaning, for example, currently accepted scientific principles. By contrast, White and Gunstone are interested in the *extent* to which a learner understands, and this seems less tied to any comparison with some external standard and rather needs to be evaluated holistically and on its own terms. Their 'richness' of understanding seems to relate to how extensive 'the set of propositions, strings, images, episodes, etc.' available to a learner is, rather than being a judgement of the correctness of those elements by comparison with some independent standard.

This actually only considers part of White and Gunstone's treatment of the topic, for it is implicit in their book – and in places explicit, for example, where they discuss scoring students' work produced in response to their probing techniques – that teachers will commonly probe understanding to find out whether student understanding is consistent with what is set out in the curriculum. However, this does raise the important point that there is a critical difference between the questions:

- Does the learner understand (apprehend the intended meaning of)…?
- *How* does the learner understand (construct meanings for)…?

It might make sense to summarise an answer to the former question with a 'yes', a 'partially' or even a 'B+' or '3/5', but that would not seem to be appropriate in considering the second question. In a significant sense these two questions refer to understanding treated in two ontologically different ways.

Another treatment of the term 'understanding', proposed in the context of discussing learning in science, offers further potential complications:

> I propose two criteria for understanding in science: *connectedness* and *usefulness in social contexts*. The first criterion deals with the structure of a person's knowledge. An idea is understood to the extent that the learner has appropriately represented it and connected it with other ideas, particularly with the learner's own prior knowledge and beliefs…The second criterion deals with the function of a person's knowledge. An idea is understood to the extent that the learner can use that idea in successfully performing significant tasks appropriate to the social context in which they occur. (Smith, 1991, p. 46)

Smith's first criterion touches upon both the issue of understanding being about the structure of knowledge ('the extent that the learner has…connected it with other ideas'), but also that the way the learner structures can be judged for correctness ('the extent that the learner has appropriately represented it'). The second of Smith's criteria potentially looks beyond what can be elicited in standard test or research conditions, to the use of knowledge in action (Driver & Erickson, 1983), and in a context-sensitive way – again as is judged by someone to be 'appropriate' to that social context.

Normative and Idiographic Approaches to Exploring Understanding

So it seems that in exploring student understanding, a decision has to be made about how understanding will itself be understood. The researcher needs to be clear about the ontological nature of what is being researched: in this case whether a learner's understanding is the kind of thing that can be compared with and evaluated against, and so be judged by an external norm, or something to be mapped out and characterised on an individual basis. This issue relates to a major distinction in research in education and the wider social sciences, which has considerable importance for science education.

Testing Student Understanding

Sometimes research is undertaken to find out the extent to which students match up to what is expected or required. For example, large-scale surveys such as TIMSS (the Trends in International Mathematics and Science Study, Mullis et al., 2005) and

PISA (the Organisation for Economic Cooperation and Development's Programme for International Student Assessment, OECD, 2007) seek to measure aspects of student performance in different countries, allowing those countries to be ranked. In such research, answers need to be considered correct or incorrect, and the nature of incorrect answers – beyond failing to match what is judged as correct – is not a primary focus.

If understanding is to be understood (sic) in this context, for example, if we set out to find out what proportion of students understand photosynthesis at some specified level or understand the relationship between force acting on a body and its acceleration, then this assumes that understanding is capable of being operationalised so as to be summarised in directly comparable statements. We might consider this type of research as 'normative' in nature, as it is concerned with norms and making evaluative judgements.

Unfortunately, the term normative itself has two somewhat different meanings (Dent, 1995) and could actually be understood to refer to either what is set out as target knowledge, 'standards', or what is typical for the population. This is clearly not the same thing, for a student who believed that an object would only continue to move if it was subject to an applied force would fall short of normative standards (the Newtonian principles set out in curriculum) but would demonstrate normative beliefs in the sense of what research suggests most students believe (Watts & Zylbersztajn, 1981).

A possible alternative term that could be applied here might be 'positivistic', as this approach to research appears to assume both that there are distinct 'right' answers, and that it is possible to judge student understanding in terms of the extent to which it matches such right answers – issues considered in more detail below. For clarity, this approach will here be labelled as *normative-positivist*.

Exploring Student Understanding

Other research, whilst perhaps ultimately motivated by a notion that such studies can inform teaching to better support learning that progresses understanding towards canonical knowledge set out in curriculum targets (Taber, 2009b), is rather different in that it sets out to explore the nuanced nature of the students' thinking, knowledge and understanding, *in its own terms*. Such research requires qualitative, some would suggest ethnographic, approaches, because of the nature of the subject matter. This second type of research could be considered *idiographic*, being related to a focus on individual learners (Gilbert & Watts, 1983).

Research that adopts the idiographic approach responds to a considerable challenge: given the ontological assumptions about the potentially unique and nuanced nature of each learner's understanding, and the indirect processes by which understanding can be elicited so it is represented in the public domain, and then interpreted by researchers (see Chap. 4). This is considered further below.

Table 6.1 Two distinct ways of considering 'understanding' in research

Approach	Normative-positivistic	Idiographic
Ontological assumption	Understanding is the kind of thing which can be judged as right or wrong/present or not	Understanding is complex and holistic, consisting of a rich array of interlinked elements
Epistemological assumption	Student understanding of a science concept area can be operationalised to produce simple statements which can be objectively compared between individuals or against specified targets for learning	Student understanding of a science concept area needs to be explored through in-depth probing using qualitative methods capable of uncovering nuances of meaning
Methodological consequences	Research starts with an analysis of target understanding, and the identification of the key elements against which student understanding is to be evaluated	Research involves a detailed exploration of the way the individual student understands the target concept/topic

These two ways of understanding 'understanding' are reflected in Table 6.1. The point here is not that one of these meanings is necessarily to be preferred but rather that these two distinct meanings are both commonly used and have very different implications in a research context.

Two Approaches to Research

The logic of research is different in these two approaches. In normative-positivistic research, the starting point is the target understanding of the topic, that is, the understanding that the student is asked to acquire, and the students are in effect being *tested against* that. Conversely, in the idiographic tradition, the research needs to be more open-ended, seeking to explore how the student understands a topic or concept, without being too channelled by the researcher's own understanding of the target understanding. Here the starting point is what the student has to tell us, not what we think a 'good' understanding should be like. However, there are quite significant problems in carrying out research through either approach.

Testing Student Understanding: Challenges of the Normative-Positivistic Approach

The challenges of this approach to researching into student understanding can be appreciated by considering the assumptions underpinning the approach (see Table 6.1). Accounts of research reporting on student understanding of some aspect of science in the normative-positivistic mode can be considered to be valid to the extent that:

1. It is possible to set out the target knowledge in an operationalised way. That is, it is assumed that target knowledge exists in an objective sense, so that different

members of the teaching/research community would agree on how students were meant to understand the content area/topic after teaching, and that it could be defined or set out in a form that allows the development of test instruments. (The nature of knowledge will be revisited in Part III.)

2. Student understanding is the kind of entity that can be considered either to match or to not match target understanding.

3. It is possible to develop test items that distinguish understanding that matches target knowledge from student understanding that does not.

This process is not unique to research, as in effect this is how formal assessment works, and reflects the processes undertaken by examination boards in setting papers and grading student scripts.

Operationalising Target Understanding

The assumption that target understanding can be identified is itself not unreasonable, as it also forms a basis of science teaching. The curriculum, or scheme of work, or examination syllabus or specification, sets out what it is that students are asked to know and will be tested on. To the extent that such documents may not offer precise specifications, it is part of the work of the teacher to interpret such documents in the process of planning teaching. (This is reflected in Fig. 1.3 in Chap. 1.)

Clearly this is not a straightforward matter. There are processes of interpretation involved in forming curriculum models of scientific knowledge and judgements to be made about the level of abstraction and complexity appropriate for students at different levels (Taber, 2000a): students starting secondary school would not normally be expected to learn about quantum-mechanical models of the atom and the most detailed models of the chemical and electronic processes involved in photosynthesis. Teachers have limited subject knowledge (Gilbert, Osborne, & Fensham, 1982) and may hold alternative conceptions in some topics (Taber & Tan, 2011), and this will influence how they interpret the curriculum documents and so the basis upon which they set out the science to be learnt. In research, there may be a process of producing an initial representation of target knowledge which is developed with, or checked by, various individuals considered to be in a position to confirm the validity of the formalism, perhaps university faculty, perhaps teachers working with students at the level concerned (Treagust, 1988).

Developing Test Items

The process of developing test items that can discriminate between a student holding or not target understanding is clearly a complex business. This is a large topic, and the issue will only be touched upon here. Objective items, such as multiple-choice items, are readily 'marked', but their production requires reducing aspects of understanding to rather specific elements that can be independently considered. They are also open to guessing. Perhaps even more problematic in research contexts, students can select correct responses, the nominal evidence of understanding, based on scientifically incorrect reasoning (Palmer, 1997), making such items limited indicators of understanding.

Questions asking for more extended responses give an opportunity for learners to demonstrate their understanding, but may:

- Be harder to 'mark' against the formalism ('marking scheme') representing the target understanding.
- Rely on higher levels of literacy and metacognitive planning skills (which may be a problem with younger students and complicates what is being tested).
- Require respondents to demarcate the topic being asked about in a way that matches what is wanted. That is, in writing about an area such as the shapes of molecules or the circulatory system, there may be things the student understands which are on the marking rubric, but not recognised as relevant by the student when producing the response.

As always, we only access the production – the representation made in the external world – not the underlying thought processes. It could be argued that if the student does not recognise the relevance of certain ideas to the topic, then that shows a deficit in understanding as in the target understanding there is a clear link such that its relevance is part of the understanding. However, this does not allow discrimination between a student with this deficit in linking or demarcating ideas and a student who simply does not have those ideas available.

So different types of test item have different strengths and weaknesses, but there is always the likelihood of any test items leading to false positive or negatives, for example, recognising understanding (that was not present) based on a guess on a multiple-choice question and/or failing to recognise understanding (that was present) due to a respondent not appreciating what is relevant to include in an extended response item. These inherent difficulties need to be considered when interpreting research that offers accounts of the proportion of students said to understand certain scientific concepts – such as the claim in Table 1.1 that 'about one-third of the pupils at the compulsory school have little understanding of chemical change' (Ahtee & Varjola, 1998, p. 310).

The Messy Nature of Student Understanding

All of this assumes that student understanding is such that in an ideal testing context the student understands things in either one way *or* another. However, considering much of the research into aspects of student understanding, this seems over simplistic. It seems more reasonable to suggest that *often students have several available ways of understanding the same phenomena.*

For example, research has shown that most students enter school science with an existing way of thinking about how and why objects move or stop moving (e.g. Gilbert & Zylbersztajn, 1985; McCloskey, 1983; Watts & Zylbersztajn, 1981). Moreover, there is usually a strong commonality in their thinking, in that it is generally assumed that a force is needed to both initiate and maintain movement (as opposed to the formal science understanding that an applied force is required to change the state of motion, but that in the absence of any forces a moving object would continue to move indefinitely). This so-called impetus or F-v (Force-velocity) framework for thinking about moving objects has been shown to be tenacious, such that even after school learning of the Newtonian account, most students still seem to understand movement in terms of the requirement for a continuously applied force. Indeed, even advanced physics students and graduates can be 'caught out' as explaining phenomena in these terms. In some senses this classic example of an alternative conceptual framework (see Chap. 11) might better be understood as an alternative formalism (Taber, 2009b) that is largely coherent and consistent with empirical experience, but for present purposes the important point is that many students, indeed most, seem to understand phenomena in these terms.

However, whilst it seems that school teaching does not undermine a person's ability to understand force and motion in terms of an impetus framework, it is still the case that many students can learn the Newtonian formalism and demonstrate understanding of this way of thinking. So Palmer (1997) reported that when students were asked to answer a series of 8 objective questions, all concerning 'the context of a freely moving body in linear motion' (p. 692), they commonly based their responses on *more than one* reasoning pattern, apparently depending upon scientifically irrelevant contextual cues in the questions. Palmer's sample included two groups of students, secondary (year 10) pupils and tertiary students who were preparing to be teachers. Impetus-type thinking was most common among the secondary students, although 'the majority of students in both groups held the alternative conception that "motion implies a force" in at least some of the questions in the survey' (p. 691). The university students were also more consistent in their reasoning (i.e. 'the average number of reasons used by the university students was 1.82 and only one student used more than three types of reasoning over the eight questions'), whereas 'the Year 10s tended to use more different types of reasoning… and almost one-third of them used four or more types of reasoning in the survey' (p. 690).

A key issue for research here is whether it is more appropriate to consider student understanding in such situations as (a) manifold or simply (b) complex enough to appear 'convoluted'; that is, do students have available resources which in effect offer them alternative ways of understanding; or is it better to consider they have *a* way of understanding phenomena that is nuanced by multiple considerations (cf. Pope & Denicolo, 1986)? In the latter case, what may seem as inconsistent reasoning might instead be considered as the application of a set of rules that indicate different approaches to a problem depending upon features that may not be considered relevant from the scientific perspective (cf. Camacho & Cazares, 1998). That is, Palmer's tertiary students may either have acquired a scientific way of thinking to supplement existing impetus ideas or they may have developed their existing understanding to incorporate new considerations, with potential incoherence avoided by having ways of determining how to respond to different contexts. So a student may maintain a belief that a moving object has motive force acting on it but in some contexts (light or slow moving objects; obvious resistive forces) does not see that as the most salient or critical feature of the situation.

This distinction, that is, between (a) several discrete ways of understanding which can be cued by features of a stimulus question or the context of questioning; and (b) a complex understanding, where some kinds of priority rules determine which of several possible principles will be considered most significant in different specific situations; may seem to be purely a matter of semantics. However, some researchers have offered accounts of students' thinking which seem to presume a distinction here. In particular, Solomon (1992) argued that alternative ways of thinking about energy available to secondary students should be understood as linked to two rather distinct domains. Solomon's model was definitely aligned with option (a) here, for she saw students being channelled to either formal scientific or everyday lifeworld ways of thinking about the same phenomena depending upon the perceived social context of a question, rather than learners making distinctions between different physical contexts represented in problem situations (see Chap. 12).

Of course, both of these possibilities may exist – either depending upon science topic or even within science topic – and researchers looking to make sense of learners' understanding of science topics should not presume complex responses are either manifold *or* convoluted prior to data collection and analysis. Rather, given the current state of knowledge in the field, this should be a question for empirical research rather than an ontological commitment to guide interpretation of data.

An Example of Manifold Understanding of a Science Topic: Student Understanding of Ionic Bonding

There are probably many situations where student understanding takes a 'manifold' form, where it may be best understood as the individual having available more than one way of making sense of a particular phenomena or concept area.

In my own research, I have found evidence of something of this nature. Interview studies with English college students (cf. 16–19-year-olds) had allowed me to identify a way of thinking about ionic bonding, which was at odds with the curricular model presented in teaching. There were strong common elements to the thinking of different students, as well as consistency with aspects of the findings of other research undertaken in a different educational context (e.g. Australian research such as Butts & Smith, 1987). This led to the construction of a model comparing the target knowledge with the alternative way of understanding ionic bonding that I had identified from my informants (Taber, 1997). This is represented in Table 6.2.

The alternative way students understood ionic bonding in NaCl (often discussed with students as a familiar archetypal ionic compound, having a 1:1 ratio of ions)

Table 6.2 Comparing two ways of understanding ionic bonding

In the alternative way of understanding ionic bonding identified among students	In the target knowledge in the curriculum
An ionic bond is an electron transfer event between atoms leading to ion formation and allowing atoms to obtain octets/full outer shells of electrons	An ionic bond is the electrical interaction between oppositely charged ions adjacent to each other in a lattice
Therefore, each ion only has bonds with those counter-ions it has donated/accepted an electron to/from, so can only form a number of bonds equivalent to its charge (Na^+ has one ionic bond to one Cl^-)	Therefore, each ion is bonded with those oppositely charged ions it is adjacent to (so in NaCl, each Na^+ is bonded to 6 Cl^- and vice versa) – the number of bonds an ion has is determined by coordination number, not electrovalency
So there is a difference between the interactions between adjacent ions that are bonded, and those that are just attracted together by forces	So there is no difference between the interactions between any one ion and any of its oppositely charged neighbouring ions (in a symmetrical lattice)
And in effect the ionic lattice contains molecules of ions bounds together	And there are no molecules present in the ionic lattice

suggested that they saw an ionic solid as being very much like a covalent solid, with molecules, or at least quasi-molecular ion pairs, with strong internal bonding, and held in the lattice by attractions to other molecules/ion pairs that were not considered to be proper bonds. Indeed, given the abstract and unfamiliar nature of the molecular models used in learning chemistry (Gilbert & Treagust, 2009; Harrison & Treagust, 2002), it has been suggested that if covalent bonding is met first, students' learning about covalent bonds may act as a template for making sense of teaching about ionic bonds, and so (as we saw in Chap. 5) existing learning may bias new learning (Taber, 2001a). This is not likely to be the only feature of teaching that may lead to students developing this alternative understanding. The molecular way of understanding ionic bonding builds upon the inappropriate idea that bonds are a means for atoms to fill shells (see Fig. 6.1), and many textbooks include misleading diagrams showing ion formation based upon interactions of single metal and non-metal atoms, so that this becomes the first substantive representation that students meet that they associate with ionic bonding.

The model of two ways of understanding ionic bonding, as represented in Table 6.2, was used as the basis of developing a diagnostic instrument to help teachers find out whether students in their classes held the alternative understanding. The instrument consisted of a figure showing a two-dimensional representation of a part of NaCl lattice (cf. Fig. 6.2), with 30 statements that students were asked to judge as true or false (with the alternative of a 'do not know' response option). The items were mostly written to represent aspects of the target knowledge (understanding ionic bonding in electrostatic terms) or the alternative framework for understanding ionic bonding in molecular terms.

Students surveyed were in classes (a) studying science at school-leaving level (i.e. 14–16-year-olds in the UK system) who had studied the topic of bonding at this level, (b) students in post-compulsory classes taking chemistry at 'advanced' level

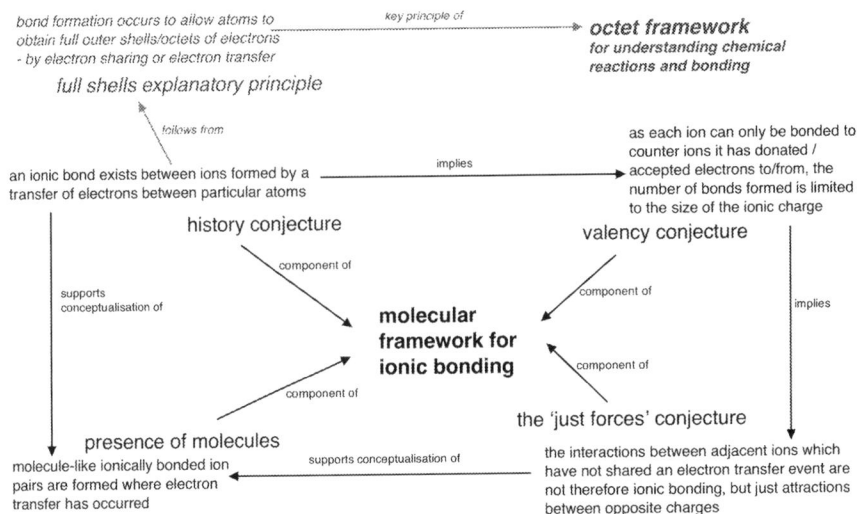

Fig. 6.1 An alternative framework for understanding chemical bonding (From Taber, Tsaparlis, & Nakiboğlu, 2012)

Fig. 6.2 NaCl, a structure with ionic bonding

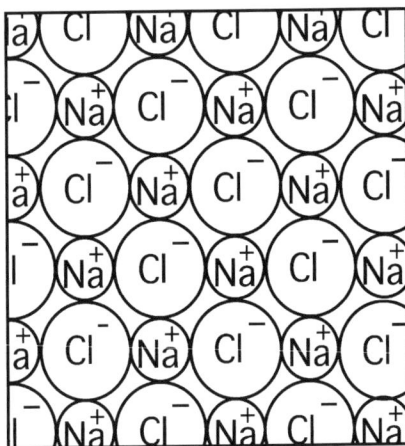

(i.e. studying for university entrance level examinations, normally 16–19-year-olds) who had not at that that point studied the topic beyond school-leaving level, and (c) students in post-compulsory classes taking chemistry at 'advanced' level who had studied bonding at this higher level (Taber, 1997).

On a simplistic notion of student understanding, it might be expected that individual students would either have learnt the target understanding of ionic bonding, and so would judge the statements accordingly, or if they understood ionic bonding in terms of the alternative framework, they would instead judge statements according to that way of understanding. However, patterns of responses at all three levels were more complex than that, with individual students often responding to items in a more nuanced way.

Overall there was evidence of progression, in the sense that response patterns better matched target knowledge in the students studying chemistry at advanced level than the younger respondents studying at school-leaving level and among those advanced students, in those who had studied the topic at that level. It should be noted that students were a convenience sample, volunteered by their teachers, and so comparisons between sub-samples should be interpreted with care. However, students generally judged as true some statements that were based on the curriculum model, *and* also some statements derived from the alternative framework. Similar findings were obtained when the diagnostic instrument was translated for use in Greece and Turkey (Taber et al., 2012). In some cases, this meant selecting as true statements that appear contradictory. So, for example, among the students studying for school-leaving examinations, most agreed that 'a chloride ion is only bonded to the sodium ion it accepted an electron' (56 %), and that 'a chlorine atom can only form one ionic bond, because it can only accept one more electron into its outer shell' (64 %), statements which reflected the alternative 'molecular' framework for understanding. Yet most of this group of students also agreed with the statement 'each chloride ion is bonded to more than one sodium ion' (58 %) as the curriculum model would suggest.

If we were to assume that students understand ionic bonding *either* in terms of the curriculum model presented as target knowledge or in terms of the common alternative conceptual framework, then the pattern of responses on the instrument is difficult to explain, as most students' profile of responses do not completely match either framework. That in itself could simply imply that the alternative conceptual framework developed from the interviews study is not a good model of the way many students understand the concept of ionic bonding, that is, that they have developed *alternative*, alternative understandings, which are not well reflected by either of the ways of understanding the topic built into the design of the instrument, that is, the curriculum target knowledge and the molecular framework. The nature of such 'conceptual frameworks' will be considered in Chap. 11. However, even if this is the case, it does not explain what kind of understanding would allow students to agree with apparently contrary statements, as were found among responses to the instrument.

It is known *both* that students may use technical scientific terms in rather different ways to their teachers and scientists (Watts & Gilbert, 1983), *and* that they may see concepts as contextually bound in ways that make their understanding of concepts idiosyncratic (Palmer, 1997) – and this has potential for making a student's responses in research seem inconsistent until we are able to develop a good model of how the student understands the concept area. This is of course part of the basis of the argument for using in-depth open-ended research techniques in this type of work, as discussed above. However, the common judgement of what appear clearly contradictory statements (presented in relation to the single context of the NaCl solid lattice) as true by respondents to the diagnostic instrument discussed here suggests that this is *not* just a matter of student understanding not being caught by either of the frameworks used to design the instrument.

What seems more feasible here is that to some extent many respondents have more than one way of understanding, and so thinking about, ionic bonding. So many respondents are capable of thinking about ionic bonding in terms of discrete tightly bound NaCl units, bonded through a process of electron transfer and which are held in the lattice by weaker interactions; and also to appreciate something of the curriculum perceptive of the bonding being no more than the electrostatic binding of the lattice because of the array of positive and negative charges. Given these two ways of understanding the concept, the presented statements may well be *judged true* if they fit one or other way of understanding, and so apparently contradictory statements come to be judged to be true. In the UK study, the results from the three groups of students suggest that the relative weighting given to these two ways of understanding ionic bonding shifts towards the curriculum model with increasing level of study (Taber, 1997). This interpretation is consistent with what was found in interviews, with students such as Tajinder (see Chap. 5), where several ways of understanding the 'same' concept were identified, and conceptual change was best understood as a shift in which ideas were cued (Taber, 2001b), rather than a switch in how a topic was understood (cf. Part IV).

Different Ways of Understanding

The presence 'within' one mind of several ways of understanding aspects of the world should not be surprising, as it is a commonplace of human experience. This is represented in Fig. 6.3. Sometimes decisions are difficult because we can mentally model a situation in different ways, exploring possible advantages and consequences of particular actions. Presumably some 'floating voters' in political elections are able to accept something of the different understandings of the world presented by more than one political party.

Even when we may feel we have a clear, preferred, understanding of a situation, this does not exclude appreciating different understandings. A mature 'theory of mind' (i.e. the ability to mentally model what might be going on in other people's minds – see Chap. 2) allows us to construct an understanding of how someone else might understand something differently to us. This is certainly essential in argumentation; for unless one can appreciate the other person's way of understanding the focus, it is not possible to do more than simply reiterate one's own position or simply disagree with other points of view: 'argument is an intellectual process [whereas] contradiction is just the automatic gainsaying of any statement the other person makes' (from the 'Argument Sketch' from the BBC Television Show, Monty Python's Flying Circus, episode 29).

The premises of the present chapter are (1) that students may understand scientific concepts differently to how they are presented in the curriculum, and (2) that researchers, and indeed teachers, may build up an understanding of how the student understands the concept. For example, a teacher may simultaneously:

• Understand the curriculum model of force and motion
• Understand Aristotle's way of thinking about force and motion

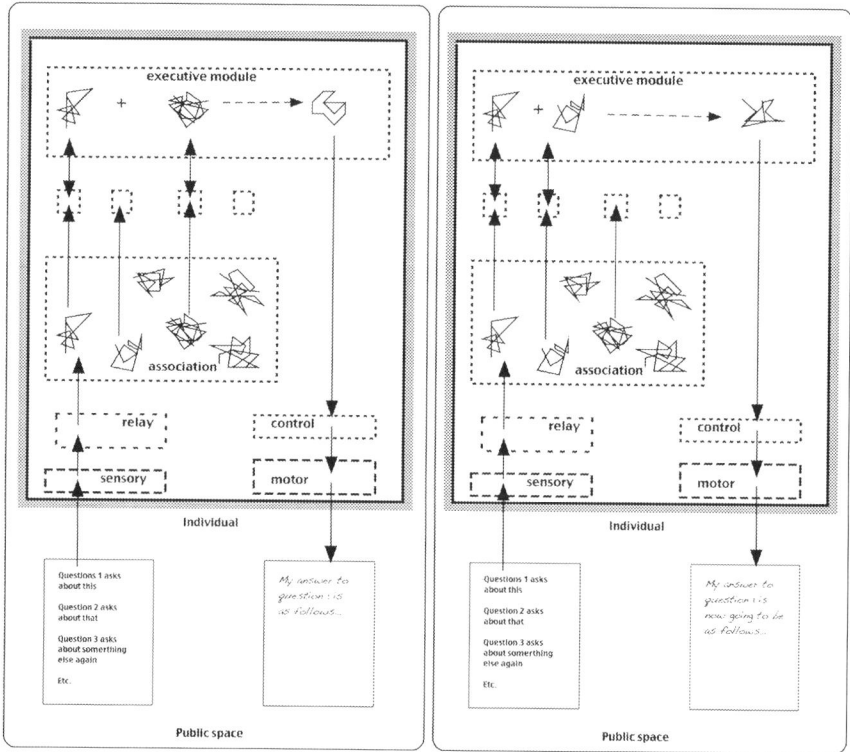

Fig. 6.3 An individual may have the mental resources to potentially understand something in more than one way

- Understand the common alternative conceptual framework for force and motion reported in the literature
- Understand the specific way a particular student thinks about force and motion

This would involve having available as 'mental resources' several different ways of making sense of (i.e. understanding) the same phenomenon. For argument's sake this particular teacher may well be committed to the first of these options, so we could say that she only understands force and motion in one way herself but that she also *understands other ways of understanding* the concept area.

Meta-understanding and Multiple Understanding

There is an important distinction here, between (a) one's own understanding, that one is committed to as the best way of making sense of some aspect of the world, and (b) an understanding of other understandings. In the latter case, what is understood is considered by oneself as something *other* than (in this example) force and motion. This teacher has a way of making sense of an aspect of the world, a way of making sense of how Aristotle made sense of this aspect of the world, a way of making sense of how (according to researchers) many students make sense of this aspect of the world and a way of making sense of how a particular individual learner makes sense of this aspect of the world.

Although the teacher has alternative ways of understanding the same phenomena available as mental 'resources', only one of these *is committed to* as the appropriate way to make sense of that aspect of the world (and the others are resources used to make sense of other people's thinking, committed to as appropriate ways to make sense of *those specific aspects of the world that are the public representations of the contents of other people's minds,* cf. Chap. 4). This would seem to be somewhat different from the situation of a learner who has two distinct ways of making sense of ionic bonding, or the relationship between force and motion, and is not strongly committed to either as the best way of understanding that aspect of the world. So there is an important distinction relating to the levels of commitment to particular ideas as representing how some aspect of the world actually is.

Understanding Distinguished from Beliefs

This important distinction can be denoted by distinguishing *meta-understanding*, that is, the understanding of a possible understanding of a phenomenon, from *multiple personal understandings,* that is, having several competing understandings for the same phenomenon. The difference between meta-understanding and personal understanding is one of commitment or belief. So the historian or physics teacher can come to understand how Aristotle understood force and motion (Toulmin & Goodfield, 1962/1999), that is, to develop meta-understanding of Aristotle's mechanics, without committing to understanding the world the same way: without believing that is the best way to make sense of force and motion.

In a similar way, a student who comes to school with a strongly committed alternative understanding of some scientific topic, perhaps an impetus like understanding of force and motion, would need to develop an understanding of the curriculum presentation of the topic before there was any possibility of committing to it as a better way of understanding that aspect of the world (Thagard, 1992). Indeed, research suggests that many students who do come to understand the curriculum presentation well enough to use it successfully to answer formal assessment questions still do not commit to it as a better way of making sense of forces and motion in their everyday lives. We might say for these students they have acquired a meta-understanding of the curriculum formalism but do not themselves personally understand the world that way.

Whilst this is a key difference, there is an important relationship between these two situations, as it is not possible to change one's way of understanding until one has available an alternative. That is, shifting commitment towards a new way of understanding is only possible once that understanding is available as a mental resource (see Chap. 15). This leaves the question of whether genuine cases of multiple understandings of which the learner is consciously aware as viable alternatives, and so not meta-understandings, and commits to, can occur. To be *firmly committed* to more than one such possibility would seem to be logically excluded; it might be expected to lead to an awareness of a difficulty, perhaps what Piaget termed disequilibrium (Piaget, 1970/1972) or what has sometimes been called cognitive dissonance (Chapanis & Chapanis, 1964). If a human's cognitive system is a self-regulating system, then such a state should motivate changes that are intended to remove the ambiguity.

Yet, by the same token, when we consider people as inherently learners, as those whose conceptual systems are still developing and are 'works in progress' – from a constructivist perspective as actors in an environment receiving constant feedback to allow them to adjust their mental models to better match expectations to experience (Glasersfeld, 1989; Kitchener, 1987) – we should not expect the 'current' state of a person's conceptual system to be fully coherent and consistent. We are all, to some extent, such 'works in progress', with an ongoing programme of making sense of the complexity of our experiences of the world, including from time to time events that seem completely incongruent with our expectations based on how we understand past experiences.

Alternative Interpretations of Perceived Manifold Conceptions

Given these considerations, it would seem that researchers need to show caution in making assumptions when interpreting evidence of learners' thinking (i.e. the public representations of their thinking; see Chap. 4) as demonstrating manifold ways of understanding the same phenomena. The discussion in this chapter suggests that there are various related possibilities here, then, which researchers should look to disentangle:

(a) The individual may have a single internally consistent way of understanding a phenomenon, which may or may not seem coherent when public representations of her thinking are interpreted from the perspective of canonical knowledge (e.g. what seem similar cases to an observer are within this individual's scheme perceived as significantly different on some characteristic imbued with salience by that individual).

(b) The individual may have available several ways of understanding the same phenomenon, and be aware of this, and is committed to one way whilst acknowledging (having meta-understanding of) the different ways others such as a science teacher understand the phenomenon.

(c) The individual may have available several ways of understanding the same phenomenon, and be aware of this, and is not sure yet which is the best way to make sense of the phenomenon.

(d) The individual may have available several ways of understanding the same phenomenon, and be aware of this, and consider that each has something to offer and so is not motivated to be restricted to one approach: which could in some cases reflect a more sophisticated epistemological stance that our ways of understanding some things are necessarily limited and imperfect, in which case commitment to one likely imperfect perspective may prove inadequate.

(e) The individual may have available several ways of understanding the same phenomenon, only being consciously aware of one of these, yet sometimes applies other implicit ways of thinking that operate at the preconscious level without noticing this.

If a researcher was only interested in whether a learner offered a 'correct' response to a question on an occasion when testing was carried out – which is in effect the stance implied by the nature of many public school examinations – then this complexity is of little relevance. Yet anyone claiming to be undertaking research exploring student understanding in a topic needs to be aware that an account that does justice to an individual's understanding is likely to need to engage with issues of the status of understandings elicited, in terms of (i) whether they are unitary or part of a manifold of available ways of understanding and (ii) of the level of commitment the learner has to the ways of understanding elicited.

Describing Student Understanding: Challenges of the Idiographic Approach

It is becoming apparent that, realistically, no research report is likely to do full justice to an individual's understanding, so all such reports will be simplifications and approximations. Therefore, it could be argued that the distinction in Table 6.1 between normative-positivistic and idiographic approaches to describing understanding offers a false dichotomy, for if idiographic research inevitably produces simplifications, then it is just a matter of degree to summarise further, by taking such research outcomes and reducing them to simple statements that are suitable for comparative evaluation – for example, to determine that 'the student does not understand photosynthesis' or 'has a good understanding of how force relates to acceleration'.

In practice, in science education, these two aspects are sometimes combined, with research to explore a students' understanding leading to a detailed description (an idiographic approach), which could then be compared with either target knowledge or what was considered typical in the wider population such as of students of that age studying in that educational system (switching to a more normative-positivistic stance).

This would however almost certainly involve some form of discontinuity in the research process. Given what White and Gunstone have suggested about understanding (see above), any hypothetical authentic account of a learners' understanding of a topic or concept in science will be a report of something which is complex and so to some extent needs to be appreciated holistically (Pope & Denicolo, 1986). Yet there is no simple way to compare accounts of such complexity (without producing equally complex evaluations), and so in general such comparisons can only concern *highly reduced features* of the account.

It might be argued by some researchers valuing the richness of idiographic research that the reduction and consequent simple evaluation/comparison of students' understanding with statements of target or 'typical' understanding is not justified as the reduction completely distorts the original account. However, if done carefully and recognising necessary limitations, it may be argued that the reduction and consequent simple evaluation/comparison of students' understanding with statements of target or 'typical' understanding can be justified as the reduction represents *key elements* of the original account which may be used as *valid summary statements* (and which can be claimed to be valid as such *because* they derive from an in-depth investigation and so take into account and acknowledge the complexity of the students' understanding).

The Researcher's Dilemma

Pope and Denicolo referred to this kind of issue in science education as a dilemma. They discussed how the essence of much research involved detailed work, because the nature of what was being investigated was complex and could not be clearly represented in pithy summary statements (Pope & Denicolo, 1986). However, journal space tends to favour brief summary reports, and teachers usually have limited time to engage with research, so a brief summary of main themes in reporting student understanding might at least inform classroom work in a way a dense ethnographic report is unlikely to.

This *is* a genuine dilemma, and there is no simple solution. However, one can take a pragmatic perspective. In principle, the reduction of detailed, nuanced accounts of students' understanding to simple statements suitable for ready comparison with other such statements will always *somewhat* distort the original account. That is inherent in the notion of reduction. However, it is less clear that this need always necessarily be an invalid process as long as one is aware of the shift in the way 'understanding' is understood. Part of the work of a researcher in analysing data is to reduce the material to produce accounts that are both authentic, yet concise enough to be of value to others.

A priori it is not possible to determine whether such reduction will produce outcomes that remain at some level 'valid' representations of student understanding and so offer authentic accounts suitable for making useful comparisons with target understanding. This is likely to depend upon a range of factors, including perceived purposes of the research. Certainly, in principle, we might expect there to be circumstances where such a process is invalid, for example, when a student's understanding is heavily context dependent, such that reduction might lead to a set of apparently contradictory statements – some of which might be judged correct and some incorrect. (Any attempt to draw a comparison taking this into account is immediately both building back in some of the complexity and requiring judgements about how to do this on an individual basis.)

In other cases, however, it may be that although understanding is complex and nuanced, it is possible to identify a key conception which appears to be at the core and which can reasonably stand for the students' understanding of some topic or concept 'to a first approximation'.

So this would suggest that reducing rich idiographic accounts to simple summary statements that can be used to make comparisons always risks oversimplification, but providing this is acknowledged and care is taken not to continue to simplify when this starts to introduce significant distortions, then sometimes such reduction can be justified and can support valid comparisons (e.g. with target understanding). Clearly such work depends upon the sensitivity of the researcher (including familiarity with the full data set) and careful judgement based on the researcher's own interpretations.

From a perspective that acknowledges the richness and subtlety often found in students' thinking, even when research aims at making such comparisons, it still depends upon an initial in-depth approach to investigating the students' understanding, as only after that work has been done is it possible to make judgements about the extent to which simple summary statements (of the sort suitable for comparison and evaluation) might be able to stand for the full description. Reports of such work should also acknowledge the issues raised here, and the inherently problematic nature of shifting from the idiographic stage in the research considered necessary because of the complex and nuanced nature of what is being studied (a person's understanding), to the normative-positivistic phase required to provide the simple summary statements needed to make comparisons and evaluative judgements.

Comprehending Language

Learning in science certainly involves students observing and interpreting phenomena. However, many phenomena of interest in science are not readily observed, at least not directly, being too small or too large, to hazardous, too slow, etc. Moreover, much of the content of science comprises of theoretical ideas that need to be explained. A good deal of science teaching and learning is dependent upon communication through language (Lemke, 1990), for example, listening to the teacher and reading. This might be considered to short-circuit the need to learn by direct experience, to allow us to take advantage of how someone else understands some aspect of the world (Karmiloff-Smith, 1996). However, such short-circuiting needs to be understood within the context of the material presented earlier in the book:

> Understanding is a very personal thing…Understanding is not something that can be passed or transmitted from one person to another. No one can make the connection for someone else. Where there are connections to be made, the mental effort has to be supplied by the learner. (Newton, 2000, p. 2)

Most humans learn a mother tongue at a young age (Chomsky, 1999), largely without explicit teaching (although learning to read is not generally acquired spontaneously), and for most of our lives we tend to communicate in that language, readily producing utterances and usually comprehending (i.e. making sense of) the speech of others. Johnson-Laird (2003a, p. 5) argued that much of the process of understanding an assertion in your native tongue 'is profoundly unconscious'. There are thought to be a series of automatic processes that precede conscious awareness of what the speech means. Johnson-Laird suggested that

> Once you have recognized the words in the sentence – no mean feat – there are three main steps in grasping the significance of its utterance…The first step is to compose the meaning of the sentence out of the meanings of its words according to the grammatical relations amongst them. The second step is to use general knowledge to modulate this composition… And the third step is to use this interpretation to construct a mental representation of the situation described in the assertion. (Johnson-Laird, 2003a, pp. 6–7)

Johnson-Laird argues (2003a, p. 10) that 'the results of the first two steps – compositional semantics and modulation of knowledge – must be an expression in an unconscious mental language'. It has been argued that the language we have evolved for social communication is essentially different from this 'machine code' used within the brain, as 'communication of mental objects is usually accomplished through the symbols of language, a heavy and cumbersome coding system, not necessarily well adapted to the "language of thought"' (Changeux, 1983/1997, p. 162).

So it is argued that thought is primarily non-verbal and in effect has to be expressed in language for the purposes of communicating to others:

> There is no translation between speech and thought. … To express a thought is not to translate it. The relationship is not between inner and outer speech; nor is there simply something in our heads that has a character open to translation. Rather, there exists something that could be more or less expressed. This statement is grounded in the idea that thought is to some extent independent of the capacity to handle a language, while at the same time it is dependent on this capacity when we have to conceptualise and express our thought in language. (Anderberg, 2000, p. 110)

An immediate objection to this is the subjective experience of thinking in language: of 'talking to ourselves'. However, if we have to learn to express thoughts in verbal language to communicate with others, it is feasible that the verbal thoughts we are aware of consciously are better considered *the expressions of our thinking* presented to consciousness, rather than 'pure thought' itself. Thinking is another of those key lay notions used to describe aspects of cognition and learning and is again a 'fuzzy' concept in everyday use.

Chapter 7
The Learner's Thinking

This chapter will explore what is to be understood by our everyday term 'thinking' in research in science education. This account will build upon the earlier chapters in this part, by discussing thinking in the context of the 'cognitive system' of an individual learner. Thinking is a term used for a *mental* process (see Chap. 3) and so according to the analysis offered earlier in the book relates to personal subjective experience ('thinking' is part of what was called the 'mental register' in the part introduction) and is not available as an object for direct 'objective examination'. Indeed, a key theme that will be stressed in this chapter is that much of what is of interest to science education researchers in terms of learners' thinking is not solely related to those conscious processes that are open to report following introspection. Establishing this general feature will be important in setting out a background for the subsequent parts on student knowledge (§3) and learning in science (§4).

A Study on 'Scientists and Scientific Thinking'

Coll, Lay and Taylor report a study on *'Scientists and Scientific Thinking'* (Coll et al., 2008). They reported that 'the interviews provide a window into scientific thinking as practiced by modern scientists, and suggest that the scientists are rather more open to alternative thinking than might be supposed' (p. 197).

The study involved an initial administration of a set of statements that participants were asked to rate in terms of whether they believed they were true or false, followed up by in-depth interviews. This is an interesting study, which would bring into question any stereotypical view of scientists *necessarily* having beliefs (commitments to how the world is; see Chap. 6 and also Chap. 15 for a discussion of worldviews) that would exclude the existence of ghosts, UFO sightings, the possibility of prayer leading to healing, or health-improving effects of crystals.

However, the interest for present purposes is in the way that the notion of scientific thinking was used. As with many studies that focus on everyday notions

K.S. Taber, *Modelling Learners and Learning in Science Education: Developing Representations of Concepts, Conceptual Structure and Conceptual Change to Inform Teaching and Research*, DOI 10.1007/978-94-007-7648-7_7,
© Springer Science+Business Media Dordrecht 2013

such as 'thinking', the authors (Coll, Lay and Taylor) do not feel it necessary to offer a technical account of what they mean by 'thinking'. The terms 'thinking' and 'scientific thinking' are treated by the authors as unproblematic (their discussion draws on the 'mental register') and are apparently assumed to have a clear meaning for readers. In this volume, I am suggesting that scientific investigation of mental 'phenomena' (such as learning and thinking) requires us to conceptualise these mental phenomena in terms of cognitive systems (Chap. 3), and that thinking is the term used at the mental level that best links to *the processing* of information in the cognitive system. Some of the uses of the term in Coll and colleagues' paper would seem to fit with this meaning:

> Some personal experiences were seen to influence the scientists thinking about beliefs, making them at least potentially believable. (p. 204)
>   It was noteworthy that some scientists 're-worked' some of the items presented in the surveys, thinking on their feet and seeking alternative explanations. (p. 209)

In these two uses, the authors seem to be referring to thinking in the sense of mental processes. However elsewhere, the term thinking refers less to the process than to the outcome of that process (with the present author's *emphasis*):

> Similarly, anecdotal evidence from "fairly stable sorts of people" was seen as a basis for *thinking that* some houses might be haunted… (p. 208)
>   Such thinking also applied to the scientists' perceptions of our understanding of the brain, with many of the scientists *thinking that* there remains much unexplained about the brain – thus they were open to alternative explanations including paranormal phenomena. (p. 210)

It is widely accepted that word meaning is partly determined by context, and the shift in how 'thinking' is used here, although perhaps not ideal in a research report is understandable given the way the term is used in everyday life.

The more specific term 'scientific thinking' seems to be used in a different way in the paper. The subtitle of the paper refers to 'understanding scientific thinking through an investigation of scientists [sic] views about superstitions and religious beliefs', which would seem to imply that 'scientific thinking' refers to an aspects of the individual scientist (who holds view and beliefs). Yet in the body of the paper, the term is used rather differently (again, with my added *emphasis*):

> The panel of experts consisted of scientists across a range of disciplines that examined each item statement in the instruments and asserted that it was *in conflict with current scientific thinking* in that discipline. (p. 201)
>   Likewise, the few that were less sceptical about astrology like Judy, thought that there were, potentially, underlying theoretical reasons *not inconsistent with current scientific thinking*… (p. 208)

Here 'scientific thinking' does not seem to refer to a process or product of cognition in individual scientists, but rather is considered to be linked to the scientific community: that is, presumably thinking that is consistent with current scientific knowledge. Here the authors are again using language in a commonly acceptable way. However, this does seem to raise a significant question of *what we might mean by the thinking of a community*. I have argued that thinking is the mental level description of the cognitive processing that occurs within an individual cognitive system (i.e. the individual, knowing subject, described at the cognitive system level), which in turn can ultimately be considered to arise from electrochemical activity within the brain of that individual. To shift to a community perspective would require considering the *overall* possessing activity in/across the network of cognitive systems.

This is a challenge that needs to be addressed. However, it is also pertinent to note that the way Coll and colleagues refer to scientific thinking here makes it clear they are not primarily referring to the process (i.e. processing) itself, but the outcomes of that process: the ideas and evaluations that are products of the process. So, for example, a view about whether some crystals might have inherent healing properties would be a product of thinking processes. For this reason, this issue will be deferred to a later chapter (Chap. 10), where the nature of scientific knowledge will be discussed.

Coll and colleagues' paper offers a useful insight into the outcomes of the thinking of some scientists about a range of topics, to test the notion that scientists would adopt a kind of 'party line', and so take on consensual positions, on certain issues. It therefore makes a useful contribution to scholarship. Some of the data presented does offer indications of aspects of the thinking processes of the participants, but generally the study is concerned not with scientific thinking (as a process) but more with the *outcomes of* scientists' thinking about focal topics.

This is not a peculiarity of the way Coll and colleagues use the term 'thinking'. A study on explanation in science classes by Braaten and Windschitl (2011) includes both of the following statements:

> The term "explanation" is also used to connote the communication of reasoning in an effort to make thinking visible or audible in science classrooms. (p. 654)
>
> In science classrooms, it can be difficult, if not impossible at times, to provide students and teachers with sufficient access to theory and evidence to allow for reasoning through alternative explanations to ultimately arrive at an understanding consistent with current scientific thinking. (p. 665)

The first of these quotations talks of making thinking visible or audible – something that it was earlier argued was problematic, and seems to refer to *the process* of coming to a view or judgement. However, the phrase 'understanding consistent with current scientific thinking' seems to refer to the outcomes of thinking, rather than the processes by which these outcomes were reached.

Establishing a Meaning for 'Thinking'

Thinking has understandably been an important concern in science education, one purpose of which is said to be to facilitate the development of 'scientific thinking' among learners (Laugksch, 2000; Lawson, 2010; Lehrer & Schauble, 2006). However, thinking is one of the terms that (as highlighted in Chap. 3) are used as an everyday label for something which is understood in a non-technical sense – that is, it is a phenomenon of the lifeworld, part of the mental register of our folk psychology.

> But what is thinking? This might seem a pointless question, since everyone knows by acquaintance what thinking is from his own first-hand experience of doing it…Very few people ever think about thinking. It is one thing to practice an activity and quite another thing to stand back and try to observe, describe, and account for that activity. It is one thing to realize that certain activities happen, but quite another thing to take special steps to show precisely what does happen and how it happens in the way it does. (Thomson, 1959, pp. 12–13)

Thomson goes on to suggest that there are at least six ways in which the term 'think' is used in normal discourse. So people may refer to daydreaming, recalling, deliberate imagining, concentrating and their opinion or reasoning when reporting what they think. Similarly, another commentator notes that the term 'thinking' 'is used to cover reasoning, conceiving, imagining, perhaps day-dreaming, though rarely dreaming proper' (Aaron, 1971, p. 91). As suggested earlier in the book, this is not a problem in everyday conversation, as context usually suggests intended meaning, and we can interrogate the speaker if unsure. However, this book is concerned with conceptualising and reporting research in science education, and if research literature is to be properly understood so it can be built and acted upon, it is important that when research results are reported, they use terms that have been explicitly operationalised.

In Chap. 3, an approach to describing cognition, which would include thinking, at three levels was presented (see Table 3.4). It was argued that these three levels of description are complementary – that mental activity such as thinking can be understood in terms of cognition as processing activity within a cognitive system, that in turn could be in principle explained in terms of electrochemical processes in the nervous system and in particular the brain.

Normal everyday conversation focuses on the mental level of description – considering 'ideas' and 'thoughts'. At this level we might consider that thinking is the activity that leads to ideas: thinking is a *process*, and ideas or thoughts are the mental *'products'*. There is nothing wrong with this level of description for some purposes, including much everyday conversation. However, as introspection only offers a very limited appreciation of the nature of thinking, this may not be sufficient for research purposes.

Thinking and Processing

In terms of the ways cognition has been represented in previous chapters in the book, there is a key issue in how we understand the relationship between the systems-level description of processing within the cognitive system and thinking as a mental phenomenon.

This is recapped in Fig. 7.1, which shows a model of the key stages of processing information within the cognitive system. Conscious thinking is considered to be a correlate of processing in some executive module of the system (see Chap. 3), usually identified with working memory (see Chap. 5). However, if we consider the

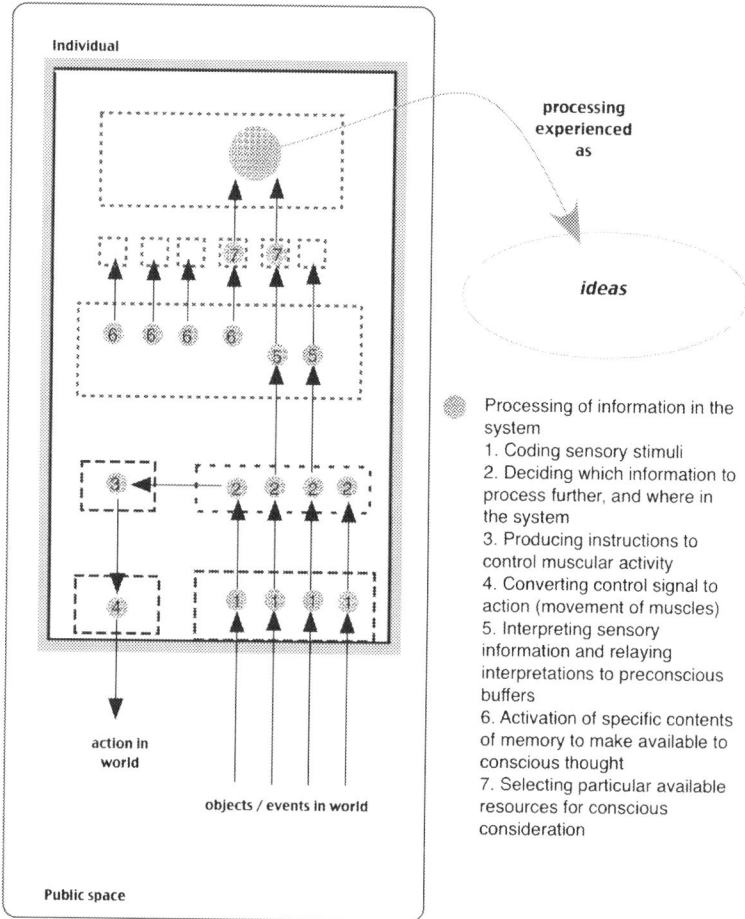

Fig. 7.1 Processing and thinking – recap of discussion of processing in the cognitive system in earlier chapters

cognitive system to be the basis of intelligent behaviour, then much processing below the conscious level contributes to this. Arguably even reflexes are a form of intelligent – certainly adaptive – behaviour, even though the processing of information, decisions about action and control of that action (blinking, moving a limb, etc.) all take place without conscious involvement. Yet, in general conversation, we would not normally call that level of processing 'thinking'.

Indeed references to such phrases as 'processing below [sic] the conscious level' impose a topological metaphor reflected in figures such as those I have used in this volume that might seem to reflect or imply a notion that what goes on at a 'lower' level is less important. However, at the very least, it is clear that conscious thought is facilitated and underpinned by cognitive processing that the individual is not conscious of. Indeed, to use another metaphor, conscious thinking is 'just the tip of the iceberg'.

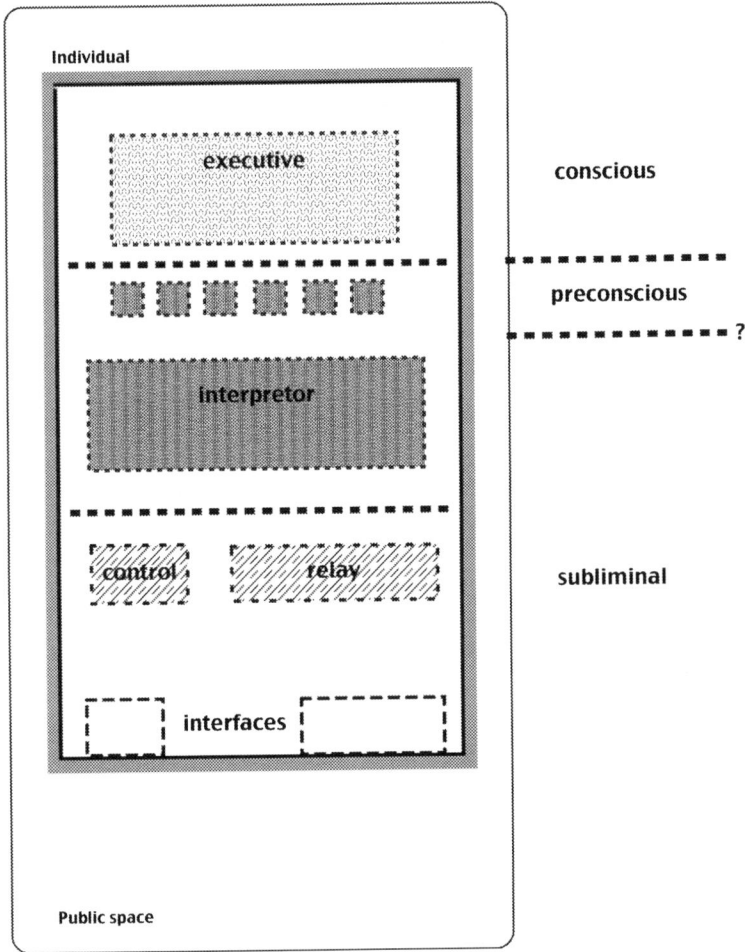

Fig. 7.2 Levels of processing in the cognitive system

If we consider the distinction between conscious, preconscious and subliminal processing (introduced in Chap. 4), then we can consider the cognitive system to be divided as in Fig. 7.2.

If we argue that the module concerned with conscious thinking (perhaps working memory; see Chap. 5) is the executive, then we could extend this organisational metaphor to other aspects of the system. We might think of the interfaces as being routine operations that are the 'unskilled' sector of the system, where work is carried out mechanically. The subliminal level at which filtering and control of information is undertaken is in this analogy a more technical level, with decisions being made as if according to an established code book, with problematic cases 'referred upstairs' for higher-level consideration. At the preconscious level is the professional/managerial work. Here – in terms of this analogy – incoming information is interpreted, and reports are presented with recommendations for possible courses of action. These reports are selectively attended to by the executive, which relies on the work of its professional part to provide accurate accounts and to suggest creative scenarios and options. Whilst such a metaphor is clearly only meant to offer some heuristic value, it does reflect how a good deal of important processing takes place prior to any conscious awareness and how the effectiveness of conscious thought is limited by the quality of the information provided by preconscious processes. *It would seem perverse to exclude this level of processing from being considered 'thinking'.*

The Significance of Preconscious Processing

So the processing which correlates to *conscious* thought may only be one stage of a more complex sequence of processes, much of which we are never aware of consciously. Were these subconscious processes limited to general physiological regulation and reflex actions, and so involved something largely unrelated to conscious thought, then we might feel it is useful to reserve the term 'thinking' for conscious processes. However, this does not seem to be the case.

I would suggest that the activity of the crossword puzzle offers a useful insight into the limits of conscious awareness during thinking. My own experience here is that some clues lead to a possible answer appearing immediately in consciousness; others do not, but I am often able to get an impression of whether I am going to be able to readily think of the answer – even though I do not at that moment have one 'in mind'. Sometimes I have the impression that I have nearly got the answer, and I am just waiting for it to appear in consciousness, although I am not quite sure how to help the process along. This is a widely reported experience, known as the 'tip-of-the-tongue' phenomenon or a 'feeling of knowing' (Parkin, 1987, p. 37) – when someone finds they cannot (yet) produce the word they are 'looking for' although they are pretty sure it is in their vocabulary and *at some level* they 'know' which word they want to use.

This tip-of-the-tongue experience could be put down to wishful thinking or some kind of cognitive error, except that it often seems to be accurate: it is usually accompanied by the production of the word or answer that we then recognise as being what we were trying to access. It would seem that at some level of the cognitive system, we are able to recognise that the target of some kind of search process has been located, before we are able to form a representation of it at the level of processing which is associated with conscious thought. Perhaps this links to the issue of thinking largely being in a form of 'machine code' that then has to be expressed into verbal language (see the previous chapter, Chap. 6). So Brown and McNeill reported that when students were asked to identify words used at low frequency in the language from definitions, about half of the words generated before finding the word they considered matched the definition shared its initial letter (Brown & McNeill, 1966/1976). This suggests that the students were not just accessing semantically similar words as might be expected working from a definition, as these would more often than not have different initial letters.

This is just one example of how 'subconscious' (preconscious) processing is important for, and seems to blend into, conscious processing of information. Freud's work showed that much of our thinking seems to be subconscious – 'a sea of unconscious ideas and emotions, upon whose surface plays the phenomenal consciousness of which we are personally aware' (Hart, 1910, p. 365) – to the extent that, it is claimed, we often act on motivations that we do not consciously recognise leading to us finding alternative rationalisations to explain and justify our actions.

Various pathological conditions also support this type of argument. For example, people exhibiting the condition of blindsight have no visual awareness and consider themselves to be blind, although no physiological damage may be detected on medical examination. A person with blindsight cannot report what is in their visual field, because as far as they are aware, they do not have vision. However, they can be very good at 'guessing' what is in their visual field. So it seems that one part of their brain has access to the visual stimuli and is able to process the sensory impressions to the degree they are interpreted into what should be precepts (objects and so forth), but the visual images themselves are not accessible to consciousness. When asked to guess about objects placed in front of them, the blindsighted person is often able to report accurately, although as they have no conscious awareness of how they could know, they consider they are just guessing (Churchland, 1980; Gazzaniga, Fendrich, & Wessinger, 1994).

Thinking as an Inclusive Term

This suggests that to reserve the term 'thinking' for conscious thought would be a rather arbitrary distinction. So instead I will use the term 'thinking' to describe cognitive processes that are not necessarily conscious (see Fig. 7.3). That is not to suggest that reflex actions or automatic adjustments of posture which require some low level of processing in the nervous system should be considered as thinking; but rather processing that is considered to be related to cognitive activities such as

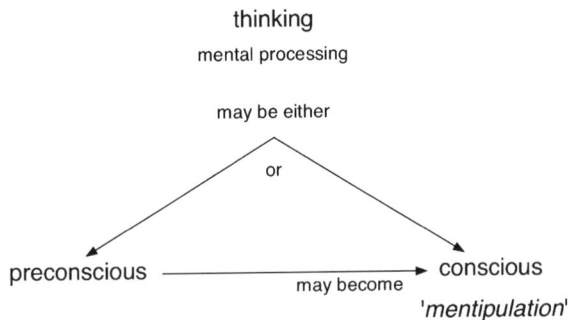

Fig. 7.3 Thinking is not necessarily conscious

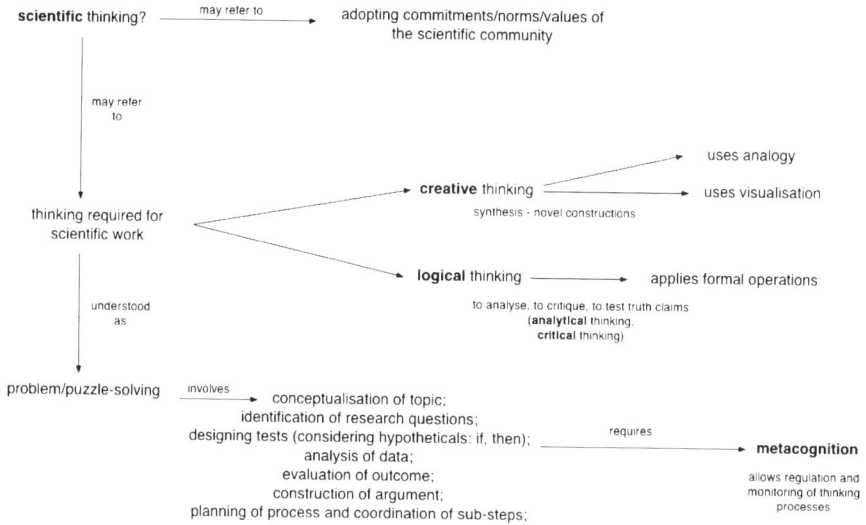

Fig. 7.4 Aspects of scientific thinking

concept learning and problem-solving will be considered as a kind of thinking, whether conscious or preconscious. Where it is important to differentiate, or to avoid being misinterpreted, the inclusive term 'thinking' can be qualified by being preceded with 'conscious' (or 'preconscious') as seems appropriate. Alternatively, the type of processing of information that is consciously experienced directly 'in the mind' could be termed mentipulation – that is, the mental analogue of manipulation (perhaps a rather obvious neologism to coin, and the term has previously been suggested by Ivić, Pešikan, & Antić, 2002).

Forms of Thinking Valued in Science Education

Traditionally, science education has been associated with the development of certain types of thinking styles or skills, and often this has been 'logical' or 'critical' rather than 'creative' thinking. However, science also involves creative thought, and creativity is important to learning in science (see Fig. 7.4).

Scientific Thinking

Science is much more than a body of knowledge. It is a way of thinking. This is central to its success. Science invites us to let the facts in, even when they don't conform to our preconceptions. It counsels us to carry alternative hypotheses in our heads and see which ones best match the facts. It urges on us a fine balance between no-holds-barred openness to new ideas, however heretical, and the most rigorous skeptical scrutiny of everything – new ideas *and* established wisdom. (Sagan, 1990, p. 265, emphasis in original)

One of the aims of science education is to help students develop 'scientific' thinking. As we saw earlier in this chapter, the term scientific thinking is sometimes used to refer to public scientific knowledge, that is, current scientific thinking about a topic.

It has also been characterised as the adoption of a scientific worldview, that is, a set of assumptions about the world that guide thinking (see Chap. 15). If thinking is for the purposes of this chapter considered a process, then these other meanings relate to either the outcomes of thinking (output) or resources to support thinking (input) and not the processing itself. Scientific thinking in these senses then lies outside the scope of this chapter, and these points are picked up in later parts of the book.

More commonly, scientific thinking can refer to the thinking processes required to undertake scientific work. Whilst it might be difficult to agree a definition of what scientific thinking is, it is recognised to be central to the practice of science. It has also been argued that 'scientific thinking is a paradigmatic example of cognition that reveals the key features of many basic cognitive processes' (Dunbar, 2001, p. 115). It has sometimes been strongly associated with the application of the 'scientific method'. In particular, it has been associated with rational, logical thinking (cf. Fig. 7.4).

Science-as-Logic: Logical Thinking

Logic has long been associated with a key form of thinking. Indeed, Bonatti has suggested that

> An old and venerable idea holds that logic is concerned with discovering or illuminating *the laws of thought*. Its psychological corollary is that a system of logic in the mind underlines our thinking processes. … In a nutshell, it holds that reasoning consists of operations on mental representations, according to logical rules implemented in procedures activated by the forms of the mental representations. (Bonatti, 1994, p. 17, present author's emphasis)

Logic is central to scientific work because designing and interpreting scientific investigations requires the application of particular types of general rules. These are involved in establishing the conditions under which it is appropriate to draw specific conclusions. So in school science, learners will be taught about control of variables and how to set up control conditions so that results will allow them (in ideal cases at least) to draw conclusions about whether a particular cause produced a particular effect (i.e. 'fair testing').

In practice, science is seldom as simple as it often tends to be represented in school science (Taber, 2008b). Rather, experimental design always depends upon both an existing conceptual framework that suggests which of the potentially infinitive range of variables might potentially be relevant (e.g. in exploring the effect of wire length on resistance, we might decide it is important to control for material and temperature; we may however consider that it is not relevant whether the wires are vertical or horizontal, whether they are aligned North–South or East–West, whether it is a Tuesday or a Wednesday, whether we have said a prayer before we collect data, whether the lab is on the ground floor, the gender or nationality of the lead researcher, what they had for breakfast and so on, ad infinitum) and a theoretical or methodological framework informing the choice of particular techniques which can collect valid and reliable data (Taber, In press).

However, whilst these considerations are of great practical importance, underlying any such investigation is a more fundamental framework (a metaphysical belief system; see Chap. 15) that says that science is possible because the world can be understood in terms of causes that bring about effects, and do so in regular ways (we assume there are 'laws of nature' that will not change from day to day or from one place to another), so that through setting up suitable combinations of conditions, we can determine necessary and sufficient causes by applying simple logical rules (Sijuwade, 2007). So, for example, we cannot claim that some factor is a necessary cause of some effect if sometimes we see the effect when that factor is not present.

The *Handbook of Child Development* refers to one image of science as being 'science-as-logic', where

> Science-as-logic emphasizes the role of domain-general forms of scientific reasoning, including formal logic, heuristics, and strategies, whose scope ranges across fields as diverse as geology and particle physics…These heuristics and skills are considered important targets for research and for education because they are assumed to be widely applicable and to reflect at least some degree of domain generality and transferability. (Lehrer & Schauble, 2006) p. 156

The authors of this review, Lehrer and Schauble, suggest that 'learning to think scientifically' is variously conceived of as:

- Acquiring strategies for coordinating theory and evidence
- Mastering counterfactual reasoning
- Distinguishing patterns of evidence that do and do not support a definitive conclusion
- Understanding the logic of experimental design (p. 156)

These rules of logic often seem self-evident to researchers, so that conclusions can be drawn from data without having to explicitly refer back to them. Yet Dunbar (2001, p. 116) notes that 'much cognitive research and research on scientific thinking has demonstrated that human beings are prone to making many different types of reasoning errors and possess numerous biases', and that 'informing subjects of these biases does little to improve performance, and even teaching subjects strategies for overcoming them does little to ameliorate these biases either'. He goes on to suggest that 'much research on human thinking and reasoning shows that thinking is so error-prone that it would appear unlikely that scientists would make any discoveries at all!' Yet clearly some of the population are either less prone to these logical errors or alternatively are able to learn to overcome them – at least in their professional work. Making the logic of scientific investigation more explicit in school science may be important here (Lawson, 2010), especially where what seems self-evident to the teacher is being missed by the learners.

Indeed, the very notion that the human brain has innate apparatus for logic has been questioned. Bonatti (1994, p. 17) suggests that 'even if the thesis loomed around for centuries, there is still little convincing psychological evidence of the existence of a mental logic'. An important difference here might be between a logical analysis suggesting how scientists ought to think; and psychology, which tells us that we fall short of the ideal (Nickerson, 1998). Presumably such imperfections are the result of natural selection pressures, because some logical corner cutting is often the pragmatically more effective approach in dealing with the everyday problems of survival. After all, when faced with food shortages or dangerous predators, a weakly supported plan of action that can be implemented now might often prove to be better than a well-researched plan of action that we are hoping to have ready at some hypothetical point in the distant future.

Creative Thinking

Creativity is a central part of doing science, and one criterion by which scientific work is judged is that it shows originality. Indeed creative and logical thinking are complementary prerequisites of scientific discovery (Taber, 2011). Arthur Koestler (1978/1979) argued that science, art and humour all relied on the same creative processes of bringing together previously unrelated ideas into a new juxtaposition. Although scientific work does require logic to devise and interpret tests of ideas, it also relies upon someone producing the idea that will be tested. How we have such novel ideas is not well understood. Whereas, in logical thinking, conclusions are in a sense already implied by the premises; creative thinking means coming up with something that goes beyond the information available and that is not logically justified. In creative thinking there is no set procedure or set of steps to follow, and often an idea just appears in consciousness. Indeed, there are many stories of how creative thinking is best supported by relaxed distraction (Taber, 2011) – taking a bath, dozing, going for a walk, etc. – albeit usually after an extended period of intense engagement with the problem area being studied. Plant geneticist and Nobel laureate Barbara McClintock talked of how her brain would 'integrate' information in the background and come up with possible solutions to scientific problems (Keller, 1983):

> I read the paper and when I put it down I said, 'This can be integrated'. My subconscious told me that. I forgot about it, and about three weeks later I went into the laboratory one morning at the office. I said 'This is the morning I'll solve this'. (Quoted in Beatty, Rasmussen, & Roll-Hansen, 2002, p. 282)

This might be described as relying on intuition or tacit knowledge which has been developed but which is not consciously available (see Chap. 11). Einstein is commonly quoted as suggesting that 'the intuitive mind is a sacred gift and the rational mind is a faithful servant'.

It seems sensible to assume that creative ideas or problem-solutions that seem to appear suddenly in mind are actually the outcome of processing within the brain outside of conscious awareness (see Fig. 7.5). That is, thinking is occurring that is

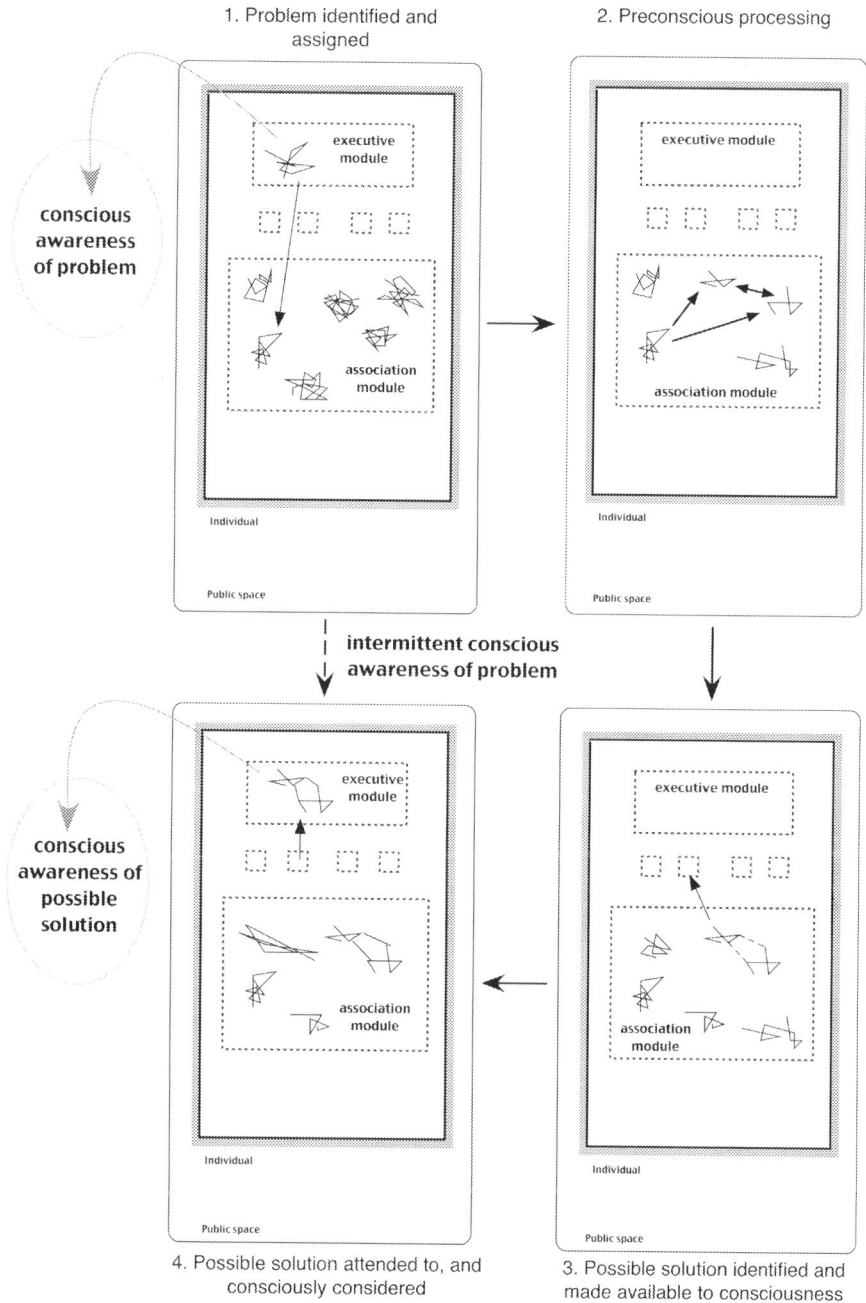

1. Problem identified and assigned

2. Preconscious processing

conscious awareness of problem

executive module

association module

Individual

Public space

executive module

association module

Individual

Public space

intermittent conscious awareness of problem

conscious awareness of possible solution

executive module

association module

Individual

Public space

executive module

association module

Individual

Public space

4. Possible solution attended to, and consciously considered

3. Possible solution identified and made available to consciousness

Fig. 7.5 We can consider the executive component of the cognitive system to assign (problem-solving, creative) tasks to preconscious processing and then later to accept reports

inaccessible to consciousness, but is still goal-directed, and often motivated by known problems and issues previously considered in conscious thought. Baddeley (2000) has suggested that the component of working memory (i.e. the executive module of the cognitive system; see Chap. 6) that he labels the episodic buffer may have a role to play here in creating new cognitive representations, that is, new syntheses from, or modifications of, existing available representations.

Extending the analogy used earlier, we might think of the executive referring an issue to some kind of working group to consider, for later report. Whilst the analogy should not be given too much weight, it does seem that this type of delegation/assignment of processing to preconscious levels is a key part of much human cognition.

Changeux has suggested how such processes may be understood in terms of physiological properties at the physical level (see Chap. 3):

> The neurons participating in assemblies of concepts will be both dispersed and multimodal, or perhaps amodal. This should bestow on them very rich 'associative' properties, allowing them to link together and above all combine… This recombining activity would represent a 'generator of hypotheses', a mechanism of diversification essential for the genesis of pre-representations and subsequent selection of new concepts. In a word, it would be the substrate of imagination. It would also account for the 'simulation' of future behaviour in the face of a new situation. (Changeux, 1983/1997, p. 169)

Analogical Thinking

Analogy has been proposed as one major source of creative ideas in science (Muldoon, 2006). Wong argues that 'analogical reasoning is one means by which experience is related to and differentiated from what is already known. Through analogies, an understanding of novel situations may be constructed by comparison to more familiar domains of knowledge' (Wong, 1993, p. 1259). According to Gentner (1983, p. 159), 'an analogy is a comparison in which relational predicates, but few or no object attributes, can be mapped from base to target'. She gives the example of the analogy that a hydrogen atom is like our solar system, where it is intended that relations are mapped from the solar system to the atom (e.g. the *electron* orbits the *nucleus*, like the *planets* orbit the *Sun*), but not object properties (e.g. not that the *Sun* is yellow, so the *nucleus* is yellow).

Analogical thinking can therefore be a component of creative thinking, as by considering that X could be like the more familiar Y, possible features of X are suggested which can then be subject to testing. Analogical thinking would seem to involve at least two separate processing tasks: first searching through representations of the familiar to find a possible analogue for the target to be better understood by recognising some form of similarity and then undertaking a formal mapping process to see what the analogy would suggest may be the case about the target. When 'teaching analogies' are presented in class (Harrison & Coll, 2008), the learners are faced only with the second (analytical) part of this process, whereas when learners are asked to generate their own analogies (Wong, 1993), they must also undertake the initial step of finding a source analogue that offers some kind of similarity to the target.

The mapping exercise is undertaken with full conscious awareness: decisions are made about how aspects of the analogue map to the target and are open to conscious evaluation before being represented in the public space to communicate the analogy to others. The search process, whilst initiated with conscious awareness, would seem to be largely undertaken by processing that occurs below the level of conscious awareness, so that we become consciously aware of a limited number of possible candidate analogues and are not aware of the vast number of other potential analogues represented in the cognitive system which are judged at a preconscious level not to offer any similarity and so not worthy of further consideration. This process would seem to be an example of the type of cognitive work that largely takes place without our being able to (consciously) access and monitor the process itself. Rather, we just have access to its 'outputs' (see Fig. 7.5).

Imagery

In Chap. 6 it was suggested that our thinking does not occur primarily in verbal language, rather that 'thought is to some extent independent of the capacity to handle a language' (Anderberg, 2000, p. 110). Clearly, as modern humans, much of our *conscious* experience is in the form of verbal language, and language acquisition certainly provides important tools for internal conscious thought (Vygotsky, 1934/1986) just as much as for communication between minds (Vygotsky, 1978). Moreover, part of the 'executive' processor associated with conscious thought (the phonological loop in working memory; see Chap. 5) seems to have evolved to facilitate this. However, this executive module is also thought to include what is labelled the visuo-spatial scratch pad, which provides representations in imagistic form. As Baddeley points out, 'there are many examples of the importance of visual or spatial imagery in scientific discovery, including Einstein's development of his general theory of relativity' (2003, p. 834).

Einstein is just one of a number of scientists who have described how much of their creative thinking was imagistic (Miller, 1986). Nersessian (2008) has described how scientists form mental models, often represented in images, which act as mental simulations that can be 'run' so that the outcomes can be compared with the target phenomenon. Kekulé famously described a kind of exploratory imagistic simulation when he claimed to have had the idea that the benzene molecule was cyclic after interpreting an image of a snake grabbing its own tail:

> I turned the chair to face the fireplace and slipped into a languorous state. Again atoms fluttered before my eyes. Smaller groups stayed mostly in the background this time. My mind's eye, sharpened by repeated visions of this sort, now distinguished larger figures in manifold shapes. Long rows, frequently linked more densely; everything in motion, winding and turning like snakes. And lo, what was that? One of the snakes grabbed its own tail and the image whirled mockingly before my eyes. (as quoted in Rothenberg, 1995, p. 425)

The ability to imagine in this way is thought to make use of the same areas of the brain that are involved in visual perception (cf. Chap. 4). That is, the cognitive system includes apparatus for producing visual images, presumably evolved initially for converting sensory information to visual percepts, which are also (i.e. during evolution, have been recruited to be) used to generate visual images from processing initiated internally within the system – when remembering or imagining (Changeux, 1983/1997; Parkin, 1993).

Given the suggestion that there may be rare individuals who as adults retain eidetic imagery ('photographic memory'), when most people have very limited visual working memory (see Chap. 5), it seems possible there may be quite significant variations in the capacity of individuals to mentipulate visual images – and this may be one area where individual differences between science students may be quite significant in determining cognitive styles and mental capabilities.

Critical Thinking

Another descriptor often associated with scientific thinking is critical thinking (Lindahl, 2010) which has been described in a consensus statement from a Delphi study as 'purposeful, self-regulatory judgment which results in interpretation, analysis, evaluation and inference, as well as explanation of the evidential, conceptual, methodological, criteriological, or contextual considerations upon which that judgement is based' (Facione, 1990, p. 3). Critical thinking has been associated with the ability to make decisions in complex situations or to find solutions to weakly structured problems (Lubben, Sadeck, Scholtz, & Braund, 2009).

That is, whereas logical thinking can be analysed in terms of following certain rules (e.g. effectively 'if/then' rules, albeit often nested in complicated ways) – such that providing the structure of rules is followed correctly and assuming the information provided was accurate, then the 'right' answer will be obtained – logic alone is insufficient when the 'ifs' remain 'iffy'. This is a more realistic scenario for most real-life decision-making as there is seldom a full, unambiguous data set to support a single assured solution to real-life problems.

Problem-Solving

Problem-solving is widely recognised as a key concern of education, and indeed Lawson and Wollman (1976, p. 413) report that according to the Educational Policies Commission, 'the central purpose of American education is the development of problem-solving processes called rational powers'. Problems, by definition, do not have ready solutions, and in an educational context, a problem is something that a learner is *not* able to solve by simply applying a familiar routine: as Jonassen (2009, p. 17) suggests '…problem solving entails a lot more cognitive activity than searching long-term memory for solutions'. What is a problem for one person who is a novice or less advanced learner may be straightforward to another who is a more advanced learner or an expert. Therefore judgements about what count as a problem have to be made relative to specific learners (Phang, 2009).

As an example of this point, consider the example of a teacher asking a secondary student to complete the following word equation:

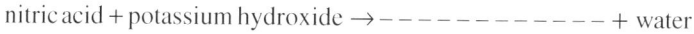

$$\text{nitric acid} + \text{potassium hydroxide} \rightarrow - - - - - - - - - - - - + \text{water}$$

This is a trivial task for chemists, chemistry teachers and advanced students. Indeed, if the author is given a task such as this, then the answer appears in consciousness without any explicit attempts to work out an answer, much quicker than I can formulate the rationale for why the answer is potassium nitrate. Yet for many secondary students, this task, which certainly has the appearance of a simple exercise to the teacher, takes on the nature of a problem, as the student has to seek out relevant knowledge and a way of coordinating that knowledge to produce a candidate answer (Taber & Bricheno, 2009).

Problem-solving, then, is a special kind of creative thinking, in that the individual has to find some new synthesis that although perhaps well known to others is a novel association for that individual. Problem-solving is believed to be an area where the limitations of working memory capacity can restrict learner performance (Tsaparlis, 1994). It is also thought that successful problem-solving is dependent upon metacognitive processes.

Metacognition

Metacognition, cognition of cognition, involves thinking about one's own thinking and has been defined as 'knowledge of the processes of thinking and learning, awareness of one's own, and the management of them' (White & Mitchell, 1994, p. 27). The idea of metacognition is closely related to that of 'self-regulated learning' and in schools may be linked to 'independent learning skills' or 'study skills'.

A key point made above is that much cognitive activity occurs without conscious awareness; metacognition is concerned with conscious thinking about one's own cognition and how this can be used to plan, monitor, evaluate and redirect one's own thinking. Clearly then, in terms of the model of learner as cognitive system developed in this book, 'metacognition is closely related to executive function, which involves the ability to monitor and control the information processing necessary to produce voluntary action' (Fernandez-Duque, Baird, & Posner, 2000, p. 288).

This does not undermine the claim that much thinking is undertaken away from conscious awareness, rather it is *because* so much cognitive processing occurs at preconscious levels that metacognition becomes so important. For example, in Fig. 7.5, it is suggested that a problem that the individual is consciously aware of can be metaphorically 'assigned' to be worked upon preconsciously, and only once a solution has been identified does it get flagged to be made available to consciousness. However, the executive module is then able to decide whether to accept the solution as a basis for action (e.g. to represent it in writing on an examination script) or whether to continue the search for a better solution. So the suggested solution presented to consciousness needs to be seen as a proposal (if a course of action is sought) or hypothesis (if an explanation is sought) that can then be evaluated. Without the metacognitive 'layer', we could do, say, or be satisfied as an explanation with, the first thing that 'came into our heads' (i.e. was presented to conscious awareness).

This also links back to the complementary roles of creativity and logic in scientific discovery (Taber, 2011). When making claims to new scientific knowledge, the context of discovery is less important in persuading others than the context of justification. It should be of no relevance to the evaluation of the quality of our ideas whether they occurred to us in the bath, or when chatting with a colleague over coffee, or through a sudden insight of an analogy with a work of art we were inspecting. However, we must make a logical case based on evidence for why the idea should be taken seriously. Our metacognition allows us to consider if there are good grounds (the justification) for considering an insight (the discovery) to be correct or at least productive as the basis for further action.

In both problem-solving and scientific discovery, the creative step seems mysterious but is essential, and the logical work concerns evaluating the output of the creative stage. Scientific discovery may be associated with the public community of scientists and their outputs (see Chap. 10), but in terms of it depending on individuals processing information in their cognitive systems, it has parallels with a school student solving a problem or suddenly making sense of what the teacher is trying to explain. In both cases progress depends upon an insight that is the outcome of thinking that largely takes place outside of conscious awareness.

The Fallacy of 'Machine Code'

Before leaving the consideration of cognitive processing, it is useful to revisit one notion that was referred to earlier in the book. In Chap. 4, the way in which sensory information had to be 'coded' so that it could be processed in the cognitive system was considered. Information available from, for example, photons – quanta of energy – being incident upon, and being absorbed by retinal cells, is represented into patterns of electrical activation in the optic nerve. So retinal cells act as transponders that convert a signal of one kind into something different. The term 'coding' seems appropriate as perception of the external world is only possible because the cognitive system's sensory interface (see Fig. 4.2) converts the patterns of, for example, illumination, into patterns of electrical activity in a non-arbitrary way. This allows the information available to the senses to be interpreted by the rest of the cognitive system.

The Limits of Computing Analogies for Cognition

The notion of 'machine code' draws upon an analogy with electronic computers. These computers are basically extensive networks of binary switches (on or off), so all processing must be in terms of signals cuing switching between on and off states to initiate or stop signals elsewhere in the system. Yet programming was traditionally undertaken in 'higher-level' languages designed for that purpose, for example, using logical operations such as 'if/then', such as COBOL, ALGOL and BASIC. The instructions written in these higher order languages therefore had to be translated into the code used by the machine by a conversion process (known as compiling). We can understand what happens when sensory information is coded into the human cognitive system as being *analogous to* this. We should however bear in mind, as suggested above, the important proviso that analogies reflect notable parallels in structure within different systems, but this does not imply the target system is in all senses like the analogue.

Whilst use of early computers required writing programmes in one of the higher-level languages, and then using the compiler to translate into the machine code, the experience of modern computer use is very different for most users who use the computer as a tool and have limited interest in programming it themselves. The operating systems of modern personal computers have inbuilt programming that allows users to undertake a wide range of operations with user-friendly interfaces. So a file can be copied into a new folder without any knowledge of programming languages, simply by using iconic representations on screen, such as dragging an icon from one location to another with a mouse, touchpad or on the screen itself.

As I tentatively suggested earlier, when we undertake these operations, we 'see' the icons as the files and folders they represent and conceptualise them as having the physical locations shown, and we can consider this in some ways akin *to* the role of consciousness in our own thinking: in using the computer the interface presents us a simple visual representation of the 'world' inside the computer that allows us to operate on (in) that world effectively because the representational system becomes that world to us. The extent to which we are aware that we are only operating with a representational interface, and understand how what we do with the icons relates to aspects of computer architecture, might be seen as akin to metacognition, allowing us to reflect upon our use of the interface and perhaps better think through problems when the systems do not seem to be doing what we want or when we wish to undertake an operation we are not familiar with. Many people, however, use computers without reflecting on these issues, lacking the (analogue to) metacognition to move beyond a kind of phenomenological experience where the desktop *is* the working space and the icons *are* the folders and files they represent.

A Different Type of Processing System

At one level, this is a strong analogy. It is clear that the processing in the brain is in a form of 'code' that is not the same as sensory input data (it clearly cannot be – light and sound do not travel into the cortex), nor is it the same as our conscious experiences, as pointed out earlier in the chapter. However, there are some significant differences between human cognition and electronic computers that we must be aware of when using such an analogy.

One important difference is the nature of the processing. An electronic computer has a fixed array of switches and works in terms of switches being on or off. The human brain has synapses which act as switches, but these can vary in connection strength: they can be more or less on.

Moreover, the actually set of 'switches' is not fixed: connections between cells can be added to or removed, and indeed a key aspect of development in young humans is an extensive pruning of most of the initial connections established, to provide a more selective network. This provides the potential for a much more flexible, and responsive, processing apparatus than is possible with an electronic computer – where the stability of the set of connections in the processor is rather important to its normal functioning. This immediately suggests that humans and computers are going to have rather different properties as processing systems and so also have rather different strengths and weaknesses.

Related to this is the type of processing undertaken in a system with the type of structure and inherent plasticity of a synaptic network. The input to one node, one neuron, can be from a range of different other nodes, and it can in turn provide output to a range of other nodes – each with changeable connection strengths. Moreover, output can through a chain of connections influence input: that is, there is potential for feedback in the system.

Systems of this type have potential for undergoing changes that can be considered as learning. Artificial networks of this kind have been 'trained' to undertake processing tasks such as, for example, distinguishing the sonar patterns obtained from shoals of fish and submarines: a kind of task that because of its complexity – due to variations in size, shape and distance of target; water conditions and other objects influencing echoes, etc. – is very difficult to achieve with binary electronic computers that have to be programmed in terms of a series of if/then decision-making steps.

Whereas programming an electronic computer for such a task requires an extensive analysis of the task and subsequent highly complex programme, synaptic networks are 'trained' by changing connections strength patterns in an iterative manner: feedback is used to see which changes produce more effective outcomes, and which is less effective, and over time the system is tuned to its function. Now as human cognition is more like the artificial synoptic networks than binary electronic computers, it seems that their processing is better understood by analogy with such networks. That is, rather than considering processing in the brain to be *as if* someone has written a compiler to translate sensory input signals into a machine code, it may be more appropriate to think of it *as if* somebody has finely tuned output patterns from available input, by considering the feedback to an extensive set of trials. It should be noted that the actual system properties of a particular brain depend upon both features which are in effect genetically programmed by the 'feedback' effects of natural selection acting over many generations (see below) and the individual's experiences, responses to those experiences and subsequent experience of effects of their responses.

The Ghost in the Machine: Who Tunes Our Processing Networks?

However, a major difference between the artificial synaptic network and a human brain is the nature of a trainer. When an artificial network is trained to distinguish fish from submarines or perhaps appendicitis from intestinal wind in a medical diagnostic system, the external trainer makes external judgements about the accuracy of outputs and uses these to decide how to proceed to tune the circuit.

But just as there is no programmer writing code for human brains, the newborn infant has no external trainer to suggest which outputs are accurate representations of input from the external world. Rather, learning has to involve modifying synaptic connection strengths depending upon whether the output of attempts to act on the world are considered more or less effective in meeting internal drives. The hit-and-miss nature of this business suggests that a great many trials might be needed before any human cognitive system could be well enough 'tuned' to offer an effective model of the outside world.

However, there is a mechanism for the learning of one generation to allow the next to have something of a 'head start' in this process. The transfer of genetic information to offspring has allowed natural selection to operate over a great many generations so that each new individual is not starting ab initio, expected to make sense of the world with a random network of neutrons that has to be moulded into a tuned processing system ex nihilo, but enters the world with both initial apparatus and inbuilt tendencies to direct development, which are the results of testing over millions of years. Certainly within the chordates, the initial state of any individual's cognitive system is highly biased. In humans this is especially so, for as we have seen there are innate tendencies to acquire verbal language, for example.

That is an especially helpful adaptation, because it means that from quite early in the young person's development, the tuning of the cognitive systems can be *supported* by external trainers (Vygotsky, 1978), who supplement the initial drives to make sense of an act in the world by providing extensive additional feedback (e.g. no that is not a doggy, that is a cat). Whilst this feedback itself needs to be interpreted within the system (see Chap. 4), it offers additional sources of information about how humans have found it helpful to understand the world.

Emergent Systems

This reiterates a point made near the start of this part of the book (in Chap. 3), about the emergence of properties. If we accept the consensual scientific model of evolution by natural selection, then the way in which human brains process information about the world is simply the result of selection pressures acting on organisms that cope differentially in their environments and which are able to pass on genetic material causing their offspring to tend to be like them.

The human sensory interface that translates sensory information into electrical patterns in the cognitive system is simply an outcomes of that process, and the 'coding' is simply what has emerged as a solution that worked. That is, it has in the past tended to allow individuals to leave offspring. It is a system that 'fits' with the need to be able to make sense of, and act on, the environment. Similarly, other aspects of cognition have presumably emerged in the same way: the nature of our percepts, the development of consciousness, the limitations of working memory, the nature of our concepts and conceptual systems, our use of logic and our creative abilities, our ability to reflect on our own cognition and so forth. If all of this relies on the self-organisational properties of synaptic networks that can be tuned through feedback, then there is no established machine code that acts as a natural language of thought that we would recognise as akin to human codes or languages: rather we have evolved cognition that offers survival value, based on coding templates that are largely our common genetic inheritance as humans, but which presumably themselves show individual variation. Unlike electronic computers that are often cloned in the millions, we each have both a somewhat unique processor and a somewhat unique 'operating system'.

Key Terms from the Mental Register

As noted in the part introduction, this part contains five chapters exploring what we understand by, and how we might investigate, such matters as learners' ideas, learners' thinking and learners' understanding. In this part I have attempted to explore what these terms might be considered to mean when used as part of a way of modelling the learner's cognition.

As suggested earlier in the book, during our childhood, we all develop an implicit 'theory of mind' that supports everyday dialogue in terms of a more or less shared 'folk psychology' supported by an informal 'mental register' of terms. So in everyday dialogue, we may commonly use a term like 'thinking' to refer to both processes and the outcomes of those processes, and it is common to talk about memory as though it is a place within the mind where we can store experiences or information that we may wish to access later. This works fine in the context of normal social dialogue but can become problematic when we use the same terms in the context of formal research and scholarship. As I suggested in the introductory chapter, too often in research in science education, the components of the mental register are adopted without any formal definition, as though they are well-defined technical terms.

Within any research programme, there will be widely shared technical terms that have been established and which do not need to be spelt out in detail each time they are used in a study. In the research programme into learning in science, there are such technical notions (such as alternative conceptions and phenomenological primitives; see Chap. 11), but commonly researchers have also borrowed terms from the mental register – thinking, understanding, remembering, etc. – as though these are also accepted technical terms. The need for using either these notions or

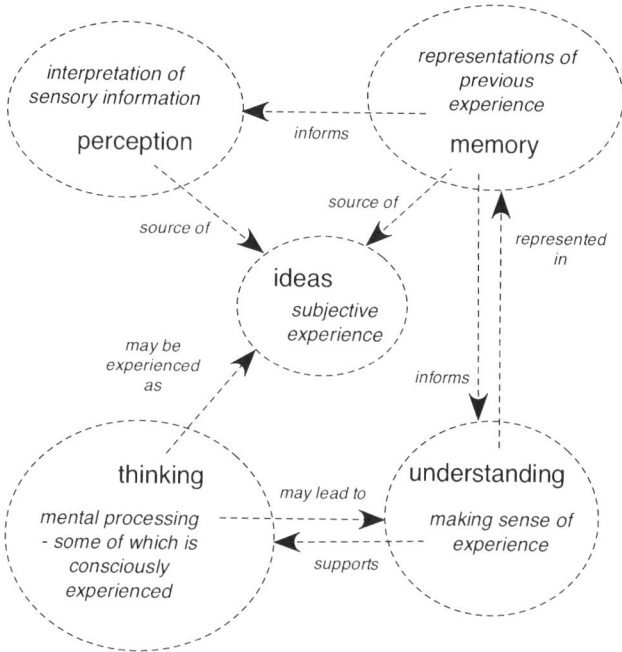

Fig. 7.6 Mental processes – a model relating some key terms used in research into learning in science

alternatives that cover the same set of phenomena in the research programme is clear, but the close familiarity with and taken-for-granted nature of the use of the mental register has led to such terms being used *as if* technical terms, without usually being operationalised in any explicit way.

In this part of the book, I have set out a way of thinking about key notions from the mental register, informed by well-accepted ideas from the cognitive sciences, that can act as a basis for reformulating these notions as constructs suitable for discussing in research. That is, I have been building up a model for how these terms can be understood to relate and to refer to observables in research. So assuming that mind is an emergent property of the central nervous system – indeed of largely the cortical areas of the brain – and is for many purposes best understood in functional terms as a system for processing information (but which leads to conscious experience), I have suggested how we could best understand notions such as thinking, understanding and remembering. Figure 7.6 summarises the meanings that have been established for five of the key terms: perception, memory, ideas, understanding and thinking.

Part III
Modelling the Science Learner's Knowledge

Chapter 8
Introduction to Part III: Knowledge in a Cognitive System Approach

> In short, the interviewer is constructing a model of the child's notions and operations. Inevitably, that model will be constructed, not out of the child's conceptual elements, but out of conceptual elements that are the interviewer's own. (Glasersfeld, 1983, p. 62)

The previous part of this book explored mental notions – such as ideas and understanding – and how these might be understood in terms that support modelling of research into aspects of learning in science (see Fig. 7.6).

Key ideas arising from this exploration of mental constructs were:

- As subjective experience is only available to the individual, any research claiming to report the ideas or understanding or beliefs, etc. of a learner must be based on interpretations made by researchers of public representations of the learner's mental experiences in the external world.
- There is a range of common terms such as thinking, memory and understanding which are widely used to discuss mental phenomena but make up a lifeworld register of signifiers of fuzzy concepts, and which when used in research reports without further clarification can compromise the precise communication expected in technical writing.
- Mental phenomena can be understood as correlates of physico-chemical processes in human nervous systems that ultimately relate to electrical activity occurring in networks of neurons.
- But mental experience is an emergent property of the complex nature of the nervous system and in particular the brain, making descriptions at the physiological level generally less helpful in research into learning in science.
- However, a more fertile approach is to model the learner as a cognitive system that processes information, but where such a model is *constrained by* what is learnt from anatomical and neuroscientific studies, as well as from psychology and science education research.

K.S. Taber, *Modelling Learners and Learning in Science Education: Developing Representations of Concepts, Conceptual Structure and Conceptual Change to Inform Teaching and Research*, DOI 10.1007/978-94-007-7648-7_8,
© Springer Science+Business Media Dordrecht 2013

The Cognitive System Approach

Figure 8.1 sets out an overview, in the form of a kind of concept map (Novak, 1990a), of how one might conceptualise the learner as an organism that is supported in the 'task' of surviving in an environment by a cognitive system that enables the individual to sense the environment and act in, and so on, it. The individual is a *learner* as he/she is able to use this feedback to modify action because of the plasticity of the system.

That is, the cognitive structure through which processing occurs when sensory information informs action is itself modified by experience (see Chap. 5). The bolder arrows in the figure constitute the basis for a feedback cycle: information about the environment and/or the internal state of the organism is processed to direct behaviour that changes the environment and/or the internal state of the organism, and then new information about the environment and/or the internal state of the organism allows that change to be detected. To support this process, our sensory/ perceptual apparatus is especially tuned to notice changes – movements in the visual field, variations of tone or volume, etc.

The overall approach taken here is hardly original and, for example, has much in common with the way that Piaget (1970/1972) thought about young children learning to make sense of their environment. Piaget approached his seminal pro- gramme of work on cognitive development with the perspective of a biologist, and recognised that it was productive for understanding children as developing people to consider that they were also biological organisms (with the constraints and

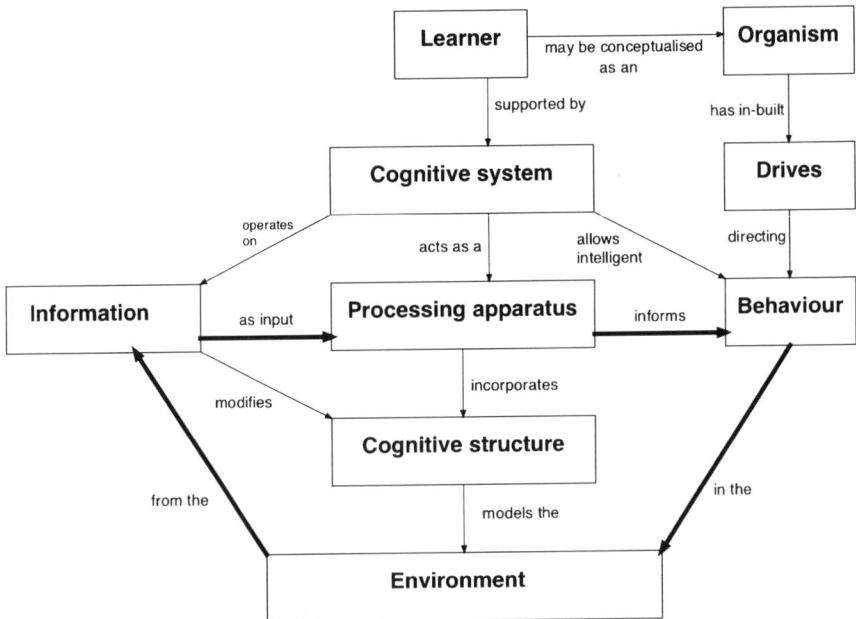

Fig. 8.1 The organism in the environment supported by a cognitive system

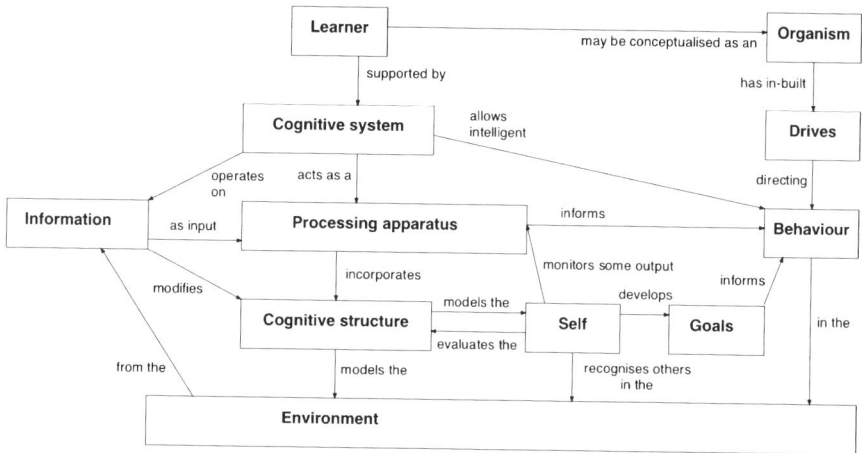

Fig. 8.2 The conscious organism can include itself as a discrete element within its model of the environment

affordances that implied), and so came to be as they were in considerable part because of natural selection acting over extended periods (Piaget, 1979).

Taking a biological perspective as his starting point may have contributed to common criticisms of Piaget as underplaying the role of socialisation in development. Indeed, the conceptualisation presented in Fig. 8.1 need not refer to a human learner, but would apply to any organism with a complex enough nervous system to move beyond purely instinctive behaviour which can only be selected upon at the generational level by natural selection, to be able to modify its own behaviour in response to feedback from the environment, when acting to meet its needs. Such an organism, by modifying (tuning) synaptic connections within its nervous system (see Chap. 7), will learn from experience and may be considered to be modelling the environment to inform future action. This need not require conscious awareness, but should be considered a form of intelligent behaviour, where intelligence is fundamentally the ability to learn from experience. Consciousness, however, enhances this system, as suggested in Fig. 8.2.

Consciousness provides awareness of self as separate from the environment, allowing deliberate goal-directed behaviour. Awareness of self also supports the development of a 'theory of mind' (see Chap. 2), which allows us to identify others as discrete elements of the environment to which we can posit needs, emotions, ideas, etc. The extent to which we are likely to be able to effectively model the minds, that is, the conscious experience, of those others depends to a large extent on how like us they are. The philosopher Thomas Nagel (1974) famously argued that although we have good reason to assume that a bat has a sophisticated enough nervous system to be conscious, and so it is meaningful to talk of *what it is like to be a bat*, there was no way that a human being with his or her very different nervous system, and in particular sensory system and cognitive apparatus to support it, could ever know what it is was like to be a bat.

Awareness of self as a conscious actor in the world also allows the development of metacognition, which it was suggested in the previous chapter, provides the facility to monitor, reflect upon and evaluate our own cognition. Thus, the metacognitive being not only learns from feedback about actions in the world, but can also learn from feedback about cognitive processes themselves.

Linking Back to the Mental Register

In the previous part a number of key themes were explored, taking terms from the lifeworld mental register and considering how they could be understood in terms of a model of the learner as a cognitive system. Terms such as perception, ideas, memory, understanding and thinking have been discussed from this perspective and can be mapped onto different parts of Fig. 8.2 – as has been done in Fig. 8.3:

Seeking to Understand 'Knowledge' Within the Cognitive System Approach

Another key term in the lifeworld mental register is 'knowledge', which – like the other terms considered in the chapters in Part II – is widely used in everyday communication, both in lay and professional educational contexts, but again proves difficult to pin down to a precise meaning.

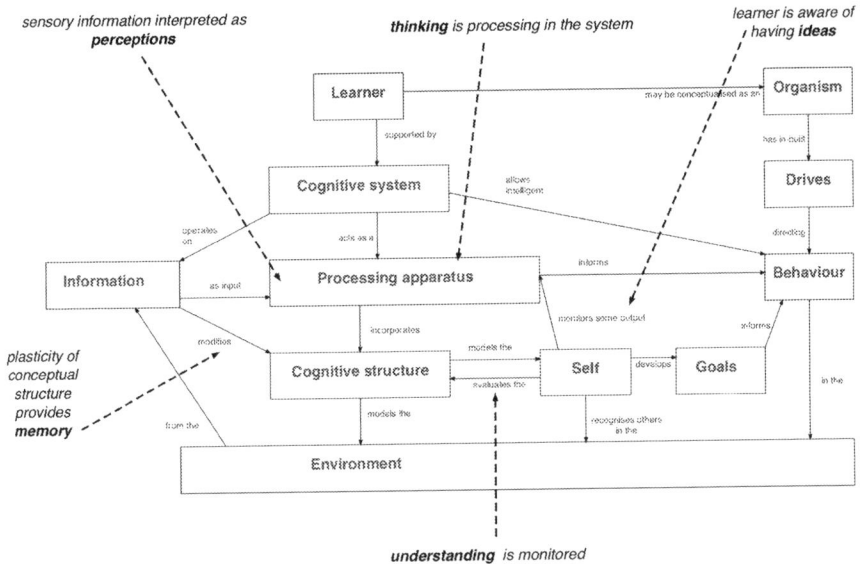

Fig. 8.3 Linking the cognitive system description to the mental register

A particular problem with the notion of knowledge, as widely used, is that the term is applied to mean:

- Something which is an attribute of individuals: the students' knowledge of science topic X
- Something which exists in the public sphere: scientific knowledge

This presents a particular difficulty for the analysis being followed in this book. An individual's knowledge can be understood in terms of the cognitive system approach, as with terms such as understanding and memory (as in Fig. 8.3), but it is more difficult to see how knowledge – if understood in these terms – can also be said to exist in the public space between individuals.

The chapters in this part seek to take forward the analysis presented in Part II. The next chapter, Chap. 9, addresses the core issue of what is the nature of a learner's knowledge, examining various meanings that have been given to the term, before suggesting how knowledge can best be understood as part of a cognitive system. Although the concern of the present book is modelling *the learner and learning*, the learner's knowledge is commonly judged in education against what is understood to be scientific knowledge, so the relationship between personal and public knowledge offers a challenge for the approach adopted here.

One sense of the term 'public' knowledge is that it refers to what is generally known, rather than being private information, but whilst this is a useful notion, it is clearly problematic. For one thing, the public is a large body, so there is likely to be little knowledge that can be considered to be known by everyone. Public knowledge is therefore better understood as knowledge that is widely known and generally accessible through being represented in the distributed system of the network of people. We can ask someone, 'look it up' in a book, or perhaps more often these days, use an Internet search engine.

If something is public knowledge, in the sense that it is widely known, then that might seem to imply that many people have *the same* knowledge. This might seem a reasonable suggestion if we are interested in 'factual' information such as the answers to questions such as who is the current president of the European Union, what do the initials NARST stand for and where is Pitcairn Island? However, there is a problem if we are interested in the more complex information needed to answer such questions as how does photosynthesis work, why did the dinosaurs become extinct and what is the molecular structure of benzene? Here knowledge depends upon understanding that we have seen (Chap. 6) is nuanced and may be quite idiosyncratic. It would not be sensible to expect the 'same' knowledge to be held by many different people in such cases. Clearly, the very notion of 'public knowledge' is a problematic one. The issue of how the personal knowledge of an individual relates to notions of public knowledge such as scientific knowledge is taken up in Chap. 10.

The final chapters of this part then shift the focus back to the individual learner. Chapter 11 tackles a long-standing issue in science education research, that is, of the nature of the different kinds of knowledge components reported in research, and sets out a model (mindful of what has been established in Part II) for making sense of these different types of knowledge element. Finally, Chap. 12 considers another key referent in some science education research, conceptual or cognitive structure, and explores how aspects of an individual's knowledge might be organised within some kind of structure.

Chapter 9
The Nature of the Learner's Knowledge

Knowledge as a Problematic Notion

Like a number of other key terms considered in this volume, 'knowledge' is a problematic term because it is widely used in everyday discourses as a term that is generally understood but without being closely defined. The notion of 'knowledge' is part of the lifeworld 'mental register' that comprises of informal, fuzzy concepts but which can suggest a veneer of technicality in professional and academic discourse. That is, studies that claim to report someone's knowledge might seem to be referring to something that we all take for granted that we understand, yet may actually fail to clearly operationalise how knowledge is defined in that research.

Public and Personal Knowledge

An important complication in the way the word knowledge is used is the distinction between what can be termed as public knowledge, as in the expression 'scientific knowledge', and what might be termed personal knowledge, as the knowledge that an individual can be considered to hold. As this chapter clearly concerns the notion of the learner's knowledge, it will only discuss personal knowledge, not public knowledge. However, public knowledge is important to our theme, because – as was seen in Chap. 6 – in education, both in research and teaching, it is common to evaluate a learner's knowledge, and so their learning, and to consider we are doing this in comparison to some system of public knowledge, as represented in a curriculum, for example. This is reasonable when it is considered that a large part of education is framed in terms of learning about a public system of knowledge, such as scientific knowledge (see Fig. 9.1).

Public knowledge therefore needs to be considered as part of any overall account of modelling learners and learning. A consideration of how public knowledge may

K.S. Taber, *Modelling Learners and Learning in Science Education: Developing Representations of Concepts, Conceptual Structure and Conceptual Change to Inform Teaching and Research*, DOI 10.1007/978-94-007-7648-7_9,
© Springer Science+Business Media Dordrecht 2013

Fig. 9.1 Education may be
seen as learning about public
systems of knowledge

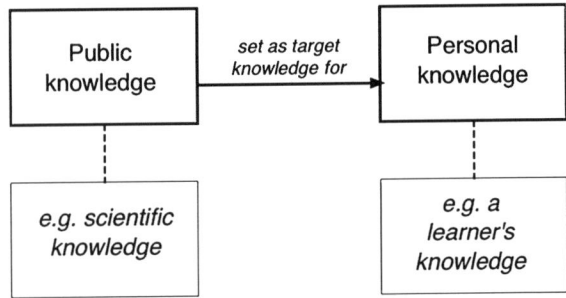

Public knowledge	set as target knowledge for →	Personal knowledge

| e.g. scientific knowledge | | e.g. a learner's knowledge |

be understood is presented in the following chapter (Chap. 10), and further
discussion of this issue will be deferred till then.

What Does It Mean to Know?

Our knowledge is what we 'know', but this way of talking may seem to imply
something unambiguous and readily demonstrated when, as the previous chapters
have suggested, this may be far from the case. For example, Aaron suggests,

> The word 'know' is used in widely different ways, but the epistemologist has always found
> his main interest in that use of it which involves a contrast between knowledge and opinion.
> In the widest sense of the term 'knowledge', opinion is itself a species of knowledge, but in
> the narrower sense knowledge is contrasted with opinion. (Aaron, 1971, p. 3)

Knowledge as True Reasoned Belief

In philosophy the term knowledge is reserved for reasoned, true belief (Bhaskar,
1981). Matthews describes how.

> Plato holds a Reasoned True Belief account of knowledge that can be systemized as
> follows, where A stands for a cognizing subject and p for some proposition or statement
> of fact purportedly known by them: "A knows p" ≡ (i) p is true – Truth condition;
> (ii) A believes p – Belief condition; (iii) A has good reasons for believing p – Evidence
> condition. (Matthews, 2002, p. 127)

Consider a learner, Jean, who

(a) Thought that *phosphorus was a metal.*
(b) Recalled being told that *hydrogen could be a metal at very high pressures* but
    found this unconvincing.
(c) Thought that *all metals conducted electricity* because on testing iron, copper
    and zinc, it was found that a lamp would come on when the metal was inserted
    into a test circuit.

By Plato's criteria, these statements could not be considered as representing Jean's knowledge. Statement (a) would be considered to be *untrue*, as phosphorus is considered in terms of accepted 'scientific knowledge' (see Chap. 10) as a non-metal, and so thinking it is a metal would not count as knowledge. Statement (b) could not be considered knowledge either, as even though scientists consider hydrogen would be a metal under the right conditions, and suspect that metallic hydrogen may even exist in our own solar system in the cores of the giant planets, Jean *does not believe* this reported fact. Finally (c) should not be considered knowledge as testing such a small sample of the metals hardly provides secure evidence for a generalisation to all metals even though Jean's experiential basis for the belief might well have been a school science 'experiment' presented as showing that metals conduct electricity. That the conclusion is considered true would not justify the statement as knowledge when the reasoning is flawed. After all, on a similar basis, Jean might just as well claim that all metals were (ferro-)magnetic, based on testing iron, nickel and cobalt – and that would be an incorrect conclusion.

Plato's system may clearly be useful in a philosophical context, although only if we consider we have the basis for determining truth. This would work in an axiomatic system, for example, within Euclidean geometry. In such a system, certain axioms are established as tenets for the system, and then logic is applied to see what else follows. True knowledge becomes possible in such mathematical contexts, because deductive logic can be applied to find what must be the case.

In science such an approach might be meaningful where definitions are involved. If the scientific community decides to measure temperature in Kelvin, and Jean believes that temperature is measured in Kelvin because (a) the science teacher told the class this, and this was corroborated by (b) the respected school text book, and (c) because Jean had seen that scientists used this convention in a documentary about space exploration, then we might well judge that this constitutes knowledge (true, reasoned, belief) in the philosophical sense.

However, there might be more difficulties in the case of Jean believing that *the universe began in a 'big bang'* – or that birds evolved from dinosaurs, or that hydrogen adsorbed into palladium cannot provide a source of 'cold fusion' power. These things cannot be considered definitely 'true' as they are scientific ideas which are – through the nature of science itself – only accepted provisionally (Gilbert & Watts, 1983).

If Jean was sophisticated enough to believe that *the big bang is the currently most widely accepted scientific notion of how the universe began* (or that *there is currently no generally accepted evidence that cold fusion can be a major potential source of power*), then this might be considered knowledge, as long as Jean's grounds for belief were considered strong enough. However, it is known that school students often fail to appreciate the provisional nature of scientific knowledge (Driver, Leach, Millar, & Scott, 1996), and so if Jean considered his beliefs about these scientific ideas to be 'facts', then these beliefs fail to pass the 'reasoned true belief' criteria of knowledge.

Whilst the reasoned true belief notion of knowledge offers criteria for what counts as knowledge, it *also* excludes most of what is normally considered as knowledge and so is not that helpful to science educators (teachers or researchers) in talking about their work. For what Jean believes he knows about phosphorus, hydrogen, metals, the origin of the universe, the evolution of birds and the failure of cold fusion is of relevance to further learning in science and so of interest in science education, regardless of whether the term knowledge is considered to technically apply to them.

Finding a More Useful Notion of Knowledge for Science Education

So whilst the 'reasoned true belief' version of knowledge may be useful in philosophical discussions, it does not seem to 'do the job' in supporting research in science education. The influence of current thinking upon new learning does not depend on how some external observer evaluates the truth of those thoughts. If a focus of science education is to be the learner's knowledge, then we need an alternative definition of what we mean by knowledge that better fits how this term can be of use to the research community.

Although the need for rational justification is not always maintained, the *Oxford Companion to Philosophy* suggests that 'virtually all theorists agree that true belief is a necessary condition for knowledge' (Goldman, 1995, p. 447). Yet when considering the nature of understanding earlier, it became clear that notions of what are believed are themselves not always entirely straightforward (see Chap. 6). An individual may have several inconsistent ways of understanding the same phenomena or concept area and be cued into thinking in one way or another in different contexts. A student who answers school physics questions using Newtonian physics, but thinks in terms of an impetus notion of motion when dealing with force and motion in the lifeworld, could be said to demonstrate 'belief' in, or commitment to, both, or neither, ways of thinking. What would be clear in this example is that a researcher interested in the students' 'knowledge' would have to take both ways of thinking into account to produce an authentic account of what the individual 'knew' about force and motion.

Student learning is considered to be influenced both by ideas that the student entertains without necessarily strongly committing to them, as well as the things they do strongly believe which would not be considered true. Indeed much of this area of research has been concerned with 'misconceptions' or 'alternative conceptions' or 'intuitive theories': entities that would fall outside of a 'true belief' model of knowledge.

Perhaps an alternative term is needed for what researchers tend to be interested in when they explore students' thinking about science – a term for 'things the student thinks might be the case'. Within that category there is scope for different levels of commitment and different levels of match to what might be considered 'true' by others. We might report that we are researching into 'notions that the learner entertains as possible mental representations of some aspect of the world' but that is a rather convoluted expression. Slightly less clumsy might be 'plausible mental constructions'.

This highlights a major ontological distinction in this area of research, which follows from the considerations earlier in the book. Research into student knowledge can be intended to explore one of two rather different things:

(a) The ideas generated by learners in response to specific stimuli
(b) The conceptual resources available to learners

These are very different in that (a) ideas relate to the outcomes of some kind of mental processing, thinking (see Chap. 7), and are transient, whereas (b) conceptual resources are more permanent although not immutable features of the mental apparatus responsible for generating the ideas, and so part of the mental 'environment' which influences those thinking processes. The issue of precisely what these resources may be is a core one for research into learning in science and will be the subject of Chap. 11. Clearly it follows from the earlier discussion in the book that all researchers that can ever access *directly* are *public representations of* (a), that is, behaviour representing what the learner is thinking at that moment, which can then be considered indicators of some aspects of (b).

So, our research at best allows us to make inferences about what learners are thinking from which we then seek to make further inferences about the stable resources (e.g. 'knowledge') that support that thinking. As has been argued earlier, the first stage of this two-phase process requires interpretations that rely on the learner offering us honest reports that are clearly communicated and even then may tell us little about the bulk of the thinking iceberg (the preconscious thinking) which does not break the surface of conscious awareness.

Those ideas experienced by the learner are the outcome of a contingent nexus that draws upon those available resources for thinking that can be accessed at that moment. Some of these resources are the internal conceptual resources (b, above), but the current environment also offers affordances (cf. Figs. 5.1 and 5.2 in Chap. 5). So, in other words, a researcher's question; some focus introduced to the research context as used in techniques such as interviews about instances/events or construct repertory test; the availability of a periodic table or a reference book or the Internet; or the presence of a peer in a group task, or a dyad interview; can all act as external resources that may be drawn upon in association with internal mental resources to produce a response in a research situation.

So, for example, consider the following tasks that might be presented to a learner:

(i) Explain the shape of a methane molecule.
(ii) Use the valence shell electron pair repulsion theory to explain the shape of a methane molecule.
(iii) Explain why the methane molecule is considered to be tetrahedral.
(iv) Use the valence shell electron pair repulsion theory to explain why the methane molecule is considered to be tetrahedral.

These different versions of the question might all be considered to be testing 'knowledge' of a common procedure for determining the shapes of molecules and applying it in a particular case. For some learners, depending upon the accessibility and 'state' of certain internal conceptual resources, this is a trivial question, as the example may be very familiar, and the distinction between the four forms of the question would not be significant in influencing the answer produced. For these students the question does not present a 'problem' (see Chap. 7), just a routine exercise in applying well-learnt principles.

By 'state' of conceptual resources here, we might consider the extent to which those resources represent a canonical understanding of the topic (itself a problematic notion, see Chap. 10) and how well organised that representation is and perhaps how well integrated with representations of other related scientific ideas (see Chap. 12), that is, the accuracy, organisation and integration of the students' knowledge. The degree of integration is likely to be closely linked to how accessible the representation is with appropriate cuing. For a student who has learnt about this topic, but has not represented the learning so effectively in terms of these criteria, task (i) may be more challenging.

However, (iii) offers the learner a rather significant piece of information – the result of the application of this knowledge. For a student who could not readily access a clear memory of the principles and how to apply them, offering the additional information in (iii, cf. i) allows an additional *strategy* in forming an answer, as the learner can think about the problem 'from both ends'. That is a general problem-solving strategy which itself would be considered an internal mental resource – perhaps an aspect of metacognitive knowledge (see Chap. 7).

Questions (ii) and (iv) tell the learner which set of ideas to access and apply in producing an answer – and that could clearly be a useful memory cue for some students, being helpful if the representation of prior learning is not well integrated with other representations of related topics, and so not so readily accessed.

Moreover, it is quite possible that some students who are familiar with aspects of atomic and molecular structure, but who have not been taught about the shapes of molecules, could actually deduce from the label 'valence shell electron pair repulsion theory' the gist of what was required. It is certainly feasible that some learners might be able to use the affordance of question (iv) as a resource and suggest that methane is tetrahedral because of the repulsion between the electron pairs in the valence shell – even if the principles are quite novel to them. By contrast, it would seem very unlikely that a student who had not previously met these ideas could generate such a response to question (i).

I recall here one interview I undertook with a college student where in responses to a series of questions she 'discovered' and explained why neutral molecules might attract each other due to what are often called van der Waals' forces. She had not yet been taught these ideas in class, and had no recollection of having come across this idea, but had sufficient conceptual resources, in terms of knowledge of the structure of molecules, for example, to build up the idea of transient dipoles leading to net forces between neutral entities. The scaffolding of the interview context allowed her to produce this plausible mental construction, and so at that point, this became part of 'her knowledge'.

Use of the Term 'Knowledge' in Science Education

Even if suitable alternative terms to 'knowledge' were available (such as 'plausible mental constructions'), making it easier for the research community to restrict the use of the term 'knowledge' to the limited meaning philosophers claim, it is clear that the label of 'knowledge' will continue to be used widely in educational discourse, where it is understood to be less restrictive:

> Much *of our knowledge* pretends to nothing more than probability. We guess, have hunches, and believe on such evidence as is available, and for the time being we take what we believe to be true without, however, claiming certainty for our beliefs. If we are wise we go on testing our beliefs, searching for further evidence that will confirm or refute them. A great deal *of our knowledge* clearly is of this kind and it has been held that all of it is so. (Aaron, 1971, p. 49, present author's emphasis)

For the purposes of this book, then, the term 'knowledge' will continue to be used, with the understanding that a learner's knowledge refers to what they believe to be the case or simply consider as a viable possibility. Their knowledge is the range of notions under current consideration as possibly reflecting some aspect of how the world is. That is not knowledge as philosophers understand it, but it better fits common usage, and refers to what is relevant to science education researchers, where true, reasoned belief is an ideal of limited practical use in understanding learners. The meaning used here fits better with that suggested by Higgs and Titchen (1995, p. 521) who 'define knowledge (of the individual) as an awareness of the individual which has current conviction for the individual, gained through the testing of acquired or self-generated understanding' and with what Park (2011, p. 1) has referred to as 'not … true justified belief. Rather… the psychological sense of an information carrying mental state'.

Personal Knowledge

All that has come before concerning what is understood about the nature of cognition (see Part II of the book) raises the issue of how we should understand the nature of personal knowledge – the knowledge of an individual person. It would seem that what science education researchers are usually interested in, and what is meant in everyday discourse, is not limited to reasoned, true beliefs but rather includes whatever is 'stored' (that is represented) in a person's cognitive system that leads them to express particular beliefs and views and indeed to entertain certain possibilities as worthy of consideration. This 'personal' knowledge may not be shared by others and indeed provides the basis for actions that may lead to outcomes that later lead us to decide it was mistaken. As Polanyi (1962, p. viii) suggested, 'personal knowledge is an intellectual commitment, and as such inherently hazardous'.

A strong behaviourist view would avoid such issues by only considering direct observables – behaviour, such as responses to the researcher's stimuli – rather than the inaccessible mental structures that are conjectured to be intermediaries between stimulus and response. Science education has long eschewed such an approach, but the behaviourist perspective offers a useful reminder that if we take a scientific view, then 'personal knowledge' is not a phenomenon open to direct inspection, but rather a theoretical term used as part of an explanatory scheme. That is, we should see 'personal knowledge' as a conjectured entity, as a construct that helps us explain the basis and patterns of our thoughts, and aspects of the behaviour of others.

So as individuals we experience a certain level of stability and continuity in our thoughts, which we can conjecture is due to a level of stability in the cognitive apparatus we have built up through our experiences in the world (as discussed in some detail earlier in the book), which 'biases' our thinking, and leads us to understand our ongoing experiences in certain ways. We might think of this as the way in which our cognitive systems have 'represented' past experience so as to allow us to organise, understand and explain ongoing experience (cf. Fig. 8.2). Through this iterative process of representing experience, an individual builds up 'content' of the cognitive system that we might choose to label as their 'personal knowledge'.

From a constructivist perspective,

> Knowledge is not passively received either through the senses or by way of communication, but it is actively built up by the cognizing subject. … What we call "knowledge", then, is the map of paths of action and thought which, at that moment in the course of our experience, have turned out to be viable for us. (Glasersfeld, 1988, pp. 1–4)

As researchers, we observe patterns of behaviour in others such as what they say, what they write and so forth and recognise both strong stable patterns from individuals when a particular person usually seems to offer similar answers to similar questions and sometimes clear individual differences where different learners in the same class may offer quite different responses to the same questions relating to topics they have studied in the same science lessons. Indeed such patterns as a tendency to stability in individual behaviour and different people behaving differently in the 'same' situation are recognised by all normal human adults and understood through their theory of mind (TOM, see Chap. 2). That is, from quite a young age, we interpret the behaviour of others in terms of them each having an individual mind, much like our own, but uniquely theirs.

As researchers we go beyond the informal TOM and look to build models of aspects of the minds of learners that help us understand aspects of behaviour (why they say this or write that). We not only assume that students have minds much like our own, but we want to understand something of the 'content' of their individual cognitive systems, the features of their cognitive systems that have developed through their experiences to facilitate them in organising, understanding and explaining their ongoing experience, leading to particular patterns of thought in their subjective world (World 2 in the sense discussed in Chap. 4) and so expressed in particular patterns of behaviour that are observable (in the objective 'World 1' – the public space referred to in the model used earlier in the book), and so are the phenomena observed in

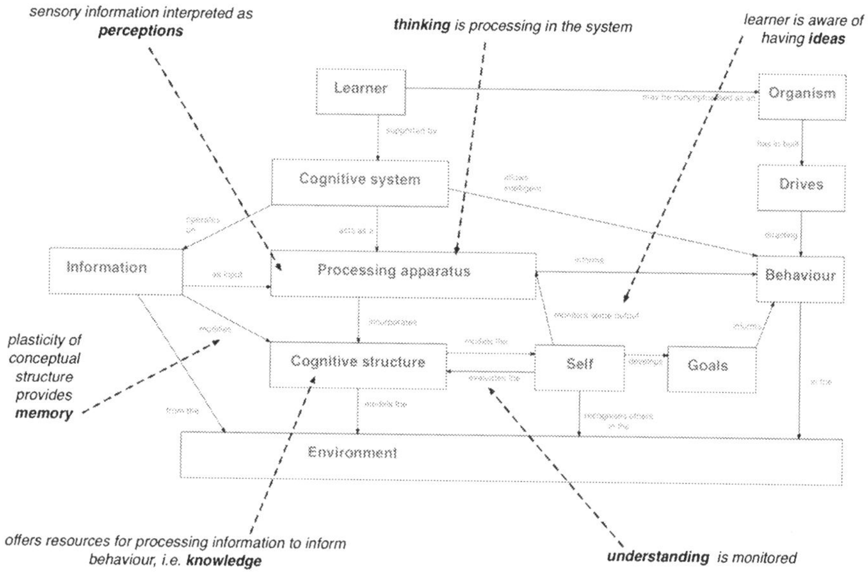

Fig. 9.2 Knowledge as resources to be drawn upon in the cognitive system

our research. It is those 'contents' of the learner's cognitive system which we are interested in when we look to explore the learner's personal knowledge. As a research community, then, we are interested in how to most usefully think about these contents, which can be considered to make up the 'components' of a learner's knowledge. It is in this sense that we understand knowledge in our model of the learner at the system level (see Fig. 9.2).

What Are the Cognitive Resources That Support Learning?

The present state of knowledge in neuroscience does not allow us to be able to identify the precise anatomic correlates of something like a concept, even if that level of detail would be useful. However, a number of ideas met earlier in the book are informative in considering how researchers should think about the cognitive resources that are called upon during cognition.

A key point is the shift that takes place during the stages of processing in perception (see Chap. 4), from fully automatic 'coding' type processes in the sensory interface, through various filtering stages that occur without consciousness, to the level of conscious thought. Karmiloff-Smith (1994, p. 694) refers to how during human development there is a process 'by which information that is in a cognitive system becomes progressively explicit knowledge to that system' (this is considered further in Chap. 15).

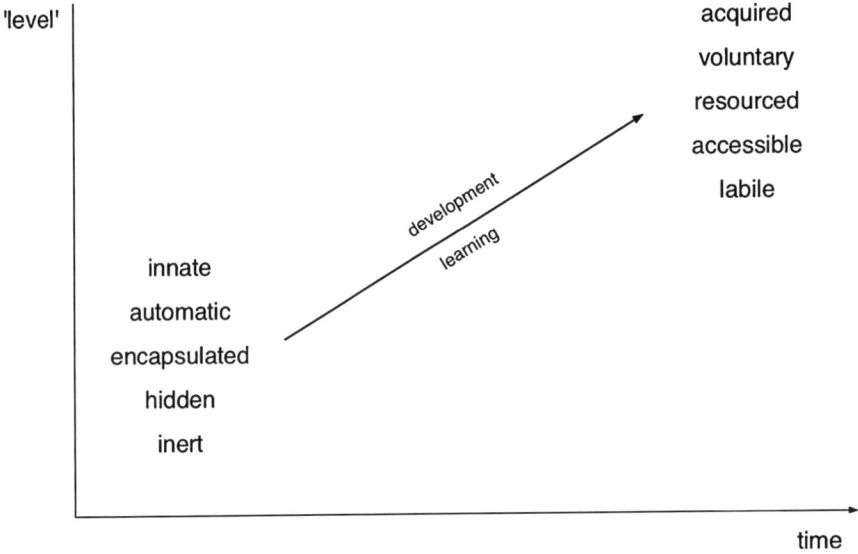

Fig. 9.3 Characteristics of cognitive resources

Indeed, it is possible to characterise cognitive resources along a number of dimensions, as shown in Fig. 9.3:

The time dimensions represented in Fig. 9.3 reflect Karmiloff-Smith's point, suggesting that resources available early in human development tend to have one set of characteristics, whilst those that develop later tend to be quite different. The other 'dimension' reflects how we tend to conceptualise aspects of cognitive processing as being 'lower' or 'higher' level (in the sense described in Chap. 4, cf. Fig. 4.7). The resources available to support cognition are shown as having clusters of properties, which reflect the discussion of the cognitive system in Part II of the book.

That is, cognitive resources may be:

Innate: hardwired into the system from very early in a child's development	or	Acquired: developed over time in response to experience
Automatic: so that processing is triggered as a reflex to certain input	or	Voluntary: able to be selected and applied under conscious control
Encapsulated: discrete modules that act independently and provide output only depending upon input	or	Resourced: able to access other resources during processing and so able to process input in relation to other 'information' represented in the system
Hidden: from conscious awareness	or	Accessible: to introspection
Inert: fixed and non-modifiable	or	Labile: able to be modified by experience (i.e. learning)

At first sight it may seem that resources with characteristics in the second cluster are much more valuable to humans as cognitive agents in the world, but it is important to bear in mind a number of key points:

(a) Processing is much faster by lower level resources; for example, the blink reflex will protect eyes much faster than the output of conscious reflection on a perceived threat.
(b) Lower level resources are 'ready to go, out of the box' and allow a neonate to start acting in its environment before there is an experiential base on which to build acquired resources.
(c) Higher resources need to be constructed from some pre-existing building blocks: the innate resources provide those starting points.
(d) The rate of input of information available to the cognitive system from the environment and proprioceptors (the body's internal sensors) is vast, so the cognitive system sequentially filters information at increasingly higher levels (see Chap. 4, cf. Fig. 4.3).

The other key point to reiterate from what was discussed earlier in the book is that there is no absolute distinction between the apparatus of cognition and the 'contents' of the cognitive system. So acquired knowledge, in effect representations of what has been learnt from experience, involves modifications of the same cognitive structure that represents the innate knowledge that is our genetic inheritance. However, the most basic levels of processing are hardwired so that they cannot be 'overwritten' by experience – as these resources are fundamental to basic processing in the system, protection from modification would seem to provide an important safeguard.

The Possibility of Distributed Knowledge

Before moving on to consider what might be meant by terms such as 'scientific knowledge' (in the next chapter), it is useful to consider how the approach adopted in this book might apply to realistic contexts. The model used here is a general one, with the cognitive system sensing and acting in 'an environment' without any further characterisation. This is only useful if it is applicable to realistic teaching and learning contexts. In science education, such environments are likely to include science classrooms with students (hopefully) listening to teachers, working in groups, talking to each other and/or a teacher, handling apparatus and materials and using texts and other learning resources. Other environments where student learning might take place would include fieldwork, home study, informal learning from museums, leisure reading, viewing television programmes and so forth.

In these various contexts, the individual's learning will depend upon particular features of the environment (the teacher, a peer, the Internet, etc.), as well as upon the internal cognitive resources within their nervous systems. The leaner is able, in these contexts, to draw upon additional (external) resources for thinking to use alongside their internal cognitive resources. This does not require a fundamental modification of the model being developed in this book. Back in Part II (see Fig. 5.1), it was suggested that sensory information and memory provided two sources for thinking, and sensory information may be derived from teacher talk, discussions with peers, reading books, watching a practical demonstration and so forth.

In the introduction to the present part (Chap. 8), the general context of human cognition was set out in terms of sensing, modelling and acting upon the environment, then sensing the new state of the environment, providing feedback for modifying the internal model of, and guiding intelligent action in, the world (see Fig. 8.1). However, it is clear that the environment cannot be considered as a static and inert context in and on which an individual learner operates, and this raises the issue of whether resources in the environment should be considered, like cognitive resources, as a form of 'knowledge'. This would be the perspective taken in some connectivist accounts of learning, where knowledge is considered to be distributed across networks – and these are not seen as limited to neural networks within a single individual – so that 'knowledge may reside in non-human appliances' (Strong & Hutchins, 2009, p. 55).

In the approach taken in this book, the focus has been very much on the individual understood as a cognitive system. This is represented in Fig. 9.4, which shows an individual who is interacting with both an object in the environment and another processor. The object could be a textbook, or some laboratory apparatus, for example, and the individual is able to both sense and act on the object. The processor can be understood as something more than a static object that can be manipulated but a special type of object able to actively process information: this could be a computer, for example, or, indeed, another person.

The same scenario is represented somewhat differently in Fig. 9.5. Here the cognitive system is seen to include not only the individual learner but also those features of the environment supporting cognition. Cognition is distributed because processing is 'shared' between more than one processor. Just as each processor will have access to internal resources (the individual's 'memories', a computer's database), the object is also used as a resource. If knowledge is understood in terms of resources that can facilitate processing, then in this system, the knowledge is distributed across both processors and the object. So, in this perspective, knowledge resides in people and in computers, and in textbooks, and indeed in anything that can be interrogated within the system, for example, a test tube of copper sulphate crystals suspended above a Bunsen burner flame and observed.

It is not sensible to ask which of these ways of understanding the scenario is correct, as both are meaningful and potentially useful ways of thinking about the situation. There is a difference of semantics here, as 'knowledge' is understood rather differently in these two ways of modelling the same situation – either as internal resources of the learner or as distributed across a network of people/things. The question is: Which is the most useful way of understanding this situation? That is likely to depend upon our purposes.

The distributed cognition perspective offers a useful way of thinking about knowledge. However, seeing knowledge as distributed across a network may lead to

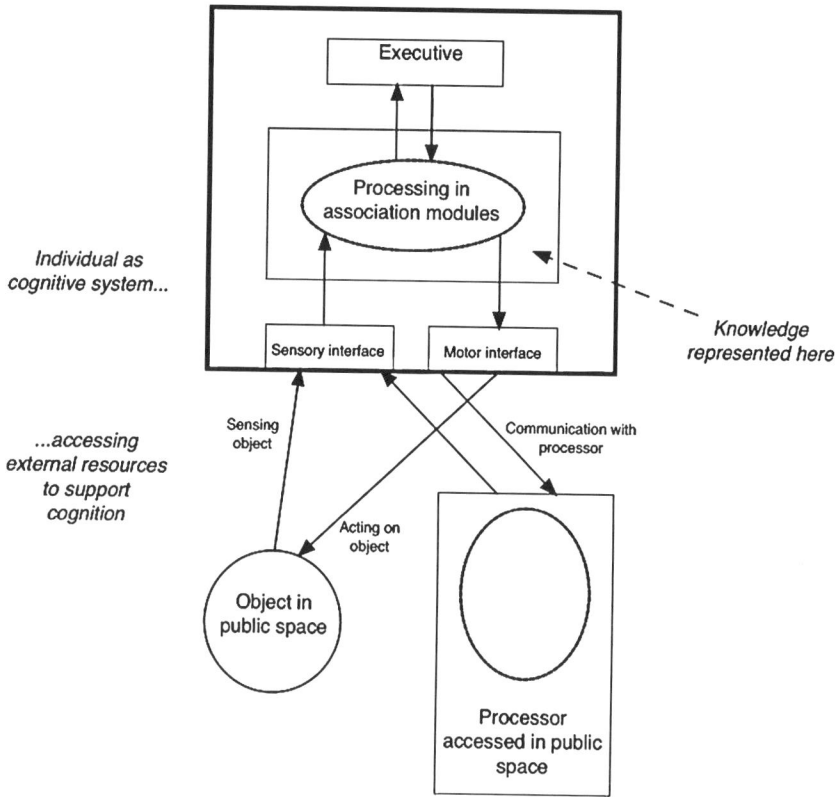

Fig. 9.4 The individual learner supports cognition by drawing upon external resources

rather convoluted notions of what the knowledge actually is in particular networks. Consider a specific example of the scenario referred to above (Fig. 9.6).

Figure 9.6 represents a very simple example of a situation we might examine through the lens of distributed cognition. Two classmates, Jean and Jerome have got together to revise for a test are looking at a diagram provided by the teacher entitled 'the structure of NaCl'. From a distributed cognition perspective, knowledge might be said to reside in the diagram. However, it is not clear that it is possible to assign any specific knowledge to the diagram in isolation (and from a distributed cognition perspective, one would not wish to, as knowledge is distributed across the system).

Now, in an ideal world, Jean and Jerome would look at the diagram, and discuss it, and come to an agreed interpretation of it. Indeed, in an *ideal* world, they would interpret the diagram in the way the teacher had intended! If Jean and Jerome appreciate the value of talking through their ideas, and if they have developed critical thinking skills in argumentation (see Chap. 7), it is quite possible that the diagram may be a useful resource to facilitate a discussion through which both learners develop their understanding of some science and so modify their own cognitive

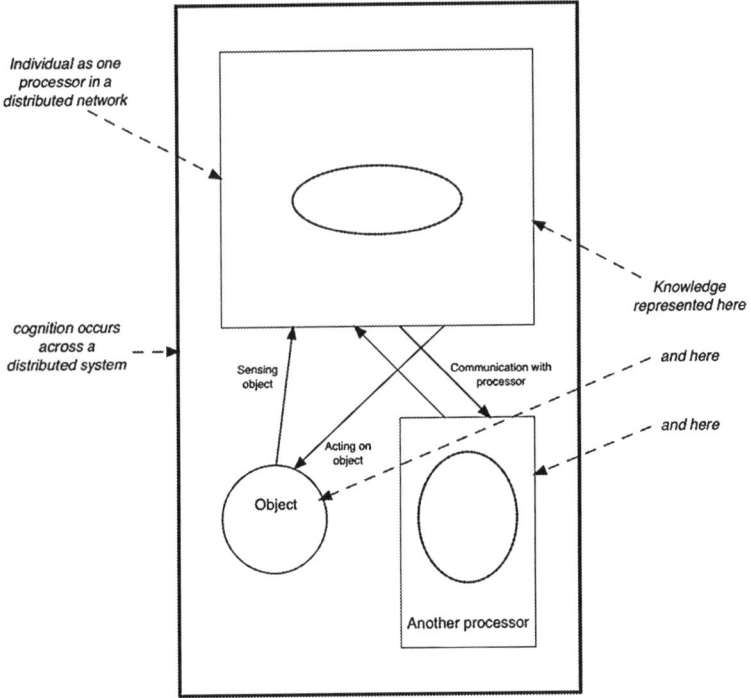

Fig. 9.5 Distributed cognition perspective: the learner is one component of a more extensive cognitive system

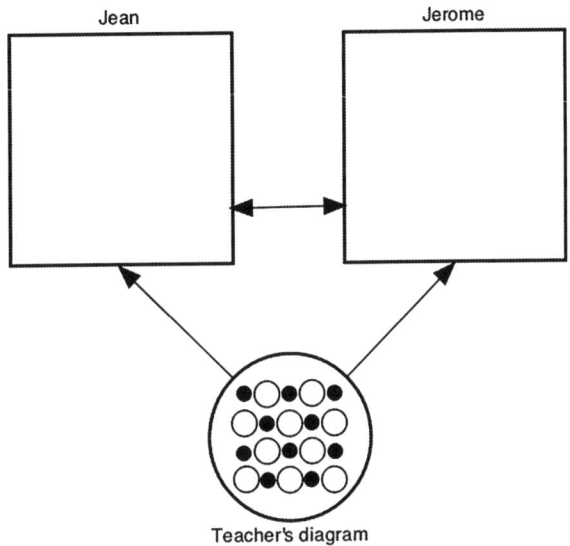

Fig. 9.6 The knowledge residing in an object

structures. It is also in principle possible that due to the interactive and iterative nature of dialogue, they may shift their views towards much the same understanding of the science – leaving aside for the moment the issue of how we could ever be sure they had the 'same' understanding (cf. Chap. 6). In this situation we might see that the two 'processing' components (i.e. Jean and Jerome) coordinated effectively, supported by the external resource of the diagram – and so indirectly drawing upon the teacher's own input – and consider this an effective example of distributed cognition. If the intention had been to write summary revision notes together, an agreed output might readily be produced.

However, without in any sense undermining the value of peer discussion and dialogue in learning, it is also clear that the ideal case is not the only, and perhaps not the most likely, version of this scenario. Both Jean and Jerome have unique internal (cognitive) resources for interpreting the diagram, and they may come to different interpretations. For example, perhaps the teacher was intending to represent a cross-sectional slice through a 6:6 coordinated crystal, showing the cubic arrangement of ions, but Jean fails to appreciate the sectioning and reads the diagram as showing that NaCl is an ionic crystal with 4:4 coordination. Jerome, however, holds some very common alternative conceptions of ionic bonding (used as an example in Chap. 6) and interprets the diagram as showing how ten molecules of NaCl, formed by electron transfer between Na and Cl atoms, are neatly packed into the crystal. Jean and Jerome might discuss their different interpretations but will not necessarily come to an agreement or even fully appreciate each other's ideas.

In this less than ideal version of the scenario, it is still possible to consider distributed cognition as it would be possible, for example, to consider a transcript of the conversation, and start to build a model of how the two students were thinking, and perhaps observe some shifts in position. However, this would not lead to a clear outcome, beyond perhaps simply an agreement to differ and move on to the next task. So the distributed processing could certainly be modelled, but as there is more than one self-directed 'processing component' involved, the model would be a messy one. In this situation, the case for preferring to see the scenario as a distributed cognitive system rather than two interacting cognitive systems may not be strong. In more complex contexts such as groups of learners working together or whole classes, the notion of what the knowledge that is distributed across the system will become even more problematic.

However, these alternative conceptualisations could both offer valuable insights, and it is not argued here that the distributed cognition model does not have value, but simply that because people each have their own goals, and the ability to moderate them, and are able to direct their own behaviour, any distributed network of people becomes a complex situation as there is no one source of executive control marshalling the distributed resources towards a common purpose, and able to make executive decisions when different processing components are unable to agree. Teachers, who might like to be in that position in regard to their classes, are well aware that the individual 'processors' (students), having their own minds, are not always prepared to overwrite the outcomes of their own internal processing on the basis of external authority! Distributed cognition offers a useful perspective to explore and analyse these issues, but because each learner is self-directed, it is considered here to often be more useful to analyse learning at the individual level and see others as providing input and feedback to that learner's personal cognition.

The question of whether knowledge resides in objects seems more straightforward. The teacher's diagram *represents* in a non-arbitrary way knowledge that the teacher has; and it has the potential to be *interpreted* in terms of the learners' existing cognitive resources in ways that may facilitate changes in their knowledge. However, it is difficult to understand what knowledge resides in the object itself. The object acts as an external resource for modifying the knowledge of Jean and of Jerome. Moreover, if they are so motivated, Jean and Jerome can share and explore their individual interpretations of this representation of an aspect of the teacher's knowledge, and by seeking to understand each other's thinking (facilitated in part by their having a theory of mind; see Chap. 2) and by evaluating their own thinking (by virtual of having metacognition, see Chap. 7), work towards a 'shared' understanding. This may lead to the point where they will tend to agree with (their own interpretations of) the other's public representations of what the diagram represents.

On reaching agreement Jean and Jerome might consider that they have acquired knowledge 'from' the diagram, and that they share the 'same' knowledge. However, failure to agree is always possible and indeed quite likely when initial understanding of a topic is very different, and so is apparent agreement based upon misinterpretations of each other's utterances – which research suggests is also very common in science education. Whatever the outcome, it seems better to consider the diagram as a resource for constructing or modifying personal knowledge than as something which, in some sense, 'contains' knowledge. I would also suggest that it might be more useful in looking to understand learning to see the two learners in terms of two cognitive systems which are separately processing information, but due to recognition of mutual interests, are seeking to coordinate their cognition, whilst ultimately each holding a veto (i.e. through retaining executive control) over what personal knowledge might be developed.

By means of analogy, learners working together in this way are often better understood as forming alliances that are contingent on appearing to give mutual benefit, rather than as components of a single corporate body. So rather than think of learners involved in cooperative learning as members of some corporate board or government cabinet, expected to come to a majority decision that is binding on all, it may be better to think of them as independent agents who choose to work together on a common project for as long as that seems productive. An organisation like *SCORE* (*Science Community Representing Education*, in the UK) comes to mind, that acts as an umbrella for several organisations with common interests (in this case the *Association for Science Education*, the *Institute of Physics*, the *Royal Society*, the *Royal Society of Chemistry* and the *Society of Biology*). SCORE produces position statements representing its constituent organisations, but only when all agree on the details of those statements. Each of the constituent organisations retains its independence and maintains its own policies and right to speak out separately.

Cooperative learning is necessarily more like that, as each learner's thinking is primarily informed by their 'internal' cognitive resources, which are only ever indirectly linked with those of other learners through the imperfect communicative processes considered earlier in the book. In the present analysis, then, it is considered that knowledge is often best considered 'personal', rather than distributed across networks. Each person acts as a cognitive agent in the world, certainly able to communicate with others to seek to access to (i.e. to form models of) their personal knowledge, but drawing upon such external resources to supplement and modify their own personal knowledge base.

However, if knowledge is considered to be personal, and located discretely in the minds of particular individuals, then this raises the question of how we might best understand what is meant by 'public' knowledge. Public systems of knowledge, such as of scientific knowledge, are often considered to act as the referents for what knowledge and understanding is considered canonical (see Chap. 6). Yet if personal knowledge is considered a property of individual minds, then the notion of 'public knowledge' may seem incongruous. This issue will be considered in the following chapter.

Chapter 10
Relating the Learner's Knowledge to Public Knowledge

The conception of knowledge discursively associated with scientific literacy is thus something that is located in every single individual and which has universal applicability. From this conception of knowledge…it is only a small step to a discursively associated conception of knowledge as something that is more or less the same in every individual and which can be observed and described as a distinct, stable body of knowledge. (Van Eijck, 2009, pp. 249–250)

In the quotation above, van Eijck describes the possibility of knowledge that:

(a) Is located within (all) individuals
(b) Is 'more or less' the same for all individuals
(c) Can be observed and characterised as a body of knowledge

Even considering this as an aspiration for full scientific literacy, rather than a description of a current state of affairs, such a characterisation seems problematic in terms of the analysis presented so far in this book.

The first point, that in a sense knowledge is located 'in' individual people, fits well with the approach taken here: each person has a cognitive apparatus which has been modified and tuned by experience and which informs intelligent behaviour in the environment – including, inter alia, answering the teacher's questions in class and writing responses to examination questions. It was suggested in the previous chapter that personal knowledge can be understood as those features of cognitive structure, the resources available to the cognitive system, which facilitate processing, and so inform the individual's actions in the environment.

The notion that knowledge might in principle be 'more or less the same' across different individuals is inherently more difficult to determine, for reasons outlined earlier in the book. The cognitive resources that are the form of knowledge 'in' the cognitive system are substantiated in the form of networks of neurons with particular connection strengths. Due to the inherently iterative nature of the development of an individual's cognitive system, during which the genetically directed neonate's brain responds to its unique experiences of the world, it seems very unlikely that the physical structure of the neural network which might be the correlate of my

K.S. Taber, *Modelling Learners and Learning in Science Education: Developing Representations of Concepts, Conceptual Structure and Conceptual Change to Inform Teaching and Research*, DOI 10.1007/978-94-007-7648-7_10,
© Springer Science+Business Media Dordrecht 2013

knowledge of ideal gases or photosynthesis, or even of something less complex such as my knowledge that 'f' is commonly used to stand for frequency, could be considered to be the same as a parallel network in another person's brain which might be said to be the basis of the 'same' knowledge in that individual.

However, it was argued earlier that to be useful in science education, our models of cognition should not be constructed at the physiological level of description, but at the cognitive systems level – that is, it should not matter how similar the pattern of neurons and synapses are, as long as that apparatus allows the 'same' processing. But here of course we run into the problems discussed in Part II of the book. We only ever have direct access to our own mental experiences, and much processing takes places 'out of mind', and so we can never have a full account of even our own thinking, much less be sure how similar 'our thinking' is to that experienced by other people.

If, however, we adopt a more behaviourist perspective, and accepting these problems decide to focus on input and output, then we might judge two individuals to have the same knowledge to the extent that they produced the same output behaviour in response to the same input stimulus. The limitations of this approach are extreme. We can never set precisely the same input conditions, and we know that the same learner will, for example, often answer the same question in different ways on different days – depending on alertness, current engagement, what they have just been thinking about, etc. The danger here is we reduce our research to relatively trite aspects of knowledge that can be expected to respond to rote learning: What do learners think the chemical formula of methane is? What is the equation defining electrical resistance? What do we call an animal that lays eggs and has feathers?

The crux of the problem is highlighted in van Eijck's third characterisation above: the extent to which knowledge is in a meaningful sense observable. Certainly behaviour can be observed and *interpreted as* demonstrating particular knowledge, but this is always a matter of interpretation by another individual. This discussion certainly underplays the special significance of those particular behaviours that are language based and so have evolved to be helpful in communication with others. We can certainly get a 'good idea' of what someone else thinks about a topic if we spend time talking to them in depth about it or if we read an extensive piece of their writing. Conversation, in particular, allows us to make iterative moves that can converge on what we judge to be agreed understandings (Bruner, 1987). However, we also know that we can never be entirely sure we fully understand another's intended meaning:

> Understanding what other speakers mean by what they have said, therefore, cannot possibly be explained by the assumption that we have managed to replicate in our heads the identical conceptual structures they intended to 'express'. At best we may come to the conclusion that our interpretation of their words and sentences seems compatible with the model of their thinking and acting that we have built up in the course of our interactions with them. (Glasersfeld, 1988, p. 6)

The stand taken in this book is certainly not that we should dismiss the possibility of meaningful intersubjectivity, but rather that we should never take for granted effective communication, as understanding another is inherently problematic and so prone to error.

The Notion of the Independent Reference Standard

In the quotation at the start of this chapter, van Eijck (2009, pp. 249–250) referred to the possibility of 'a distinct, stable body of knowledge'. As suggested in the previous chapter (see Fig. 9.1), we commonly think of education in these terms, as largely concerned with the enculturation into formal bodies of knowledge set out in curriculum, which act as the targets for learning – see Fig. 10.1.

This process was represented in further detail at the start of the book (see Fig. 1.3) where in the context of science education, curriculum knowledge was seen as being a designed body of target knowledge based upon 'scientific knowledge'. Yet this assumes there is such a thing as 'scientific knowledge' that can be accessed as a reference for developing a curriculum.

Scientific Knowledge as Public Knowledge

The assumption that there is something that can be considered as a public body of scientific knowledge is a widespread one: certainly, it is common for people to speak and write as if this is the case. This is similar to the use of the term 'scientific thinking' when used to mean the thinking of the scientific community rather than the thinking of an individual about science (see Chap. 7).

There is also, as distinct from scientific knowledge itself, the notion of the public understanding of science, where – similar to the notion of 'scientific literacy' as discussed by van Eijck – the public is discussed as if in some sense a body that embodies some kind of communal mind that may or may not understand scientific concepts. The discussion in Chap. 6 about the difference between interpretative and normative-positivist approaches to research is relevant here. Statements about the level of knowledge of some scientific topic among the public only makes sense from the latter perspective. People also talk of 'the scientific attitude' (Osborne, Simon, & Collins, 2003) or of adopting a 'scientific worldview' (see Chap. 15).

The intention here is not to suggest such terms should not be admitted into academic discourse. All of these terms clearly do useful intellectual work, so to speak, in some contexts. The purpose of examining such terms relates to the

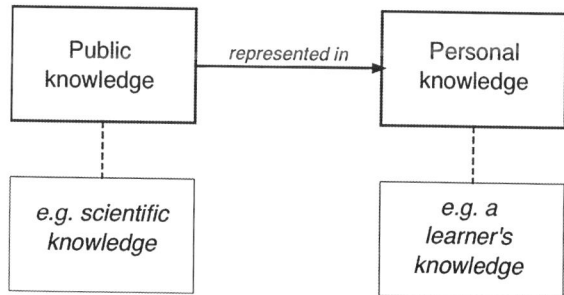

Fig. 10.1 Formal education as referenced to a formal body of public knowledge

fundamental thesis being developed in the book. For if we are to develop models of learners and learning which are to be suitable both for informing practice and for further progressing a coherent programme of research, then we need to ensure that terms that are used *as if* technical terms are operationalised such that they can actually support the development of the programme. As has been suggested throughout the book, this is often not the case, because so many of the terms which we use *as if* technical terms in our research are drawn from a familiar life-world mental register that is taken for granted. These terms are generally well understood but tend to describe fuzzy concepts.

A Taken-for-Granted Notion of Scientific Knowledge as Public Knowledge

Although it is easy to find references to 'scientific knowledge' in academic writing, the notion seems to be generally assumed and taken for granted, rather than defined or explained. Brossard, Lewenstein and Bonney wrote a paper on 'Scientific knowledge and attitude change' that was published in the *International Journal of Science Education* (Brossard, Lewenstein, & Bonney, 2005). The word 'knowledge' was used over 40 times in the main text of the paper, although largely in relation to personal knowledge (the 'participants' knowledge'). Despite the paper being in part about 'scientific knowledge', this term was not explained in the paper: indeed it only explicitly appeared once in the main body of the text, viz., 'although various evaluations of the impact of informal science projects on scientific knowledge and attitudes toward science have been published … most literature is devoted to learning in the context of science museums…' (p. 1100).

In this context the authors *seem* to be referring to personal knowledge in terms of the impact of projects on the scientific knowledge of participants, rather than some general notion of scientific knowledge in the abstract. When these authors refer to how 'good baseline data exist at the national and international levels for documenting public knowledge and attitudes toward science' (p. 1100), their use of the term 'public knowledge' seems to be in the sense of *levels of* knowledge among members of the public, that is, a kind of *aggregate* evaluation of the knowledge of individuals in a population.

In the context of the analysis in this volume, 'knowledge' is considered a viable notion when referring to individuals. Although notions of 'public knowledge' in the sense of levels of knowledge among the public present epistemological challenges for researchers, they are not problematic ontologically. Given that individuals hold personal knowledge (see Chap. 9), which can in principle be explored and evaluated (cf. Chap. 6), the idea that one can posit a construct meaning some kind of aggregate or typical level of knowledge of some topic seems reasonable, even if operationalisation may be difficult. (As discussed in Chap. 6, such aggregate or overall measures involve considerable reduction and simplification of what at the individual level may be something quite complex.)

However, other authors use the term 'scientific knowledge' quite differently. In a paper exploring the 'the relation between scientific knowledge, risk consciousness and public trust', Lidskog (1996) uses the term 'scientific knowledge' more liberally than Brossard and colleagues and in a rather different sense. Although (again) Lidskog takes this term for granted and does not define it for readers, it is possible to infer some qualities that this author associates with 'scientific knowledge' and so identify something of its ontological nature as used in this paper. In Chap. 1 I used a kind of concordance to show how one could deduce assumed ontological characteristics by examining the different contexts in which a term was used in a research paper. In that case it was possible to infer how the term 'misconception' was understood by Banerjee as used in a paper about 'Misconceptions of students and teachers in chemical equilibrium' (Banerjee, 1991), drawing upon the instances of its occurrence in the paper (as listed in Table 1.2). Table 10.1 presents a similar concordance for how Lidskog uses the term 'scientific knowledge'.

Table 10.1 suggests that for Lidskog, scientific knowledge is something that:

- Is produced but is different to other forms of produced knowledge
- Can be understood to have a role
- Can be spread among the public
- Be more or less adequate
- Be required (for individuals) to gain knowledge
- Be seen to have authority
- … but is competing with other forms of knowledge
- …yet seen to have priority

When discussed in this way, 'scientific knowledge' appears to be reified as something *other than* knowledge of individual people such as distinct scientists.

Lidskog's notion of scientific knowledge as able to be spread among the public is reflected in a paper by McInerney, Bird and Nucci (2004) who report a 'study of how scientific knowledge about genetically modified (GM) food flows to the American public' (p. 44). In their study, McInerney and colleagues do specify what they mean by scientific knowledge:

> Scientific knowledge and research generally make their way into popular literature when there are risk factors that might affect the general public or when controversies arise. We are interpreting the term *scientific knowledge* to mean those reports of experimental research that appear in the peer-reviewed journals read by scientists, subscribed to by university libraries and by industrial special libraries that cater to research scientists. (McInerney et al., 2004, p. 49)

So according to this statement, for these authors, scientific knowledge *is* the reports presented in the scientific literature. In terms of the analysis developed in this book, such reports would not be considered to be knowledge per se, but rather the *representations of* the knowledge of individual scientists moderated through team discussion, review processes, etc. As suggested in the previous chapter, it is problematic to see knowledge as residing in such representations. Rather such representations may be seen as resources that can be interpreted by readers to develop their own knowledge in ways that may or may not be considered to match

Table 10.1 A concordance for the use of the term 'scientific knowledge' in Lidskog's (1996)

Reference to 'scientific knowledge'	Page
Characteristic of present-day risks is their increasing remoteness from lay people's perception and that *scientific knowledge* is required to gain knowledge of them	31 (abstract)
This article critically discusses the role of *scientific knowledge* and experts in trust-building, and investigates factors of importance for the creation of risk consciousness as well as trust	(31 abstract)
However, when approaching concrete events, people's reactions to risks seem not to be guided by *scientific knowledge*	37
And *scientific knowledge* may not self-evidently take precedence over this local and practical knowledge	38
Noticeable is that the role of *scientific knowledge* and expertise is not given any particular attention, except that non-comprehension of the threat is one factor that impels people to stay in hazard-prone areas.	38–39
To have social authority means for science that *scientific knowledge* is perceived as the kind of knowledge that has predominance over other forms of knowledge	42
Traditional academies such as universities and official research institutions are no longer alone in their claim to be legitimate producers of *scientific knowledge*	42
Scientific knowledge claims appear in different organisational forms, and in many cases are given legitimacy both by lay people and by governmental agencies	42
In a basic sense the lay people produce knowledge too, albeit of another kind than *scientific knowledge*	48
The stress on the role of *scientific knowledge* and experts causes these theories to heavily emphasise the role of science as the producer of public risk consciousness as well as public trust	49
To sum up, conflicts over risk issues are rooted in deep cultural, social and political soil, they touch on matters of accountability, the legitimacy of government and, not least, the adequacy and authority of *scientific knowledge*	49
The invisibility of modern risks does not necessarily mean that lay people's dependence on *scientific knowledge* will increase	49
Moreover, when *scientific knowledge* is to be spread amongst the public, this closed construction has to be re-opened, deconstructed and negotiated in a new social context, a context which does not necessarily share the view of reality of the scientific community	49

the knowledge that was being represented in the reports. So, if McInerney and colleagues mean 'scientific knowledge' to be reports in the literature, then:

(a) These can indeed be understood to be entities with objective existence, and so 'scientific knowledge' in the sense of these authors has an objective referent in the public space that individuals have access to.

(b) However, from the present analysis, what McInerney consider knowledge would be better considered *representations of* the personal knowledge of some individuals and therefore potentially resources for developing the personal knowledge of other individuals.

Fig. 10.2 A representation
of McInerney et al.'s
description of the flow of
scientific to public knowledge

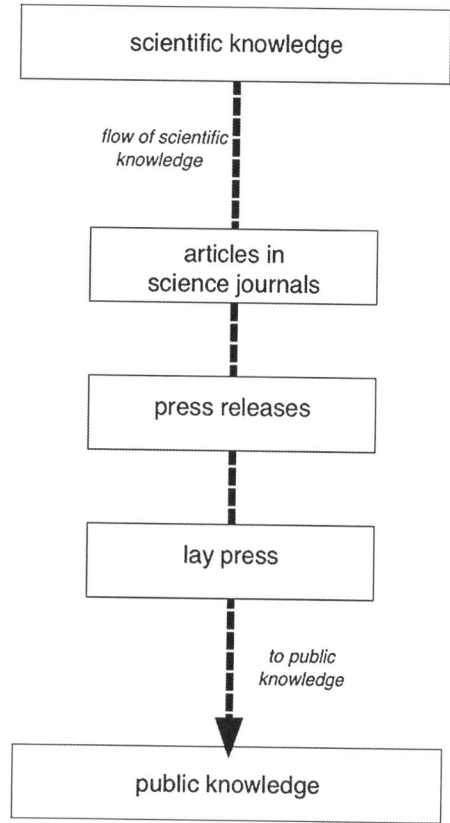

This is all very well, except in the quotation above these authors suggest that scientific knowledge 'makes its way' into popular literature, and elsewhere they refer to 'the flow of scientific knowledge to public knowledge' (p. 50), that is, that there is a 'flow of scientific knowledge to the public through the popular press' (p. 62). Indeed there appears to be a staged transfer here (as shown in Fig. 10.2), as 'scientific knowledge flows to the public through published articles in science journals, press releases, and the lay press' (p. 48), a process that can lead to 'incomplete scientific knowledge' (p. 69).

There seems to be a 'sleight of mind' here, as the problem of what scientific knowledge is is dealt with by *identifying* it with the reports in scientific literature, but that same scientific knowledge is said to *flow* to public knowledge through different forms of literature. This suggests that although McInerney and colleagues write as though research reports *are* scientific knowledge, they actually seem to consider that scientific knowledge exists independently of those reports, which act as part of a pathway for its dissemination. So scientific knowledge is actually treated as something that can be 'captured' in primary literature and is then able to be transferred through other literary forms – becoming degraded and transformed into something else in the process.

Evolving Notions of Scientific Knowledge

The notion that there might be a public body of scientific knowledge, for which peer-reviewed journals acts as the repository, seems problematic given the dispersed and sometimes inconsistent nature of scientific literature. This seems a concern even when leaving aside the more fundamental issue raised above of whether something such as a published report can ever be more than an imperfect representation of someone's knowledge and so only able to be 'reconstituted' as knowledge when it is interpreted by other 'knowers' through their own cognitive systems. Yet perhaps this is in part a reflection of the sociological turn in science studies that rejects notions of science as a fully objective activity.

That is, a naive view might be that:

> Nature is assumed to hold a unique truth and the current state of scientific knowledge is assumed to be the best available approximation to that truth. There is no need to examine why scientists believe what they believe, because there are assumed to be no social factors intervening between nature and scientific truth. (Martin & Richards, 1995, p. 5)

From this perspective, the self-correcting nature of science will ensure that when there are conflicting scientific knowledge claims about some aspect of the world, the scientific community will carefully and objectively interrogate that aspect of nature closely to quickly tidy up scientific knowledge. Such a view might well reflect the ideal, but there is now a considerable literature on the influence of psychological and social factors that inevitably complicate the supposedly objective process of science given that scientists are human. Martin and Richards report that:

> analysts have accumulated an impressive array of empirical studies of scientific controversies that have compelled attention to their central programmatic claim that scientific knowledge is socially created or constructed … According to their revised view of scientific knowledge, where closure of a controversy has been achieved, it has resulted not from rigorous testing, but from the pressures and constraints exerted by the adjudicating community. These pressures and constraints include not only the accepted knowledge of the community (the elements of its paradigm), but also the vested interests and social objectives that they embody. (Martin & Richards, 1995, p. 10)

Although Martin and Richards suggest that 'within the terms of this "constructivist" approach, the "truth" or "falsity" of scientific claims is considered as deriving from the interpretations, actions, and practices of scientists rather than as residing in nature' (p. 10), one does not have to give up notions of nature *constraining* the interpretations which are viable (Glasersfeld, 1990), to recognise that individuals will not all make the same interpretations from available data. As Part II of this book argued, interpretations can only be made in terms of the unique set of internal cognitive resources that each individual has available for making sense of the world.

The Problem of Reifying Scientific Knowledge

It seems then that the term 'scientific knowledge', when used to refer to 'public' rather than an individual's knowledge, may refer to the 'system of knowledge' that it is recorded in public forms (the primary literature being journal articles and research monographs). No individual person could ever be familiar with the entire corpus of scientific literature. It is also recognised that although scientific knowledge is 'public' in the very real sense that *publication* is considered a key aspect of the scientific process and making a scientific breakthrough implies that it is communicated, checked and accepted by the scientific community, it is not determined by simple majority voting. The scientific community is in principle open to anyone but is a self-regulating body: that is, those who are currently recognised as qualified scientists act as the gatekeepers for who can become scientists. Within the sciences, there is a complex structure of formal and informal mechanisms which determine who among the world's scientists are considered particular experts in particular areas and so are asked to evaluate the worth of new knowledge claims that are submitted for inclusion in the public scientific record.

The idea of treating a number of people as one body, that is, incorporation, is a useful legal device for ensuring rights and responsibilities when dealing with an organisation rather than individuals. Organisations that are formally incorporated in this way, such as universities, are normally required to have clear rules of governance (the limits of executive power, what can be delegated and to whom, when committee votes are needed, the extent to which 'chair's action' is possible, etc.), so that there can be unambiguous decision-making that is understood within and beyond the organisation.

If a university decides to establish a new chair, or close a department, or promote a lecturer, then it has to be absolutely clear that such a decision has been taken, and this is only possible because of clear guidelines about who makes that decision and how it is made. Such decisions will actually be made with the input of various people making arguments and offering opinions, which are considered by other people, who perhaps make recommendations to others, until decisions are formally made, which even then are often then required to be ratified by further groups within the organisation. Because of such rules, we can meaningfully say that the university, as a corporate body (and so able to be treated for such purposes as if a person), has made a decision.

Scientific knowledge can certainly be *said to be* what is currently accepted to be so by the scientific community, but like any complex body, much of the work of that community is in effect done by 'committees' (editorial boards, international scientific committees of conferences) and ad hoc working parties (reviewers appointed to consider specific submissions), which are loosely structured. The scientific community is certainly not a corporation in any formal sense. Arguably, the Soviet system attempted to treat science as if it could be incorporated, at least within the USSR, by enforcing a party line and removing dissidents from the community (Frolov, 1991). However, most scientists would consider such an approach to be entirely inconsistent with the sceptical and critical attitude necessary for progressive scientific work.

This caricature of how the scientific community works is not intended to be critical of the process, but rather to raise the issue of the potential difficulty of identifying *what counts as scientific knowledge*, when most scientists are inevitably ignorant of most of what might be claimed as scientific knowledge, and those that are 'knowledgeable' inevitably hold idiosyncratic understandings given the nature of the human cognitive system, as explored earlier in the book.

One response to this observation might be the suggestion that because science is a public system of knowledge, it is inappropriate to focus on the knowledge of the scientists themselves, with all their inevitable variations and nuances, but rather to recognise that scientific knowledge is found in the published literature. This, however, cannot be satisfactory. As considered earlier in this volume (see Chap. 4), written accounts are only a representation of the ideas of the writer and will be individually interpreted by different readers including journal referees and editors. It is also the case that the scientific literature is never going to be entirely consistent or stable, with new studies making claims to support, refute or modify earlier claims. Published claims may themselves be presented as tentative, or of uncertain range of application, or with various other caveats – and indeed scientific knowledge claims are in any case generally accepted to be necessarily provisional and to always rely upon other (uncertain) claims that make up the theoretical and methodological assumptions of a study.

Therefore, individual judgement is always needed in distilling a view of what the current status of scientific knowledge on a particular topic actually is (Latour & Woolgar, 1986). This work is undertaken by authors of review articles and textbooks, but in doing this they are *interpreting* and the work of others, that is, interpreting the public representations of the thoughts of the original authors and reporting these by themselves *representing* their thoughts in a public form. As was suggested in Chap. 4, these processes of interpreting and representing prevent there ever being a straightforward transfer of ideas between minds.

In one sense then, although 'scientific knowledge' could be understood in the abstract as what is currently taken to be the case by the scientific community and represented in the scientific literature, this can only ever be an *ideal* referent, as in practice there is no simple way to identify the 'content' of scientific knowledge. If scientific knowledge exists in this sense, it is a 'World 3' object (as discussed in Chap. 4) that does not exist as a tangible object in the material world ('World 1') and can only be imperfectly glimpsed in our subjective worlds of mental experience (World 2). Ultimately, individuals – whether scientists, reviewers, textbook writers, curriculum developers, teachers, students or members of the general public – interpret public representations of personal understandings and form their own personal understandings. This is summarised in Table 10.2.

Table 10.2 Notions of scientific knowledge as public knowledge

Notion of scientific knowledge	Complications	Conclusion
The knowledge of the scientific community	The scientific community is a large, diverse and ever-changing set of people Each scientist will hold personal knowledge that will be idiosyncratic Each scientist will only claim expertise within a narrow range of science The knowledge of the individual is represented within that person's cognitive system and cannot be directly accessed by anyone else: it can only be *represented* indirectly in a public space	A useful referent for the purposes of general discussion but too limited to be used as a technical term without considerable further specification of intended meaning
The knowledge claims made in the scientific literature	The scientific literature is vast and being added to at a great rate The literature is diverse in forms: e.g. primary research reports, reviews, communications, critiques The literature is reported in a wide range of languages (although English is increasingly dominant in most fields) Knowledge claims in different reports may often be inconsistent Individual contributions to the literature are assigned different status both through formal mechanisms (e.g. citation indices) and the individual judgement of community members Reports are public inscriptions that represent the thinking of authors but *need to be interpreted* through the idiosyncratic cognitive resources of readers to be understood	A useful referent for the purposes of general discussion but too limited to be used as a technical term without considerable further specification of intended meaning

The Analogy Between Science and the Individual

It has been common in science education to consider the nature of the learner's discovery of scientific ideas by analogy with scientific discovery, as in the common phrase, the child, pupil or student 'as' scientist (Driver, 1983). However, what seems less well noted is how it is common to consider the scientific community by analogy with the individual. So in popular idioms, scientific thinking is the thinking of the scientific community, and scientific knowledge is the knowledge produced by that community, and these terms are often used as though the community can be treated as a unified entity (despite, as suggested above, not being in any formal sense a corporation).

We have seen in the previous chapter (Chap. 9) that knowledge of an individual can be understood in the form of (cognitive) resources that support thinking and can be considered to have two sources. Some knowledge is innate, because it is part of our genetic inheritance and has been 'learnt' through the interaction of generations of our precursors with the environment. We have inherited a certain cognitive apparatus and particular processing biases, because these have been selected through evolution. We then acquire new knowledge by processing information from the environment through that apparatus – and in the process develop the apparatus itself. Thinking was earlier characterised (see Chap. 7) as the processing of information – drawing upon existing conceptual resources, and in the process developing the resource base for subsequent processing.

Earlier (in Chap. 7), I discussed Coll, Lay and Taylor's (2008) research report from a study on 'Scientists and Scientific Thinking'. The paper subtitle, 'Understanding scientific thinking through an investigation of scientists['] views about superstitions and religious beliefs', would seem to imply that 'scientific thinking' refers to an aspect of the individual scientist who holds view and beliefs. Yet in the body of the paper, the term is used rather differently (again, with my added *emphasis*):

> The panel of experts consisted of scientists across a range of disciplines that examined each item statement in the instruments and asserted that it was *in conflict with current scientific thinking* in that discipline. (p. 201)
>
> Likewise, the few that were less sceptical about astrology like Judy, thought that there were, potentially, underlying theoretical reasons *not inconsistent with current scientific thinking*… (p. 208)

Here 'scientific thinking' does not seem to refer to a process or product of cognition in individual scientists but rather is considered to derive from the scientific community: raising the questions of *what we might mean by the thinking of a community*.

I have argued that thinking is the mental level description of the cognitive processing that occurs within an individual cognitive system, that is, the individual, epistemic subject, described at the cognitive system level, which in turn can be understood to be a correlate of electrochemical activity within the brain of that individual. To shift to a community perspective would require considering the *overall* processing activity in/across the network of cognitive systems. The notion of distributed cognition across a network of people was introduced in the previous chapter (Chap. 9). In that treatment, the question of what it might mean for knowledge to be distributed across such a network was considered, and it was concluded that knowledge as understood in this volume could only meaningfully be said to reside in entities which monitored and acted in the environment (such as people), rather than passive objects (such as books) which need to be acted on and interrogated by people. However, the notion that processing itself could be distributed across a network of communicating processors (people, perhaps working with computers, etc.) was acknowledged as being reasonable. What needs to be noted in shifting focus from the individual 'thinker' (an epistemic subject) to a distributed network of thinkers is how individual cognition is subject to some form of executive controller (identified in this volume with working memory and linked to the experience of consciousness) that is able to monitor and direct at least some aspects of thinking, in relation to personal goals and motivations.

If a distributed network consists of a person working with computers, then although some, potentially quite involved, processing is undertaken by the computer, this can be considered as indirectly under the executive control of the person. For example, I have very little understanding of the processing that goes on within my computer when I enter a bibliographic reference into the manuscript of this book, but I do understand what needs to be achieved, and through using my cognitive system's motor interface – the motor nerves going to my arms and fingers – I interact with the computer's user interface (the keyboard, the mouse or trackpad) to instruct the computer to undertake some processing. Then by monitoring the screen through my sensory interface, by viewing the screen, I can check if the intended processing seems to have occurred.

Sometimes the outcome is not what I had intended, as perhaps the wrong citation is inserted or the citation is inserted at the wrong point in the text. When this happens I tend to blame either my manual incompetence or software bugs. I have no reason to interpret these errors as the computer making an executive decision that an alternative citation is more appropriate at that point in the text. Distributed cognition involving several people gets more 'messy', as each person has some degree of autonomy in their contribution to the project – albeit sometimes modified by social pressures due to employment, seniority hierarchies, etc.

Science is by its nature a communal activity: it may be carried out by individuals or groups, but an individual who is isolated from the community and so does not communicate findings and subject them to critical review is not able to contribute to science. However, the processes by which something becomes provisionally accepted in science is a complex matter (Ziman, 1978/1991). So the notion of 'scientific thinking', when used in this community context, is inherently problematic.

The approach to identifying what should be considered 'scientific thinking' taken by Coll and colleagues in their study (Coll et al., 2008) is to refer to a panel to give a view on how statements should be judged by scientific thinking. This involves sampling the scientific community – raising questions of who is a member of that community and how to produce some kind of representative sample of such a diverse community – and then converting the individual responses to a 'community' response, for example, aggregating responses. Yet clearly the way ideas become accepted in science is actually rather different from that and is probably a multilevelled process. There will often be responses to primary literature generating further primary literature supporting, criticising or developing claims; then secondary literature, such as reviews written by 'experts' in a field, setting out an overview; then text books distilling the complex muddle of research into an account suitable for teaching.

An alternative perspective might better fit the approach to 'thinking' argued here, by considering scientific thinking to indeed be thinking in the sense of mental processing, but considering the system to be the scientific community envisioned as a vast set of parallel processors, connected together with various connection strengths. *In principle*, it might be possible to model the scientific community in this way,

Fig. 10.3 Modelling the
scientific community as a set
of parallel processors

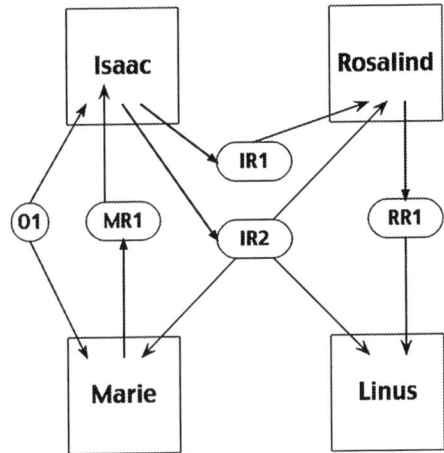

bearing in mind that the processing system of each scientist is encapsulated, in the sense that communication between these processing units – the members of the scientific community – requires the type of interfacing described earlier in this book (see Fig. 4.14).

Figure 10.3 represents this in schematic form, showing how the scientific community may be conceptualised as a set of parallel processors that can sense and interpret objects/events (e.g. O1) individually but can only communicate by representing the outputs of processing in the external world (e.g. MR1 and IR1). This places such representations where they are available in the shared 'public' space for other processors (scientists) to select as the basis of input to be coded for further processing (i.e. read, conceptualised, critiqued and evaluated). So in the simple scheme shown, both Marie and Isaac make direct observations of phenomenon O1. Marie represents her thinking about O1 in a public representation (e.g. a scientific paper) denoted here as MR1, and Isaac produces two separate public representations of his thinking (IR1, IR2). As these representations are in the shared public space, they can also provide input for further processing – so, for example, perhaps Isaac's second production on this topic might have in part been influenced by Marie's representation if he read MR1 before writing IR2. Neither Rosalind nor Linus undertook direct empirical work on this phenomenon, but both accessed at least one of Isaac's public representations of his thinking about O1, and Rosalind represented the outcomes of her own thinking in the shared public space – RR1 (e.g. this might have been a review article in the field in which O1 is considered a phenomenon), and this was also accessed by Linus. Clearly the real scientific community gets *much* more complex than this, but Fig. 10.3 represents the general principle. Even in this hypothetical example of a tiny subset of a scientific field, it is clear that whilst a parallel processing model is certainly viable, it is also describing something extremely complicated.

So each processing unit – that is, each person, each scientist – needs to access the information that is to be processed through its (i.e. his or her) sensory interface, convert it into a form suitable for processing within their cognitive system, process it; then represent its output in the external world in the form of inscriptions, gestures, vocalisations, etc. In this model (as caricatured in simplified form in Fig. 10.3), the scientific community comprises of this vast set of parallel processors, each representing information in the external world, much of which is largely ignored by other processors, but some of which is 'read' and processed by many other processors.

The language here seems very impersonal, but of course this is just because the argument made in this book is that we need to operationalise what we understand by everyday terms such as knowledge, and that it is helpful to adopt the system level of description to think about which is going on in cognition. None of this is in anyway meant to depersonalise the human aspects of knowing. Rather as explained earlier (see Chap. 2), the point is to adopt a description appropriate to a particular level of analysis. Interestingly, when considering the parallel processing of different people in a community, the image conjured is less the electronic computer, with its binary switches, than the brain itself, with its myriad neurons; each neuron has a great many connections, but with thresholds that select particular outputs as activating other neurons.

Whilst this foray into the notion of scientists as parallel processors may seem rather fanciful, it is argued in this volume that just this kind of analysis is necessarily to clarify our understanding of concepts that are discussed in research. This, in turn, is necessary if we are to recognise the extent to which research designs are adequate for producing data and analysis that can answer their research questions. Only by setting out a model of how the scientific community works as a processing system is it possible to judge what a concept such as 'scientific thinking' might mean in Coll et al.'s (2008) second sense of 'thinking of the scientific community'. Such clarity is needed if we are to evaluate the extent to which knowledge claims such as those made in their paper (e.g. assertions about which item statements in the instruments used are '*in conflict with current scientific thinking*') can be considered to be supported by the evidence offered.

Whilst the approach taken here, focusing on the individual as the unit of analysis, seems the most sensible in terms of the analysis being developed in this book, it is not the only approach that may be taken. The sociologist Harry Collins (2010) offers a very different view where individuals access knowledge by being embedded in society. In Collins' account there is collective (tacit) knowledge that is a 'property of society, rather than the individual' (p. 11). Indeed, in terms of knowledge acquisition, Collins sees the individual as having a parasitic relationship with society. Needless to say, such a view seems quite at odds with the analysis here. From the perspective adopted in the present volume, knowledge is represented in individual minds, and that forms the only viable starting point for any possible notion of public knowledge. Rather than people being parasites on public knowledge, they are the essential components of any kind of distributed knowledge network (Taber, 2013b).

Target Knowledge in School and College Science

So whilst the notion of 'scientific knowledge' as a form of 'public' knowledge has proved a useful referent, it is a problematic notion to operationalise into a technical term in research. To reiterate:

- If we think of scientific knowledge as the knowledge of the scientific community, the outcome of the thinking of that community, then we are concerned with what is represented (within minds) as personal cognitive resources for thinking in a vast and loosely connected network of separate minds.
- If we think of scientific knowledge as what is reported in the peer-reviewed scientific literature, then we have representations which have to be interpreted by a reader and which are often going to be interpreted to suggest inconsistencies and contradictions that will require judgements to be made in order to reach a view on what the current state of scientific knowledge actually is.

This is certainly *not* to suggest that something which is complex and requires careful interpretation should be excluded as a suitable topic for research: that would not be a sensible position to adopt in a book discussing research into learning. However, a consideration of the ontology of scientific knowledge highlights the epistemological challenges in making claims for what scientific knowledge in a particular topic currently is, and suggests that such claims need to be understood accordingly.

It seems clear that there is not an unequivocal process that allows us to determine what current 'scientific knowledge' about any particular focus is. In most formal educational contexts, students and teachers will be presented with some form of curriculum statement of what should be taught and hopefully learnt. So for someone studying science in such a context, there will be a formal attempt to specify what scientific knowledge and skills are the desirable outcome of the learning process. Clearly, any such curriculum statement will represent the outcome of someone's, or some committee's, deliberations on some aspect of scientific knowledge. This will be a 'public' representation of that understanding, which as the earlier chapters suggest is necessarily itself one step removed from that understanding (see Fig. 4.12). In some cases, for example, a research scientist teaching a course in her/his research area to university students, the identification of target knowledge to provide a curriculum may primarily be a matter of selection: Which parts of the experts' knowledge of his specialist topic is it most important to attempt to teach to the students in the limited teaching time available?

In most teaching contexts, however, the curriculum is not set by an individual teacher and involves more than selection of material. Rather committees of people that usually do not include most of the actual teachers of the course make decisions about not only the selection of topics but the level of treatment. This invariably involves decisions about how to simplify what might be considered current scientific knowledge to produce a form of target knowledge that is viable for the students given their existing levels of knowledge and intellectual development (Taber, 2000a).

Usually the people making these judgements are not themselves research scientists working on the topics being specified. So people removed from the current research activity use their own understandings of the science, to set out a version of scientific knowledge as target knowledge, to be represented in the curriculum. The curriculum therefore comprises of educational models of scientific knowledge that are considered by the curriculum developers (who may include government civil servants as well as educationalists) to be educationally viable and sufficiently authentic simplifications of what they understand to be scientific knowledge.

Unlike scientific knowledge, which as we have seen is often represented in fragmentary ways across a wide literature, a curriculum is usually specified in a single document. That document will however include limited detail and will need to be interpreted by readers (e.g. the teachers) through their own existing understanding of the topics specified. Again, then, target knowledge is an ideal, as knowledge is only ever represented in a curriculum document and has to be interpreted before it can be understood (cf. Fig. 1.3, in Chap. 1). When students are assessed in terms of whether their knowledge matches target knowledge, this actually involves a teacher or examiner or researcher making a comparison between their own interpretation of the target knowledge, and their own interpretation of what can be inferred about student knowledge from the representations they produce as oral statements, inscriptions, etc.

An Alternative, Idealist Notion of Public Knowledge

The account presented above, consistent with the approach taken earlier in the book, considers that our research needs to be focused on operationalised constructs that can be clearly defined. In this sense public knowledge seems problematic as the attempts above to analyse it from a cognitive perspective led to notions of scientific knowledge *as public knowledge* that seem to be of minimal value as technical concepts (see Table 10.2). Whether scientific knowledge is understood to be represented in a great many heads or in a great deal of literature, it is not likely to be readily and unequivocally determined.

In common use, however, these complications tend to be ignored, and in part this may be because of a tendency to think of knowledge in an idealist sense, for example, in terms of the ontological model of there being three worlds, as used by the philosopher of science Karl Popper (see Chap. 4). This model, deriving from ancient Greek thought, distinguishes between the (first) objective physical world, our individual (second) subjective mental worlds, and a third platonic world of ideals (Popper, 1979).

From this perspective, the notion of a sphere is an ideal. Many objects in the (first) world are considered spherical because they are judged to sufficiently match the essential properties of the ideal sphere – but it only exists as an idea. Arguably (i.e. for those who do not accept an idealist account), the ideal does not really exist at all: rather many people have mental images of a sphere, and because we are able

World 1 - world of objective objects	World 2 - world of subjective experience	World 3 - world of ideals
Evidence of student knowledge - e.g. scripts containing students' written answers to examination questions, presenting their representations of their knowledge and understanding available for objective analysis, but only a representation of what is actually being evaluated	Evaluation of a learner's knowledge is made, based on the interpretation of World 1 representations of their thinking being compared against the examiner's interpretation of target knowledge - which is a World 3 entity (i.e. only exists as an ideal)	Target knowledge - curriculum models of scientific knowledge Considered to be public knowledge but not directly accessible: only found represented in public documents (World 1) and as understood through individual minds (World 2s)

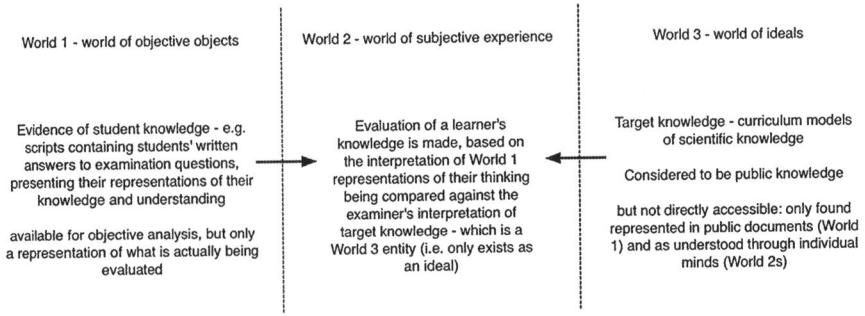

Fig. 10.4 An idealist notion of evaluating students' knowledge

to communicate essential features of the idea of a sphere, we assume that all these mental (World 2) versions have the same (idea, World 3) referent. It was suggested earlier in the book that we need to keep in mind that there are actually as many 'second worlds' as there are minds to represent and think about the first world, as by definition World 2s are subjective unlike the objective World 1.

Certainly the ideal sphere does not exist in the same sense as a spherical object, or a mental representation of a sphere, so justifying its assignment to a different ontological world. If knowledge is seen as something that actually exists in World 3, and that we humans are able to develop limited impressions of, then our common ways of thinking about scientific knowledge as public knowledge and assessment as comparing student knowledge with some 'target' curriculum knowledge may seem less problematic. However, even if we accept the usefulness of this way of thinking, we have the problem that World 3 is not directly accessible, and so it still seems that assessing and examining student knowledge in science or indeed any other curriculum subject is a rather indirect process (see Fig. 10.4).

As we saw in Chap. 6, the evaluation of a person's knowledge is an indirect process, and so knowledge claims in research reports to the effect that a learners' knowledge matches scientific knowledge to some greater or lesser extent should always be understood as carrying quite important caveats.

Chapter 11
Components of Personal Knowledge: Characterising the Learner's Conceptual Resources

It was the HC [Handsome Cognitivist]'s view that almost nothing reduces to almost anything else. To say that the world is so full of a number of things was, he thought, putting it mildly; for the HC, every day was like Christmas in Dickens, ontologically speaking. In fact, far from wishing to throw old things out, he was mainly interested in turning new things up. "Only collect", the HC was often heard to say. (Fodor, 1985, p. 1)

This book is about how we should go about modelling learners and learning in science education, and earlier in the book (Chap. 3) it was suggested that a wide range of entities have been posited as components of human minds – and so potentially components of our models of learning. As Fodor mischievously suggests, there is a sense in which the cognitive perspective invites the inclusion of a wide range of types of 'things' in minds. These entities include, inter alia, concepts, conceptions, schemata, mental models, etc. As I have previously observed, the challenge for the research programme is 'to develop models which are capable of explaining all the existing empirical content of the research area (which seems to require a multilevel, diversely populated cognitive system) but which are still able to offer useful falsifiable predictions to allow empirical testing' (Taber, 2009b, p. 318).

Who Ordered That? An Analogy with Particle Physics

Indeed the situation seems somewhat analogous to the situation in physics as the twentieth century proceeded, and newly discovered subatomic particles were regularly added to the physicists' 'particle zoo'. The simple model of protons, electrons and neutrons became supplemented by neutrinos, muons, quarks, etc. that physicists sought to 'tame' by finding a subsuming pattern reflecting a simpler underlying order:

The muon … was a particle beyond the standard model of physics at the time and …The central question "Who ordered that" was raised by I. I. Rabi when in 1947 the nature of

K.S. Taber, *Modelling Learners and Learning in Science Education: Developing Representations of Concepts, Conceptual Structure and Conceptual Change to Inform Teaching and Research*, DOI 10.1007/978-94-007-7648-7_11,
© Springer Science+Business Media Dordrecht 2013

the muon as a lepton became known – a particle which differs in all its behavior from the electron only by its mass. Up to now, this basic question why there is a second (and third) generation of particles is a strong driving force behind all modern (particle) physics. (Jungmann, 2001, p. 463)

A fundamental commitment to expecting nature to be at some level ordered and simple (see Chap. 15 for a consideration of scientific commitments and worldviews) directed scientists to develop testable models of what that assumed order might be. Whilst this programme is still active, it is widely thought that considerable progress has been made through following (what Lakatos, 1970 might have described as) the positive heuristic developed from the hard-core assumption that the messy diversity of the particle zoo reflected a simpler underlying order.

Indeed it is possible to see modern physics as one source for development of a form of realism, critical realism (Bhaskar, 1975/2008), that considered the experienced world to be real, but having an underlying nature that is only experienced indirectly through intermediate levels, and where science should be interested in the underlying level with its potentials and tendencies which are not always actualised in experience. Patomäki and Wight (2000) refer to the analogy of finding out about a nuclear arsenal, in that although the arsenal might (one would certainly hope) remain in its inert state, it is not fully understood unless the changes brought about by its potential use are considered. Critical realism suggests that approaches to science that ignore the nature of this underlying level of potentials tend to conflate two distinct levels – what is actually experienced and the underlying level of tendencies that are sometimes but not always expressed – and misjudge the nature of reality.

Finding Order in the Mental Zoo: Classifying the Cognitivist's Collection

The purpose of introducing analogy is to offer potentially fruitful comparisons. The mental zoo of concepts, and schemata, and mental models, and intuitive theories and the like, represents a level of description that is useful for many purposes. However, throughout the book I have argued that many of these notions are problematic when we use them in research in science education, because they have not been carefully operationalised for use within a research programme. Therefore, it is often possible to find research reports in the peer-reviewed literature which use the same terms in apparently inconsistent ways (Taber, 2009b).

This is perhaps to be expected given the indirect and sometimes uncertain nature of much of our understanding of human cognition, as suggested by the analysis earlier in the book. However, this also means that any attempt to set out a clear account of the distinctions and similarities between these terms is unlikely to be consistent with all uses in the research literature. My approach here will be to seek to identify the major distinctions that underpin the range of terms that have been employed, and to suggest a model for how terms might best be used in consideration of those distinctions.

I do not intend here, then, to review all shades of meaning that have been given to these different terms by various authors within science education and beyond; but rather to suggest an approach to using terms which reflects much common usage, yet gives the terms different intellectual work to do in relation to what seem important distinctions we should make.

Ultimately, however, we always face the problems outlined earlier: the tendency to talk about cognition in the taken-for-granted lifeworld mental register and the difficulty of deciding how something as abstract as knowledge can best be described at the cognitive system level (most useful for describing research into learning that can inform science education) when the underlying level of structure actually occurs at the physiological level (i.e. networks of connected neurons).

Key Distinctions

The first distinction to emphasise is one that has already been established in the book, which is between knowledge *represented in* the cognitive system, with our *experiences of* the output of cognitive processing. This is always going to be a difficult distinction in practice because of two factors explored earlier:

- The processes of thinking may themselves become represented in the underlying physical structures through which knowledge is represented, that is, our ideas both reflect and modify our knowledge.
- Conscious awareness does not have direct access to all our knowledge and at any one time is only aware of a small part of our 'explicit' knowledge.

Terms Excluded as Not Representing Knowledge Elements

So from this way of thinking, an *idea* is best understood as the output of processing drawing upon knowledge represented in the cognitive system, but not in itself a knowledge component. 'Having' an idea, perhaps as a novel juxtaposition of different existing knowledge elements, and evaluating it as fruitful, is likely to lead to that idea itself becoming represented in cognitive structure (i.e. at the physical level certain links are established or strengthened in the association cortex) in the sense that it becomes more likely that the combination of elements giving rise to that idea will be activated in the future (i.e. activation of one of what were discrete elements will more readily activate the whole new 'association'). At the mental level of description, we would say that we are likely to later recall the idea in certain contexts. However, the representation is not the idea, but a modification to cognitive structure that makes it more likely the same, or a very similar, idea will be generated again.

Similarly, the term *gestalt* is probably best not considered a knowledge element, but as the outcome of processing through such elements. The term gestalt was originally largely associated with perception (Koffka, 1967), relating to consideration of how in perception we are usually aware of whole patterns, not discrete sensations (see Chap. 4). That is, processing of sensory information involves pattern-detecting apparatus that is able to discriminate figures from their background and to associate patches of colour and edges, for example, as being discrete objects in our environment. This apparatus therefore represents a form of knowledge in the system. However, the term 'gestalt' referred to the output of that processing, and it would seem useful to use terminology that refers to the knowledge elements, the processing and the conscious experience of its output, separately.

So from this perspective the terms 'ideas' and 'gestalt' would certainly not be excluded from scientific discourse within the research programme into learning in science but would not be used to describe knowledge elements represented in the cognitive system. Rather, they would 'do intellectual work' in describing the learner's subjective experience of cognition.

Concepts as Knowledge

A key term used in relation to a learner's knowledge is that of *concepts*, and indeed key issues in science education relate to a learner's *conceptual development* (discussed in Chap. 14), and how teaching can influence *conceptual change* (considered in Chap. 15). Moreover, research into student knowledge and understanding is sometimes understood as investigating a learner's *conceptual structure*.

A problematic aspect to our understanding of concepts has been revealed by the work undertaken in psychology and cognitive science about the nature of conceptual knowledge. Much research in psychology has concerned the ability of learners to acquire artificial concepts (along the lines of being given (i) examples of different shapes in different colours and (ii) feedback on which are, and which are not, examples of plaks to test questions such as can the learner acquire the concept *plak = a blue or green shape with no curved surfaces and less than five sides*). Such artificial concepts have strict rules for membership (Seger & Miller, 2010). Yet many concepts used in everyday life are not defined through a small set of clear rules. Concepts, or categories (Ashby & Maddox, 2005), may be formed through perceptual similarity and linguistic cues in the talk of others (Gelman, 2009).

Children learn the concepts of tree, car, chair, etc., and neither are they taught these concepts through sets of membership rules nor do they apply these concepts in such a way (concept learning will be discussed in more detail in next part). We recognise an object as a tree without going through a mental checklist of attributes.

Most such concepts are 'fuzzy' in that they have somewhat blurred boundaries, and it has been shown that for some concepts we distinguish between examples which seem more typical and those which are seen as somehow less good examples of the concept. For example, perhaps a child, or an adult for that matter, knows that eels and sea horses are both types of fish, but is very unlikely to suggest them when asked for a few examples of fish, rather than perhaps salmon, cod, trout or goldfish.

However, in science classes, students can also learn about concepts that are tightly defined and do have strict membership rules. For example, the alkali metals do not comprise a fuzzy set, and there are clear criteria for whether or not something should be considered an alkali metal.

Two Types of Conceptual Knowledge

This would suggest that our conceptual knowledge is not all of the same form. Some of it is of the kind of lifeworld everyday concepts, reflecting 'the natural attitude' (Schutz & Luckmann, 1973), that was highlighted earlier in the book as being typical of how we commonly talk about thinking, learning, memory, etc. However, we also learn what Vygotsky called academic (Vygotsky, 1934/1994) or scientific concepts, which are often definition and rule based. That is the kind of thing I referred to earlier as being understood in 'technical' terms rather than everyday terms (e.g. see Table 3.1).

The term 'concept' therefore seems to have a broad referent and to relate to more than one kind of knowledge element. In particular it refers to both knowledge that is accessible to introspection and often readily represented in propositional form, and that tacit knowledge that is not directly accessible, but which operates at preconscious levels in the cognitive system.

Implicit and Explicit Knowledge Elements

This seems to be an important distinction to make, as clearly the way we use our knowledge is quite different when we are able to consciously act upon it, than when we have to rely on tacit knowledge that we only become aware of, if at all, after the event. In many aspects of our lives, such tacit knowledge is extremely valuable as it leads to quick processing and decision-making without committing of executive resources that can therefore be invested elsewhere.

However, in the sphere of academic learning, tacit knowledge can be deficient as it is inflexible and not open to justification and critique. In crossing a busy road, we need to make the right decision quickly, but in a formal academic assessment we need to be able to explain and justify *why* we suggest the answers we do. It seems useful therefore if in our research into student learning, we distinguish between these two basic types of knowledge element contributing to the learner's conceptual understanding of science topics.

The Notion of Intuitive Theories

One of the terms that have been used to describe aspects of science learners' knowledge is *intuitive theories*. This term actually has at least two meanings in the research literature. So, for example, in the context of electron diffraction in crystals, it has been claimed in a natural science context that 'there is need for a simple intuitive theory that is valid for larger crystal thicknesses' (Van Dyck & Op de Beeck, 1996, p. 99). In this context the term seems to mean a formal theory, but one that *fits with* the intuitions about the process developed by scientists working in that field.

However, in the context of science education, the term intuitive theory has been used in a somewhat different way (Pope & Denicolo, 1986). So, for example, Kaiser, McCloskey and Proffitt (1986, p. 67) refer to how, through frequent experience of moving objects, 'people develop from these encounters a systematic intuitive theory of motion'. A key feature of this 'intuitive theory' is that it is inconsistent with the scientific models. The scientific models are based around the Newtonian idea of inertia, where force brings about a change in the state of motion. However, the common intuitive theory is based around an impetus notion, something that is imparted by a force, but which somehow gets 'used up', causing motion to naturally diminish (Gilbert & Zylbersztajn, 1985). The use of the term 'systematic' by Kaiser and colleagues is quite important, as the adoption of the label 'theory' implies more than just a hunch or intuition. As McCloskey explained in another publication,

> Recent studies on the nature, development and application of knowledge about motion indicate that many people have striking misconceptions about the motion of objects in apparently simple circumstances. The misconceptions appear to be grounded in a systematic, intuitive theory of motion that is inconsistent with fundamental principles of Newtonian mechanics. Curiously, the intuitive theory resembles a theory of mechanics that was widely held by philosophers in the three centuries before Newton. (McCloskey, 1983, p. 114)

Carey and Spelke (1996), in discussing theories, whether labelled scientific or intuitive, suggest 'theories are central knowledge systems widely available to guide reasoning and action', as well as being 'open to revision' (p. 519). In this regard such 'theories' do not seem to be implicit knowledge structures, and indeed Carey and Spelke suggests that intuitive theories are distinct from what they term 'core knowledge structures' on these and other characteristics. For these commentators such core knowledge structures are 'theory-like in some, but not all, important ways' (p. 515). Carey and Spelke suggest that such core systems are largely genetically endowed and develop naturally in the child and should be considered quite different from intuitive theories:

> core systems are conceptual and provide a foundation for the growth of knowledge. Unlike later developing theories, however, core systems are largely innate, encapsulated, and unchanging, arising from phylogenetically old systems built upon the output of innate perceptual analyzers. These differences make it unlikely that the development of core systems engage the same processes as the development of intuitive theories in childhood or the development of scientific theories in the history of science. (Carey & Spelke, 1996, p. 520)

The question of whether or not children's informal ideas should be considered to be based on theory-like knowledge has been debated in the literature, and I have previously suggested that the research evidence based on students at different ages asked about various science topics suggests that the real issue is *the extent* to which such knowledge can be considered theory-like in particular cases (Taber, 2009b). The literature suggests this varies a great deal. This would seem to be what we should expect if our knowledge is partly based on implicit knowledge structures and partly on explicit representation of propositional knowledge that is available to conscious inspection and development.

The term 'intuitive theories' is itself potentially unhelpful, as it would seem knowledge must be *either* intuitive *or* theoretical but cannot really be simultaneously both. Yet if intuitive theories are understood as theory-like knowledge components that are *developed from* intuitive knowledge, then this looks less of an oxymoron. Nevertheless, it is not clear that 'intuitive theories' earn the status of being a basic category of knowledge component.

Personal Constructs

The theory of personal constructs was developed by George Kelly and was very influential in early constructivist research in science education (Pope & Gilbert, 1983). Kelly devised his system for use in therapy and suggested that it tended 'to have its focus of convenience in the area of human readjustment to stress' (Kelly, 1963, p. 12). However, Kelly considered that people modelled and understood the world through a system 'composed of a finite number of dichotomous constructs' (p. 59). That is, Kelly considered that people understood the world by making discriminations based on a set of bipolar constructs that were organised into some form of system.

Kelly thought that although we could often give labels to our constructs after the event, the process of making discriminations was not conscious or based on verbalisation. His clinical method of exploring clients' construct systems involved asking them to make discriminations by suggesting the odd one out when shown triads of 'elements', so there was no requirement to initially label the basis of the discrimination, or to rationalise why they selected a particular elements as being the one which did not fit. This was an idiographic method (see Chap. 6): there was no assumption of a right response, but rather the aim was to work through enough examples to be able to infer the constructs that were operating. Personal constructs were then envisaged as largely implicit knowledge elements that allow us to parse the world without the need for conscious deliberation or verbal labels and definitions.

Kelly believed the system of personal constructs encompassed knowledge that was primarily perceptual, as well as that which would normally be thought of as conceptual. That is, he saw continuity in the cognitive system that operated with knowledge elements at different levels: so that for Kelly the same *type of operations* would be involved in making discriminations of tone as making discriminations in the quality of doctoral theses. From this perspective, verbal description and rationalisation of judgements would seem to be considered almost as like a veneer placed on the outputs of the implicit but potentially quite sophisticated system of personal constructs.

The perspective offered earlier in this book considered a great deal of cognitive processing to take place 'out of mind', and much of that to be largely automatic, but did leave room for the executive to direct some preconscious processing (cf. Fig. 7.5). These two descriptions could be seen as consistent, depending on precisely how one interprets Kelly's distinction between the construct system and the verbal reporting that occurs after discriminations are made. Kelly would certainly have accepted that a client could censor a particular discrimination made from being reported to the therapist but saw the role of the constructs as central to how the world was understood.

Kelly included in his system discriminations that were not obviously bipolar, giving the example of discriminating red from 'the non-redness of white, yellow, brown or black. Our language has no special word for this non-redness, but we have little difficulty in knowing what the contrast to red hair actually is' (p. 63). This suggests that personal constructs may be linked to knowledge elements that can identify particular features: that is, small processing units that recognise red (or not). Whilst Kelly's notion of personal constructs is not universally adopted, it would seem to reflect important aspects of the way knowledge is represented in the human cognitive system.

Phenomenological Primitives

A slightly different type of intuitive knowledge element that has been mooted as a key part of the cognitive system is the phenomenological primitive, or p-prim. This idea has been developed in particular by Andrea diSessa, who published an extended (if intended to be somewhat provisional) account of intuitive physics based on this notion in the journal *Cognition and Instruction* (diSessa, 1993). The term phenomenological primitive is a fairly accurate label for these entities, as they relate to our implicit interpretations of the world based on abstractions from direct experience of the world. From extensive interviews with physics students, diSessa set out the case for a wide range of these primitives. Each p-prim could be understood as abstracted from common experience, and then used as part of the interpretive apparatus for making sense of the world at a preconscious level, which then feeds into our conscious thinking. In other words, although diSessa's data was largely based on elicitation of college students' explanations about physics problems, that is, an advanced academic context, he considered that much of their thinking was built upon very simple primitive discriminations that matched what was perceived with common general patterns that had been abstracted from prior experience.

So, for example, young children may come to realise that a lot of phenomena fit a pattern that might be labelled 'dying away', that is the magnitude of some qualities seem to diminish with time. The significance is that the abstraction becomes part of the intuitive model of how the world is, and the basis of implicit explanations. That is, if a novel phenomenon is understood to fit the 'dying away' pattern, then it does not pose a 'problem' for the cognitive system, as it fits within the existing model of how the world is. Dying away is treated as a natural effect, that is, one that does not need more explanation. That of course represents the 'natural attitude' (Schutz & Luckmann, 1973), not the scientific attitude, and in learning science students have to learn to question the natural mechanisms of the world that lead to these patterns. Yet many phenomena make sense to us intuitively since they are recognised as matching patterns that we have come to accept as common to experience.

A key problem with p-prims from the perspective of learning science is that they seem to only work to discriminate what fits prior patterns from novel phenomena, and so contrasting phenomena can equally fit (different) p-prims, making them of limited explanatory value. So if a person has a p-prim that we might label 'dying away' and another we might label 'building up', then both these patterns would intuitively seem natural and needing no further explanation. Simply *recognising* that something diminishing is dying away, or that something increasing is building up, would 'satisfy' this level of the cognitive system as what was being observed made sense in terms of existing expectations of how the world is. Students asked to explain phenomena will often respond that certain things are just 'natural', just the way things are, reflecting how in everyday life we do not see many familiar events as inviting explanation as we have become comfortable in accepting them as how the world is (Watts & Taber, 1996).

Research exploring school learners' thinking about chemical phenomena identified a set of potential intuitive knowledge elements that partially fitted with diSessa's scheme (Taber & García Franco, 2010) but also having some distinct features – suggesting research across different domains may help refine an account of commonly acquired p-prims. P-prims seem very similar to what Vygotsky labelled as a 'potential' concept which 'is an embodiment of a rule that situations having some features in common will produce similar impressions' and 'result from a series of isolating abstractions of such a primitive nature that they are present in some degree not only in very young children but even in animals' (Vygotsky, 1934/1986, p. 137).

Intuitive Rules

Stavy and Tirosh have suggested that one source of many of the reported student 'alternative conceptions, preconceptions, and misconceptions in science and mathematics' may be the application by the student of what they term 'intuitive rules' (Stavy & Tirosh, 2000, p. vii), which they consider to be 'expressions of the natural tendency of our cognitive systems to extrapolate' (p. 87).

Stavy and Tirosh (2000) report three examples of intuitive rules that they identify as being found in students' reasoning across a wide range of contexts: 'more A – more B', 'same A – same B' and 'Everything can be divided'. These types of general intuitive rules would seem to be the kind of primitive cognitive element that diSessa has described as p-prims and will here be assumed to be subsumed into the same class of knowledge element in the cognitive system.

P-Prims and Gestalts

Sometimes the term gestalt is used in a way quite similar to diSessa's notion of p-prims. So the 'experiential gestalt of causation' proposed by Lakoff and Johnson (1980a), and applied in the context of science learning by Andersson (1986), set out how causality in the world can often be understood in terms of a common pattern or 'a "prototypical" or "paradigmatic" case of direct causation' (Lakoff & Johnson, p. 479) which involves an 'agent' acting on a 'patient' to bring about some change in it. This would seem to be the kind of pattern recognition assigned to p-prims, and in keeping with the use of 'gestalt' elsewhere, it may make sense to consider the 'gestalt' to be *the perceived pattern*, due to the operation of an underling implicit knowledge element that is part of cognitive structure (i.e. the p-prim). That is, the gestalt is experienced due to the activation of the p-prim.

Watts and Taber (1996) used the idea of an 'explanatory gestalt of essence' to describe how it is that often, when asked for explanations in interviews, students would soon reach a point where they replied that something was 'just natural' – that is the way things were. Watts and Taber found that students varied in the extent to which they would offer layers of explanation before reaching this point, but sometimes students were clearly satisfied with recognising something as being naturally the way things were and so not needing further explanation before they exhausted the depth of explanation expected in the school or college science curriculum.

Ultimately science aims to find out the ways things naturally are, and so there is nothing wrong in principle in reaching such a point in a succession of explanations. However, science looks for underlying patterns that have explanatory value across a wide range of phenomena, whereas the natural attitude is to simply accept as natural anything that fits one of the available familiar patterns (i.e. p-prims). The explanatory gestalt of essence, the recognition that that is just the way things are, would again seem to be a way of describing the learners' subjective experience, which *draws upon* implicit knowledge elements, such as p-prims. So these mooted gestalts would seem to be related to, but ontologically different to, p-prims.

Explicit Knowledge

Whereas implicit knowledge elements are considered to do their work out of the purview and control of consciousness, explicit knowledge is directly accessible and open to deliberation. Earlier in the book, when considering memory, it was suggested that there is declarative memory, and non-declarative memory that includes both procedural memory and 'implicit' learning that takes place without conscious awareness. Procedural memory is associated with motor function and allows us to build up routines of motor actions to carry out complex tasks such as tying shoelaces or focusing a microscope. Some of this is at the level where it is open to conscious awareness and control. These elements are probably not the smallest 'grain size' and draw upon more primary, encapsulated knowledge elements, which we consciously build up into routines.

So there is parallel within this branch of cognition with declarative knowledge discussed below, in that it has both implicit and explicit components. However, the focus here is on conceptual learning, so the nature of procedural knowledge will not be developed in any detail here.

Declarative memory refers to representation of factual information that is accessible to consciousness and includes both episodic and more generalised semantic memory. By definition declarative memory refers to representations of past experience that can be reported verbally as they are consciously accessible, although that does not mean that these declarative memories are themselves representations of verbal information. So one's memory of a significant past event may well include imagery, for example. However, as one is able to access the memory leading to a conscious experience, that experience can be reported verbally.

Imagistic memory has been given most attention in the science education literature, but of course other sensory modes may provide experiences of memories that we can verbalise. We may hear the voices of others not present (e.g. in sleep), and Proust used memory evoked by a smell as a key device in for his novel *À la recherche du temps perdu* (translated as *In search of lost time* or *Remembrance of things past*). However, the visual mode would seem to be of particular importance in conceptual learning, as suggested by the incidence of eidetic memory in children and the conjectured visuo-spatial scratch pad as a major adjuvant of the cognitive system's executive module, that is, working memory (see Chap. 5).

However, the ability to rote learn passages or prose, or technical definitions, demonstrates that some knowledge representation of verbal material can and does take place. Therefore, it would seem that explicit knowledge elements within cognitive structure that are of interest in learning science can be considered to be of at least two different types. This links to Bruner's (1964, p. 2) notion of three modes of representation:

- Enactive: 'a mode of representing past events through appropriate motor response'.
- Iconic: summarising events 'by the selective organisation of percepts and of images, by the spatial, temporal and qualitative structures of the perceptual field and their transformed images'.
- Symbolic: represents 'things by design features that include remoteness and arbitrariness' (i.e. words are associated with objects and events by convention).

Enactive representation supports what has been called here procedural knowledge; imagistic memory is a form of iconic representation, and much of the knowledge of interest in science education concerns propositional knowledge represented symbolically.

Propositional Knowledge Elements

One type of knowledge represented in the cognitive system is propositional knowledge that allows us to 'know' such things as:

*   Atoms are very small.
*   Horses are mammals.
*   Energy is conserved.
*   Potassium is more reactive than calcium.
*   Humans have 23 pairs of chromosomes.
*   Electromagnetic radiation is a transverse wave.

This type of knowledge element is often a key focus of research given the central role played by language in communication and formal learning.

Conceptions

The term 'conception', and the variants 'misconception', 'alternative conception', has been widely used in science education when describing aspects of students' (inferred, assumed) personal knowledge. The term is widely used in phenomenography, research which looks to describe, analyse and understand experiences (Marton, 1981). In this context different conceptions are 'qualitatively distinct ways' in which what is objectively the same referent is understood (Anderberg, 2000, p. 94).

Gilbert and Watts recommended that the term conception should be used in science education to focus on 'the personalised theorising and hypothesising of individuals' (Gilbert & Watts, 1983, p. 69), as one way to distinguish between personal and public systems of knowledge (as discussed above, see Chap. 10). This distinction is shown in Table 11.1.

As suggested above (see Chap. 10), public knowledge is a problematic notion and indeed is arguably in some ways a fiction, but nonetheless remains a useful fiction as a referent. So following Gilbert and Watts, a learner may be said to have *a conception of* energy, or photosynthesis, or oxidation, which can be evaluated against (someone's, e.g. the researcher's) understanding of *the scientific concept* or of some curriculum model of that *concept* (cf. Fig. 1.3).

Table 11.1 Recommended use of 'concept' versus 'conception' following Gilbert and Watts (1983)

Term	Recommended use – to describe	Notes
Concept	Formal meanings as part of public knowledge systems	'World 3' objects: ideals as represented in public knowledge systems
Conception	Personal understandings	'World 2' objects: understandings as personally experienced in thinking

terms used to describe knowledge elements

knowledge elements in
cognitive structure

terms for experience of
accessing knowledge

conceptual
knowledge

procedural
knowledge

*ideas,
gestalts*

motor schema

implicit

explicit

to make
discriminations

to identify
patterns

propositional

iconic

discrete

extended

static

dynamic

*personal
constructs*

p-prims

conceptions

schemata

images

mental models

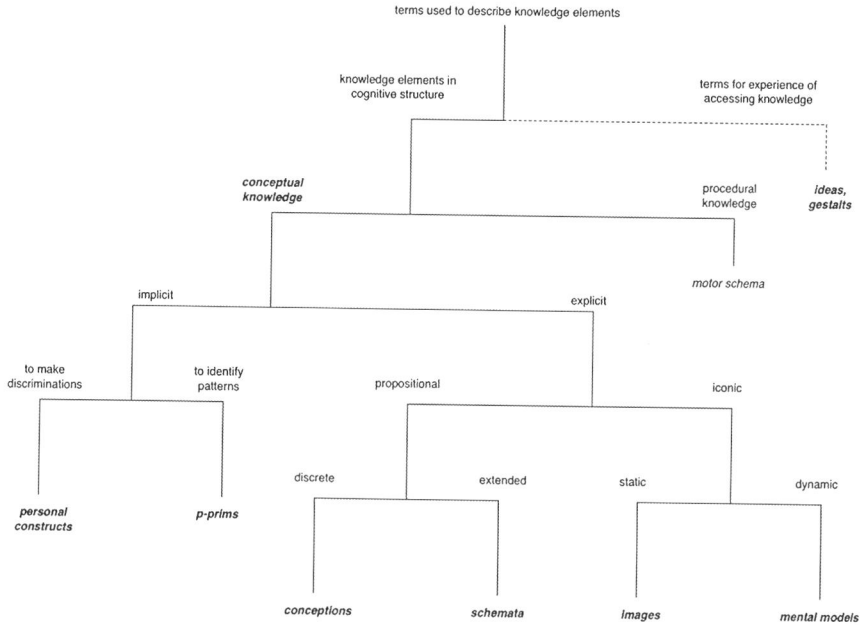

Fig. 11.1 A model typology of the main types of knowledge components represented in cognitive structure

Maintaining a distinction between the formal concept that is part of a public system of knowledge and an individual's personal conceptions might help clarify reports in science education. However, in practice there is widespread use of the term concept to refer to both formal concepts and the versions of those concepts formed by individuals, that is, their conceptions. The literature includes many examples of references to concept formation and acquisition (i.e. the appearance of new conceptions in individuals) and conceptual (rather than conception*al*) development, conceptual (rather than conception*al*) change and conceptual (rather than conception*al*) structure. These topics will be discussed in Chap. 15.

Ezcurdia (1998) suggests that 'concept'/'conception' can refer to the distinction between possession and mastery of a concept. That is, for Ezcurdia, 'one can possess a concept without having an appropriate conception, without mastering it' (p. 188). This approach may be especially helpful for the conceptions that learners have that are considered to be versions of normative concepts. So, as an example, in a secondary science class it could be said that all the students had acquired the concept of a metal, but that their specific conceptions varied considerably, or that a learner acquired a concept of a metal, and that same concept developed as his conception of metal changed (see Chap. 15). In the present chapter, the term *conception* is used to refer to aspects of the learner's personal system of knowledge representation (see Fig. 11.1), following Gilbert and Watts.

Schemata

A term that has not been so widely used in science education research but is commonly used to describe aspects of an individual's knowledge in psychological and cognitive science is schemata. A schema refers to a knowledge *structure* represented in memory: for example, 'the information that is required if a learner is to be able to solve problems [such that] if the required information (knowledge components) and the relationships among these knowledge components is incomplete then the learner will not be able to efficiently and effectively solve problems requiring this knowledge' (Merrill, 2000, p. 245). Schemata, then, are envisaged to be more complex knowledge representations than individual conceptions and indeed are perhaps not best understood as knowledge 'elements' but more, if we draw on an analogy from chemistry, as 'compounds' of knowledge elements.

Problem-solving involves more than just applying routine knowledge as in completing exercises and requires some novelty in task response by the problem-solver (see Chap. 7). So genuine problem-solving requires the learner to coordinate existing knowledge components into a more complex structure, that is, to *construct* a schema. However, the way the term schema is often used, it is also applied to schemata that have previously been compiled and therefore have some permanent 'structural integrity' within cognitive structure: so, in effect, once a schema has been constructed, that construction can be retained if it is then applied sufficiently to develop strong associations between the component elements (see Chap. 5).

So, for example, if a secondary school student is asked to complete a word equation, such as (the example used in Chap. 7)

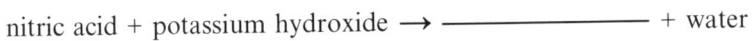

nitric acid + potassium hydroxide $\rightarrow$ —————————— + water

One way that a learner might be able to correctly respond to such an item would be if they had rote learnt sets of word equations, including this one, and so were able to access the correct word equation represented in memory by matching with the information presented and then fill in the gap by comparing the incomplete word equation with the learnt correct one. Certainly much learning of this type goes on, but such learning does not require, or demonstrate, understanding of chemistry and is only effective for specific reactions where the word equations have been (correctly) represented in memory.

More likely, this task will require the learner to coordinate a range of knowledge elements to find a solution, and for many secondary students such a question presents a genuine problem (Taber & Bricheno, 2009). More advanced and successful students might well have developed an effective strategy for answering a question of this type (see Table 11.2), which they can routinely call upon (Taber, 2002a).

The approach shown in Table 11.2 is not the only approach to attempting this task, and if students do not know the general equations, they may rely on the conservation principle (that the same elements must be represented before and after a reaction) to see what was 'missing' on the product side (Taber & Bricheno, 2009): although the coordination of other knowledge would still be needed to ensure a

Table 11.2 Suggested components of a schema to identify an unknown reagent in a word equation

Step in strategy	Note
Identify the type of reaction represented: neutralisation, acid plus alkali	Draws upon knowledge that chemical reactions are commonly classified into particular types, deepening upon the categories of reactants
	Identifies the reactants given as an acid and an alkali (i.e. classifies type/identifies set membership)
	Identifies specific knowledge that one such type involves the reaction between an acid and an alkali
Write out the general reaction: $acid + alkali \rightarrow salt + water$	Applies knowledge that each type of reaction can be represented by a general equation, where the *class of substance* stands for particular reactants and products that vary in different specific reactions of the type
	Recalls general form of equation for this class of reaction
Identifies the missing term as a salt	Compares the general equation recalled with the presented example
	Maps
	Nitric acid: the *acid*
	Potassium hydroxide: the *alkali*
	Missing term: the *salt*
	Redefines task as identifying the particular salt
Identifies the salt as potassium nitrate	Recalls/applies knowledge that salts have a two-part name, reflecting the cation and the acid radical
	Identifies the cation as potassium from the alkali
	Identifies the acid radical as nitrate from the nitric acid

correct solution, as that principle by itself underdetermines the answer. For example, potassium nitrite and potassium nitride would be possible alternative answers.

Although schemata are composed by the coordination of other existing knowledge elements, they should be considered as separate components of cognitive structure because they can be retained as long-term associations and so in effect unitary components in their own right. Merill (2000, p. 246) argues that 'solving a problem requires the learner to not only have the appropriate knowledge representation (schema or knowledge structure) but he or she must also have algorithms or heuristics for manipulating these knowledge components in order to solve problems'. He argues that,

> If the learner knows the knowledge components and knowledge structure for a conceptual network, then he or she has a meta mental model for acquiring a conceptual network in a specific area. This meta mental model allows the learner to seek information for slots in the model. It provides a way for the learner to know if they have all the necessary knowledge components to instantiate their mental model (Merrill, 2000, p. 246)

The example of completing a simple word equation in chemistry supports Merill's assertion that such problem-solving, for those learners at a stage where this task can still be considered a problem, does require knowledge of operations – of what kind of knowledge to access and coordinate – as well as knowledge of the base domain (in this case knowledge of reactions types, reagent types, etc.). However, it would seem that this 'how to' knowledge is also propositional (see Table 11.2), whereas it will be suggested below that the term 'mental model' is often reserved for something rather different. Arguably, in an example such as that used here, knowledge of how to carry out the stages of problem-solving are a part of the schema as much as knowledge of the chemical substances and reactions that need to be operated on to solve the problem.

Visual Representations in Cognitive Structure

Although there is often a focus on verbal representation when discussing [sic] students' science knowledge, it is clear that we are also able to recall images that we do not seen to construct ab initio from other kinds of representation on recall. We are also able to *form* images – from verbal descriptions, for example – but some of our memories seen to be accessed as images: the representation in cognitive structures when activated leads us to experience an image.

As suggested earlier (eidetic memory, see Chap. 5), it is considered that visual memory plays an important role in the memory of children but usually diminishes during development. However, some adults seem to retain strong visual 'photographic memories', and we all have some ability to represent visual information in cognitive structure.

Imagery as a Form of Knowledge

Images contain information, and so representing imagery in cognitive structure amounts to a form of knowledge in the system. Earlier in the book it was suggested that it was easier for a person to remember and reconstruct an image such as that showing the resonance between two canonical forms of benzene (e.g. Fig. 5.11) if they understood what the image meant, and we might imagine that in trying to reconstruct such an image some learners might draw upon propositional knowledge: I know the formula is C_6H_6, I know carbon has valency 4, I know it is described as a cyclic compound, etc.

However, it is equally the case that recalled images can support verbal recollections: mental inspection of a recalled image of, for example, the experimental set-up for measuring Young's modulus of a piece of wire, could provide information to support recall of the formula for Young's modulus, or recollection of an image of a beetle might be the source of recalling how many legs beetles have. In general, recall is supported by being able to access and coordinate both representations of images, and propositional knowledge, from memory (Cheng, 2011). Images are static, although they can be mentipulated in the mind. Moving beyond static images, there is the possibility of visual models that can act as mental simulations that can be 'run' in the mind, that is a form of mental model that is dynamic and visualisable.

Mental Models

Whilst there has been relatively limited attention in the science education literature to students' imagistic representations compared with their propositional knowledge, there has been a wide use of the term 'mental models' in the literature. Once again the point needs to be made that often such terminology is used without definition, and it is not *always* clear what researchers' reports referring to mental models (as opposed to say, student conceptions) in science education contexts are meant to refer to.

There is quite a developed literature about mental models, which can inform the use of the term within science education, although even here different authors do not seem to agree on quite how mental models should be understood (Johnson-Laird, 2003b; Merrill, 2000; Norman, 1983): as 'a consensus view about issues such as the format of the mental models and the process involved in using them has not been reached among different research camps' (McClary & Talanquer, 2011, p. 397). So, as reported above, Merrill (p. 244) suggests that 'a mental model consists of two major components: knowledge structures (schemata) and processes for using this knowledge (mental operations)', but commonly mental models are understood to represent knowledge in non-propositional form.

The notion of mental models was popularised by Norman who described how:

> In interacting with the environment, with others, and with the artifacts of technology, people form internal, mental models of themselves and of the things with which they are interacting. These modes provide predictive and explanatory power for understanding the interaction. (Norman, 1983, p. 7)

Norman describes mental models as 'naturally evolving models that must be 'functional', in that people will continue to modify the mental model in order to get a workable result'.

Johnson-Laird suggests that people construct mental models 'from perception, from imagination, and from the comprehension of language' (2003b, p. 42) and argues that a key feature of mental models is 'iconicity' in that 'a mental model has a structure that corresponds to the known structure of what it represents' (2003a, p. 11). He suggests that 'mental models can represent spatial relations, events and processes, and the operations of complex systems' (2003a, p. 19), yet he also argues that 'visual images are a special case of mental models, and many mental models do not yield images' (2003a, p. 12) and that 'many mental models cannot be visualized' (2003a, p. 11). This is an interesting attribute for an iconic form of representation and perhaps suggests that for Johnson-Laird not all mental models are part of explicit knowledge.

McClary and Talanquer (2011, p. 397) suggest that a mental model is 'a structural, behavioral, or functional analog of a real or imaginary object, process, event, or situation [that can] support understanding, reasoning, and prediction', and they use the term to mean 'dynamic internal representations that may be constructed on the spot to deal with the demands of a given problem or situation, although it is possible that in some cases mental models may be stored in long-term memory'. For these authors the construction and/or application of a mental model is guided and constrained by the explicit and implicit cognitive resources available to any given individual (e.g., prior knowledge, ontological presuppositions, intuitive heuristics), as well as by the most salient features of the task at hand… (2011, p. 397)

This sense of the moment-by-moment construction of mental models is something reflected by Shepardson and colleagues who refer to mental models as 'always under construction and based on new knowledge, ideas, conceptions, and experiences' (Shepardson, Wee, Priddy, & Harbor, 2007, p. 330).

It would seem there is no clear consensus on exactly how 'mental models' should be understood, but it is suggested here that what is useful about the idea is the notion of knowledge structures that are non-propositional, and more extensive than single images, and so 'runnable' in the sense of allowing the individual to set up an 'input' state, run the model mentally and observe a simulated process suggesting the output state that arises from running those initial conditions through that model.

In a sense, mental models seem to have a similar function to computer simulations of complex processes (e.g. ecological interactions) that allow learners to change initial conditions (e.g. population sizes) and then observe how a situation unfolds. In the case of the computer simulation, the outcome is observed on the computer monitor screen. In the case of a mental model, the simulation is imagined inside the mind.

A Model of the Ontology of Knowledge in Cognitive Structure

This analysis has considered the main types of entities that have been proposed as allowing knowledge to be represented in cognitive structure, and the key distinctions between them, as well as examining how some of the terms that have been mooted might do useful work within science education to describe distinct aspects of a learner's personal knowledge. The project here is to consider if a model can be offered which includes the main types of knowledge elements that are assumed in the literature and provide clear labels for the different components of the model.

Given the lack of consistency in how terms are used in different literature, the analyst has two options in proposing such a model: either to suggest a completely new set of terms with no history and so no semantic baggage or to draw upon the available terms and use them to do work within the model with the best fit that seems possible. Given that the field is already heavily populated with terms that although often poorly defined are widely used, I have chosen the latter course, and the outcome of the analysis developed above is represented in Fig. 11.1.

So Fig. 11.1 shows the main distinctions discussed above, with conceptual knowledge being either implicit or explicit, and explicit knowledge being proposition or iconic. The model is set up in the form of a taxonomic dichotomous key, and this almost certainly simplifies the actual complexity of knowledge structures in cognitive structure. This is considered justified in order to offer a basic system with the use of a limited number of categories (and terms) to describe the knowledge elements that may be invoked in research on learning in science. One purpose

knowledge elements in cognitive structure

terms for experience of accessing knowledge

conceptual knowledge — — — — — — — — — — — — — — —▶ *ideas,*

implicit

explicit

gestalts

to make discriminations

to identify patterns

propositional

iconic

personal constructs

p-prims

discrete

extended

static

dynamic

conceptions

schemata

images

mental models

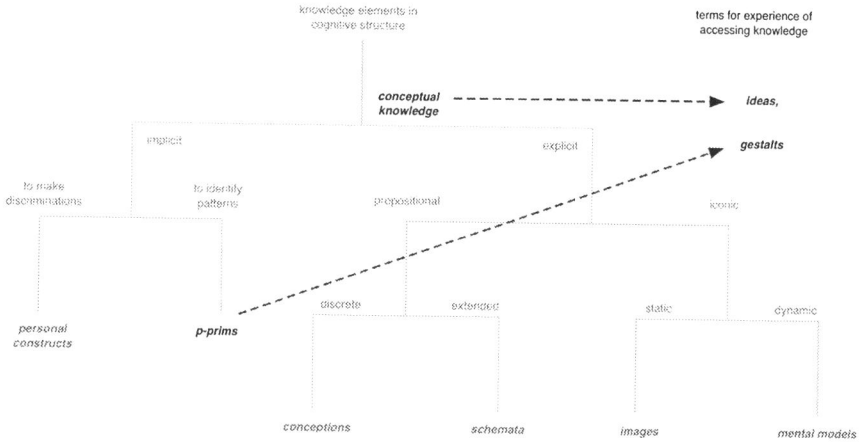

Fig. 11.2 Ideas and gestalts seen as experience of the outcomes of processing through available knowledge structures

of a model is to offer a simplified account that still reflects key features of the complexity being modelled. The model offered here is intended to include key discriminations identified in research, whilst being simple enough to be of value to those working in science education.

As suggested above, 'ideas' (or 'thoughts') and 'gestalts' may be best understood not as knowledge elements but as the experience of the outcomes of the operation of those knowledge elements: gestalts very much referring to perceptual experience and ideas more generally to the output of processing through the cognitive system (see Fig. 11.2).

Figure 11.1 neither directly represents the level of the cognitive system at which different components occur nor how they might be related, but simply the classes of knowledge component in the system. These are important issues that will be addressed separately.

The scheme presented in Fig. 11.1 excludes some terms that are commonly used in the field. So misconceptions are not included as a category, as this term inherently combines a reference to a type of knowledge element with a judgement about understanding in terms of someone else's knowledge (see Chap. 6). In terms of its nature as *a type of knowledge element* within a cognitive system, a conception has the same status whether it is judged as mistaken or not by a teacher or researcher. So some student conceptions may be judged 'alternative', whilst others with similar status within the cognitive system will be considered canonical. In the present analysis all the learners' conceptions are seen as part of a personal knowledge system which can only be labelled as misconceptions or alternative conceptions by someone making judgements from outside the system (see Chap. 10).

The scheme presented here does not distinguish between conceptions that the individual is strongly committed to, those that he/she has learnt but finds unconvincing and those that are recently formulated and are being entertained as potentially fruitful. Rather, it is assumed that these are all the same basic kind of knowledge element (see Chap. 6) yet given different weightings as representations of the external world within the system. This reminds us that although we might think of discrete knowledge components, this is certainly a simplification as our conceptions are linked into an extended conceptual structure (considered further in the next chapter).

A key feature of explicit knowledge is that it can be accessed and considered, and a judgement made about its relevance to the problem in hand, whereas implicit knowledge presents conclusions and recommendations to consciousness without providing access to the basis on which they were reached. A well-known human trait is to reach a tentative decision based upon implicit knowledge and then seek justifications for that decision based on available explicit knowledge. Arguably, the scientific attitude differs not in the exclusion of the role of implicit knowledge (Polanyi, 1970) but in the extent to which explicit knowledge is used to view one's 'hunches' and intuitions critically.

One common term used in the field is that of a conceptual framework, and I have not seen the need to include that in Fig. 11.1. The term framework is used in at least three different ways in science education research (Taber, 2009b, pp. 188–189): (1) as a synonym for a conception, (2) as a more extended conceptual structure (most like a schema in the present analysis) or (3) a technical term used to label the *abstractions* developed by researchers to describe common patterns in student conceptualisations and so distinguish these models from the personal conceptions of the individual learners.

The first sense of framework is already covered here, and the third is inherently excluded from being part of an individual's knowledge structure – at least, apart from that of the researcher, whose personal conception it is.

Conceptual Frameworks and Common Alternative Conceptions

Yet it is important not to ignore this notion of 'conceptual framework', as a key area of research has been based around identifying, and quantifying, 'common' alternative conceptions – those conceptions that learners commonly hold which are considered at odds with canonical science (Duit, 2009). In this sense, a number of alternative conceptual frameworks have been referred to in this volume: such as that motion naturally dies away, that atoms form bonds to fill electron shells, that heat is a kind of fluid substance.

It should be clear from the analysis in this volume that there are a number of problems that face researchers who make claims that some proportion of a population share a particular conception. The issue of knowing what can be taken to be canonical knowledge, given the elusive nature of public knowledge (Chap. 10) should warn researchers that definitive statements about what a scientific concept actually is should be made with caution. That is not to suggest that researchers should avoid seeking to compare student knowledge with canonical knowledge as such research is directly useful and relevant to teaching. Rather researchers need to be aware that at best they can have *a model of* canonical knowledge for comparison, and that in reporting their work they should be explicit about what they take to be scientific and/or curriculum target knowledge and on what they base this judgement (e.g. Treagust, 1988).

However, it should be clear from the discussion in this volume that when we explore students' conceptions of any topic in depth, we tend to find nuanced, often complex patterns, often with idiosyncratic ranges of application, and with evolving levels of commitment. This means both that any simple statements about student conceptions (such as the examples above about motion, bonding, heat) are likely to considerably simplify what are often actually nuanced conceptualisations.

Yet such gross simplifications are often needed when we want to produce information of direct use in the classroom. They also allow us to categorise large numbers of students into a small number of categories – a range of 'alternative conceptual frameworks' (Gilbert & Watts, 1983). This certainly has 'headline' value – so, for example, if we inform teachers that something like 80–85 % of students are likely to hold impetus-like ideas of force and motion (Watts & Zylbersztajn, 1981), then this gives a clear indication of the extent of the problem – if at the cost of loosing much of the richness of what our research can tell us (see Chap. 6). It certainly does not mean that these students are all drawing upon precisely the same cognitive resources and so would always interpret and answer different questions in the same way.

In particular, where research relies upon written instruments informed by research reports of particular conceptions, then response patterns will often vary with wording, question sequence, examples used, etc. So where instruments include a range of items about the same conception, it is quite likely that the outcome will need to be reported as a range, suggesting that more respondents applied 'the' alternative conception on some items than others.

A particular issue links to the understanding of knowledge we have adopted here (see Chap. 9): as the range of notions a person has under current consideration as possibly reflecting some aspect of how the world is, rather than only what is strongly committed to. Offered a range of statements reflecting apparently contradictory conceptions, learners will commonly agree with logically inconsistent statements (see Chap. 6) because of the tendency to agree with different positions that seem feasibly convincing. So it is sometimes possible to show students agree with both canonical positions, and also contrary alternative positions, and if we do not bear this in mind then instruments designed only to find level of support for one conception are likely to offer a distorted view.

The process of selecting particular positions from different students' conceptions as sufficiently distinct to be considered alternative conceptual frameworks is a matter of forming a model of the elicited conceptual 'phase space' which is in some way akin to factor analysis but is not supported by the statistical apparatus employed for that type of work. Designing instruments that can be used to survey populations in order to 'assign' student positions to the different alternative frameworks draws directly on these 'metal models'. This should be borne in mind when reading and writing about this kind of research, which can otherwise appear to be suggesting that given proportions of a population share *the same* cognitive components (the same conceptions).

The analogy here with statistical methods is that when quantitative analysis identifies different clusters (of schools, of students, of teachers, etc. depending upon the study), this suggests that those within the same cluster tend to be more similar than those in different clusters. It certainly does not mean that those in a particular cluster are the same in terms of what is being measured. Research looking at common alternative frameworks tends to rely on qualitative analysis, but the same caveat applies. When different students are classed as demonstrating the same conceptual framework (i.e. a particular category of elicited conceptions created by the researcher), this should be understood to be a statement about similarities in conceptual knowledge of some topic and not identify.

It is necessary and useful to look for general patterns of thinking that will be common across large number of learners in particular groups (English upper secondary students, Australasian chemistry undergraduates, etc.), but important to recognise that for such research to be meaningful, it requires careful consideration of the best ways to form models of the clusters of commonalities among what are likely idiosyncratic ways of makings sense of scientific topics. If this is so when it comes to thinking about student conceptions relating to particular topics, it is even more the case when we move to consider the next level of complication: how learners structure their knowledge elements into broader systems.

Chapter 12
The Structure of the Learner's Knowledge

Concepts do not lie in the child's mind like peas in a bag, without any bonds between them. If that were the case, no intellectual operation requiring coordination of thoughts would be possible, nor would any general conception of the world. Not even separate concepts as such could exist; their very nature presupposes a system. (Vygotsky, 1934/1986, p. 197)

As Vygotsky suggested long ago, the individual's conceptual knowledge cannot be considered to comprise a collection of completely discrete entities. Even those elements that are largely encapsulated, and so only interact holistically (i.e. given this input; that output) are connected into networks so they act as part of a processing system. So, for example, an individual is not aware of the functioning of a phenomeno-logical primitive (see the previous chapter), but that p-prim is activated by sensory information, and presents its conclusion to the 'higher' levels of the cognitive system (cf. Fig. 4.5).

When considering explicit knowledge, it would seem there is even less basis for considering knowledge to be represented in discrete units, certainly when considering how memory is represented in the association cortex (see Chap. 5). It would seem that that there is a high level of interconnectivity between different representations which, depending upon the activation patterns across the system, could become components of a variety of dynamic extended networks. That provides fluidity of thought processes and a considerable challenge for the researcher wishing to model a learner's knowledge. Part of the task of the researcher who is interested in developing models of learning, then, is to consider how the learner's knowledge is structured.

Camacho and Cazares (1998, p. 16) argue for the need to 'construct models or schemes that indicate the existence of hierarchies among the intuitive ideas and how they guide the predictions, explanations, and interpretations given by the students confronting different physical explanations'. They proposed adopting

K.S. Taber, *Modelling Learners and Learning in Science Education: Developing Representations of Concepts, Conceptual Structure and Conceptual Change to Inform Teaching and Research*, DOI 10.1007/978-94-007-7648-7_12,
© Springer Science+Business Media Dordrecht 2013

the notion of 'possible partial models', which makes a distinction between propositions (p. 17):

- 'That correspond to knowledge that students elaborate as abstract representations' (known as constrictor concepts)
- 'That correspond to students' conceptual constructions in which explicit relations among phenomenological variables are established or in which particular conditions that students attribute to physical processes are specified' (referred to rules of correspondence)

This approach, due to Sneed, is based on work designed to explore the logical structure of scientific theories (Przełęcki, 1974), that is, public knowledge (see Chap. 10), rather than personal knowledge, but Camacho and Cazares reported it was valuable in exploring high school students' ideas about pressure and floatation.

The Nature of a Conception

A key focus of much research in science education has been of student conceptions relating to different scientific topics. In the previous chapter it was suggested that 'conceptions' should be understood as the elementary level of explicit, propositional knowledge represented in an individual's cognitive system. It was also suggested that conceptions were sometimes understood as parallel to concepts but with the two terms relating to personal and public knowledge (Gilbert & Watts, 1983). So individual learners might form *conceptions* relating to such scientific *concepts* as energy, molecule, photosynthesis, species, acid and so forth.

Concepts, as suggested earlier, are often found to be quite fuzzy entities, which are difficult to clearly describe. This not only applies to natural kinds where a number of attributes may be used in defining concept membership, with different strengths, and perhaps priority rules but also to some extent to scientific concepts. As Kuhn (1977) has argued, scientific concepts are seldom fully communicated simply by providing definitions.

Moreover, as Vygotsky implies in the chapter motto above, concepts take their meanings from being embedded in a network of other concepts. So, for example, Fig. 12.1 offers a representation of the formal scientific concept of *the hydrogen bond*.

It should be clear from what has been discussed earlier in the book (see Chap. 10) that this representation cannot be said to be of *the* concept of hydrogen bond in the sense of unproblematically representing public knowledge. This is necessarily a representation of an individual's understanding of the scientific concept, and indeed this particular conception relates to teaching the concept at the upper secondary ('sixth form') level – that is, this is a public representation of an individual teacher's conception of the target knowledge set out in the curriculum. This image was prepared to support a discussion of the kind of conceptual analysis that teachers should carry out before planning the teaching of a new topic (Taber, 2002a).

It is clear that the concept of the hydrogen bond, at the level represented here, only makes sense in the context of an understanding of other concepts. Conceptual

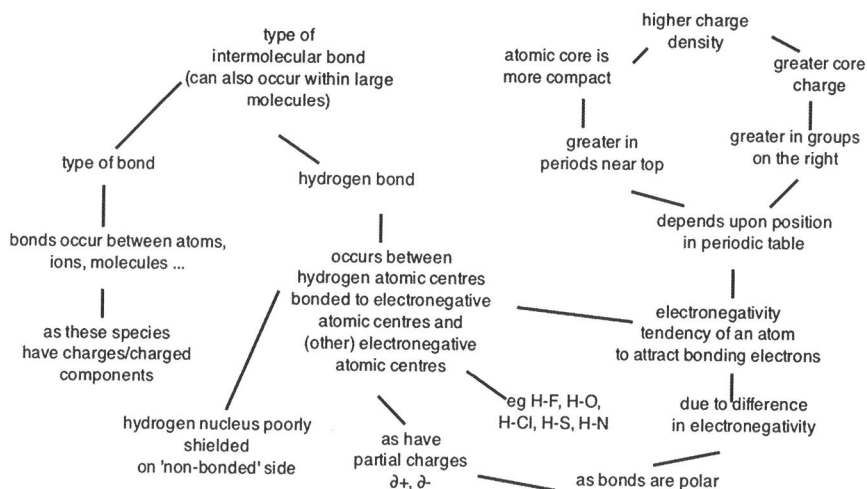

Fig. 12.1 A representation of the scientific concept of 'hydrogen bond'

analysis in science education has sometimes been seen in terms of identifying the hierarchical nature of concepts (Herron et al., 1977). However, as Gilbert and Watts (1983) have pointed out, even if it were possible for formal systems of public knowledge to be represented in terms of 'hierarchical layers which can be decomposed into smaller parts and studied independently' (p. 65), it would still be an error to see such formal representations of public knowledge and the representation of personal knowledge in a cognitive structure 'as isomorphic and by implication … part of a static, logical and organised system' (p. 66). Given that human cognition is based at the level of the physical substrate on networks, a better starting point would seem to be to model students' knowledge structures in 'network' forms, such as concept maps, which can show those propositional links that are elicited from learners as relevant to a particular topic.

Modelling Student Conceptions with Concept Maps

Figure 12.1 includes, as well as 'hydrogen bond', the concepts 'intermolecular bond', 'bond', 'charge density' and 'partial charges', which are considered linked by propositions (e.g. a hydrogen bond *is a type of* intermolecular bond). Representing this kind of linkage is the basis of concept maps. Novak (1990b, p. 29) has described concept maps as 'a representation of meaning or ideational frameworks specific to a domain of knowledge, for a given context of meaning' and has suggested that propositions can be seen as 'the units of psychological meaning'.

A concept map comprises of nodes, representing the labels given to concepts, connected by lines standing for the understood linkage between nodes that is, propositions. A concept map would therefore reflect the proposition that a hydrogen bond is a type of intermolecular bond in terms of a linkage between two nodes (Fig. 12.2):

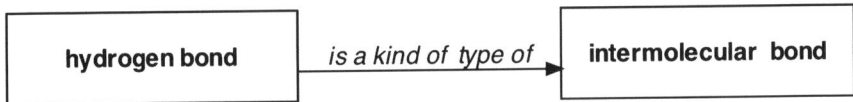

| hydrogen bond | is a kind of type of → | intermolecular bond |

Fig. 12.2   Representing one proposition within a concept map

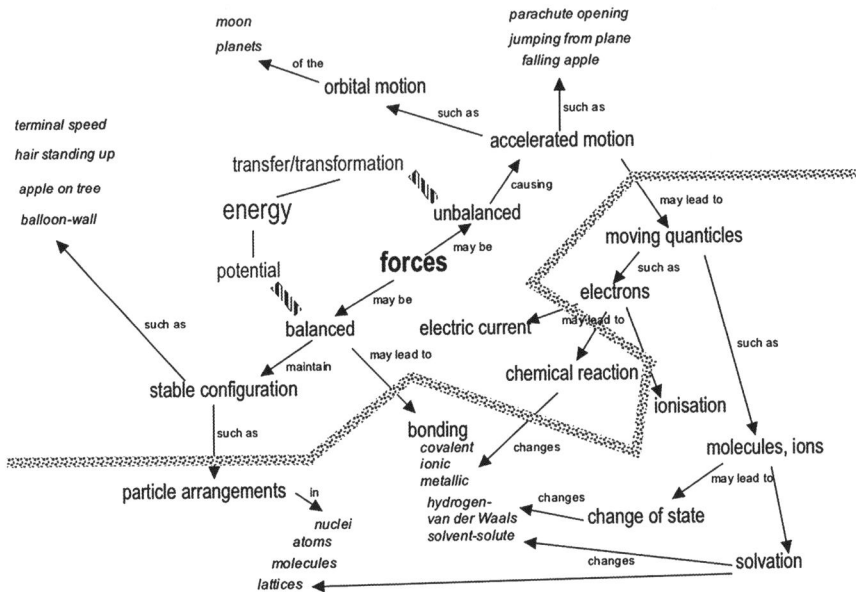

Fig. 12.3   A concept map prepared to guide an interview schedule exploring student conceptual integration across chemistry and physics topics (Taber, 2008a)

A concept map representing an individual's conception of a topic can become quite complex, as in principle 'the meaning of any concept for a person would be represented by all of the propositional linkages the person could construct that include that concept' (Novak, 1990b, p. 29). Figure 12.3 shows a concept map used to conceptualise an interview schedule exploring the extent to which college students studying chemistry and physics integrated their understanding across different concept areas (Taber, 2008a).

Concepts maps can take various forms, and in recent years software has been developed to allow concept mapping with computers that can facilitate ready modification, archiving and analysis of the maps produced (Bruillard & Baron, 2000). However, sometimes such approaches may restrict maps to a certain form, when one of the great advantages of the concept map as a tool for modelling student knowledge is its openness. Given the fluidity of personal knowledge systems, open-ended representational tools may be most suitable for modelling structure within such systems. As Novak points out, 'since individuals have unique sequences of experiences leading to unique total sets of propositions, all concept meanings are to some extent idiosyncratic' (Novak, 1990b, pp. 29–31). So concept maps, which represent the idiosyncratic nature of an individual's conception, are suitable for idiographic research that is motivated by an interest in student thinking and understanding in its own terms (rather than normative research simply concerned with the extent to which student knowledge appears to match some canonical standard, see Chap. 6).

This is not to suggest that more pre-structured representational forms may not be useful for specific purposes. For example, a particular format of representation, Vee maps (Ault, Novak, & Gowin, 1984) were designed 'to represent the structure of knowledge and the epistemological elements that are involved in new knowledge construction' (Novak, 1990b, p. 31). However, concept maps allow students themselves to represent their own knowledge in a format easily learnt and relatively unrestricted – and so can act as a learning tool that supports metacognition (see Chap. 7) by encouraging learners to focus on the current state of their own understanding (Taber, 1994). They also allow researchers to develop representations of their own models of student conceptions, to display and summarise inferences drawn from data sources such as interview transcripts.

Limitations of Concept Maps as Models of Student Conceptions

Concept maps have the useful quality of representing how an individual's conception takes its meaning as a node in a semantic network showing propositional relationships and so seem to offer the ability to reflect important features of a learner's knowledge. However, there are clear limitations in that basic forms of concept map do not offer a ready visual representation of the strengths of linkages, that is, which propositions are more central to the individual's understanding and which are more tenuous. There are certainly ways such information could be represented; colour coding, the topology of the map, thickness of connecting lines, size of concept boxes, etc. could all be used as indicators, but at the cost of complicating the production of the map.

A practical limitation is the extent of the map – where a boundary is drawn – given that potentially everything represented in cognitive structure is connected, if often indirectly through a number of 'degrees of separation'. For practical purposes a concept map has to be allowed to stop at a point where a judgement is made that further layers of links are *not central* to the meaning that is being represented. Perhaps more problematic is organising the presentation of maps which may have many interconnections between nodes, which become difficult to show on a two-dimensional representation.

That concept maps are based around propositional knowledge is also a limitation, for not all a person's knowledge is in this form (see Fig. 11.1). In a person's cognitive structure, there are links between, for example, representations of semantic propositional information and representations of images, such that when I think about 'birds', associated images (wings, beaks, different body shapes) are brought to mind as well as propositions ('birds lay eggs'). Such images can be sources for elicited propositions to include in concept maps, but such reports are less direct representations of the knowledge element itself, which is activated and so experienced by the learner as a mental image (Cheng, 2011).

Indeed, it will always be the case that attempts to model underlying cognitive structure will necessarily be mediated by thinking processes that may not involve the activation of all relevant representations. A concept map, or any other kind of representation produced in the research process, must always be recognised as a contingent model depending on the effectiveness of the elicitation process in accessing all the knowledge a person has represented which might be relevant.

A potentially seductive property of concept maps is how they may seem to reflect the underlying physiological substrate of cognition, in that such a map mimics the structure of the brain with its highly interconnected neurons; however, it is important not to consider individual nerve cells as 'storing' concepts and individual synaptic connections as representing propositions in cognitive structure. Another important complication relates to the dynamic nature of cognition (e.g. see Fig. 5.3). The selective and changing activation of representations in memory, which itself can modify connection strengths, can at any one time only offer a distorted reflection of the underlying representational structure.

A concept map produced by eliciting student knowledge at one point in time is a 'snapshot' to some extent influenced by incidental features of mental context, representing what was brought to mind at that point, and so possibly somewhat different from a concept map based on an elicitation of the 'same' conception at another time. A concept map based on data collected on a number of occasions can start to provide a more thorough representation but the knowledge being represented may not have been static in the meantime and can have actually been modified by the student's cognitive activity during the elicitation process.

Therefore, there is a trade-off between the completeness of the model of a student's conception developed and the temporal resolution obtained. If concept maps are required to model student knowledge at a particular time then there is a risk of bias towards knowledge components activated and exclusion of other relevant knowledge not activated by that elicitation. We should say the map reflects active knowledge during the time the data was collected rather than the full repertoire of knowledge that could have been accessed. This clearly becomes important if we wish to study student learning, as we seek models of student knowledge that allows us to identify changes which relate to *modifications of the underlying representations*, not just artefacts of which pre-existing representations were activated on particular occasions.

This is not a limitation of concept maps per se, although it needs to be borne in mind given their superficial similarity to neuronal networks, but of the underlying nature of the object of research. That is, if we are interested in the knowledge represented in a learners' cognitive structure, which may be vast, we can only indirectly access it by asking the learner to access it herself. That in turn means the learner interrogating the vast contents of memory through the sequential application of the limited capacity executive, working memory (see Chap. 5), based on selective activation of some of the myriad connections between the stored representations.

The extent to which this is a substantive, rather than a potential, limitation is perhaps an area where more research would be useful. If we acknowledge that a person's knowledge is an extensive network of interlinked components and that any representation of their conception of a scientific concept needs to be restricted to core aspects, then we might – at least in some circumstances – reasonably expect concept mapping, or similar elicitation activities, to represent core knowledge about a research focus reliably, even if peripheral knowledge representation may be contingent and appear more haphazard. However, if we are exploring a learner's knowledge structures during learning processes, and so are interested in less, as well as more, robust aspects of knowledge, then it may become problematic if the inclusion of more recent, 'fragile' learning, is dependent upon contingencies that are due to unknown and uncontrollable features of 'context' relating to the idiosyncrasies of the learner's recent experiences channelling the informant's 'stream of consciousness'.

This discussion highlights a major challenge for research in this area. Many papers that report on learners' knowledge of particular topics seem to largely ignore these issues. This brings us back to the central motivation and core thesis of this book: the need to think about research in a more technical sense, and not be seduced by the familiarity of the mental register (everyday talk about knowing, thinking, learning, etc.) and the tendency to assume we can unproblematically read minds. As suggested earlier in the book, we acquire a theory of mind in childhood that is essential for us in making sense of the social world but therefore makes the issue of knowing what another person thinks or knows seem unproblematic, as in everyday life we often seem to do well enough in 'reading' each other for social purposes.

Many research reports fail to include what would seem necessary caveats about the dynamic and inaccessible nature of much of a person's knowledge, and so the limitations of the models and representation presented as results. Perhaps this is sometimes because authors consider such problems should be obvious to all and so are taken for granted, but the definitive way in which results are often presented suggests that often it is not the caveats that are being taken for granted, but the researcher's ability to access and model aspects of other people's minds.

In the present book the research process is being problematised, to suggest to others working in this area, or reading research reports, that there are complications and challenges in modelling another's knowledge that have consequences for the status of reported findings. These complications need to be acknowledged by researchers and made clear in reporting findings, so that the nature of results – as the researcher's models and representations – and the likely limited fidelity of these models and representations to what is being studied are explicit.

The Importance of Conceptual Integration

If individual concepts take their very meanings through their associations, then the extent of connectivity of a learner's conception of some scientific topic is a key aspect of the quality of that conception. A conception that is relatively 'isolated', with limited associations is less readily accessed (see Chap. 5) and can only do restricted intellectual work for the individual. Certainly, creativity requires the ability to recognise potential links across different knowledge representations (see Chap. 7). Moreover, conceptual integration seems especially important in science.

The Significance of Conceptual Integration and Coherence in Science

One of the most powerful features of science is the way it provides a theoretical net, a complex web of concepts that are interlinked and offer, substantially at least, a consistent body of knowledge. Science produces knowledge of the world that has been labelled 'reliable' (Ziman, 1978/1991). That knowledge is expected to be largely coherent. In particular, scientific theories and models are usually expected to be internally consistent.

Moreover, the highest status is often ascribed to those ideas that are thought to be fundamental or unifying – seen as having very broad application, or showing how different areas of science can be brought together within a single framework. Science is generally guided by attempts to develop new understanding that fits with existing ideas. The Nobel Laureate Abdus Salam was reported as noting that 'the whole history…of physics, is one of getting down the number of concepts to as few as possible' (Wolpert & Richards, 1988, p. 17). 'Grand unification' of different areas of physics is an aim for some physicists and cosmologists, and the term 'Grand Unified Theory' is now widely used in an aspirational sense within and across many disciplines.

Even the most counter-intuitive ideas (relativity, the common origin of species through descent) can be accepted in science *if* they can be fitted into existing conceptual schemes and are strongly supported by interpretation of empirical evidence – but lack of coherence is seen as a severe problem. Historically, much of the debate about the acceptance of quantum theory involved arguments over how to find a coherent interpretation (Petruccioli, 1993). Finding a coherent synthesis of general relativity and quantum mechanics is seen as an important aim in physics.

In one commonly discussed description of science (Kuhn, 1996), anomalies are seen as often being crucial to major developments: as they can indicate that new scientific ideas are needed to bring back coherence. The recognition of an anomaly is the recognition of something not fitting – of something being 'wrong'. Results that contradict our theories need to be explained or to be explained away. To 'save the phenomenon' (Kosso, 2010) is really to preserve the coherence of our understanding. Perhaps this need for coherence also explains why so many apparent anomalies are put into a kind of 'quarantine' (Lakatos, 1970) by scientists when they do not seem to fit into available interpretative schemes.

Conceptual Integration as a Demarcation Criterion for Science Education

The central importance of conceptual coherence and integration within science suggests that this feature, or value, should be reflected in science teaching. I have gone as far as to moot this as 'a demarcation criterion for science education: teaching that does not show the links between topics and between the sciences does not provide an authentic science education' (Taber, 2006b, p. 287). Research in science education suggests that the links that may seem obvious to scientists and science teachers are not always spotted by students or even seen as helpful when pointed out to them (Taber, 1998b).

Degrees of Integration of Students' Science Knowledge

From this perspective, one key feature of a learner's scientific knowledge is the extent to which it is integrated into a coherent network of related conceptions, that is, a conceptual structure. In general, progression in science might be expected to involve an ongoing increase in these dimensions: the more advanced the student, the more coherent and integrated their knowledge would be.

Logically we might expect that integration and coherence would be necessarily related: only when different knowledge elements are seen as having a potential relationship does the issue of coherence arise. A student who does not recognise the molecules discussed in chemistry as intended to be related to the notion of gas particles discussed in elementary kinetic theory in physics has no logical imperative to expect the (interacting) molecules in chemistry and the (inert) particles in physics to show consistency in their behaviour and properties.

However, this is not a simple matter, as sometimes the tendency to integrate new knowledge by subsuming it within existing knowledge structures produces integration by distorting information. In terms made popular by Piaget, new information may be assimilated, without sufficient accommodation of existing knowledge. To borrow a common idiom, the square peg is forced into the round hole – or, in the present context, perhaps the square peg is forced to become part of the round whole.

An Example from Learning About Atoms

As an example, when students are introduced to the notion of atomic orbitals, as quite a different model of the electronic structure of atoms than electron orbits, the tendency to simply identify orbitals *as* orbits and transfer learnt properties of the orbit concept to the new orbital conception (Taber, 2002b) can undermine intended learning. Here the new learning is intended to be about how one model relates to another and how the incoherence between the models is allowed within an understanding of the nature and role of models. That is, electron orbits are not consistent with electron orbitals, but the student's conception of electron orbits as an aspect of one model can be coherent with their conception of electron orbitals as components of a different model, because there is nothing logically inconsistent in having *knowledge of* inconsistent models of the same target object (see Chap. 6), providing the epistemological status of the models qua models is recognised. In terms suggested earlier in the book, the learner can acquire a meta-understanding of different models of the same target concept, rather than holding inconsistent manifold conceptions reflecting multiple personal understandings.

In Chap. 6, I drew a distinction between meta-understanding and multiple understanding – to suggest, for example, that the historian who understood different historical models or a teacher who understood different students' alternative conceptions of a topic should be considered to demonstrate meta-understanding rather than multiple understanding. That is, the historian or teacher may have available multiple understandings of the 'same' concept area, but actually understands these as the understandings of others of the target, rather than having alternative understandings of the target themselves. In effect they have an understanding of concept X, plus a (model of the) understanding of scientist Y's, or student Z's, understanding of that concept.

Something very similar may be understood to potentially be the case here in student learning of different scientific models. A student who considers electrons to be in orbits around atoms and is then taught that electrons are located in atomic orbitals and who does *not* appreciate the status of these ideas as components of models is logically able to respond in a number of ways apart other than completely dismissing the new information:

- Simply subsuming references to orbitals under the existing conception of electron orbits – in effect seeing this as an alternative label for the existing idea
- Adding the new orbital concept as a supplementary feature of atoms so that they have both electron orbits *and* electronic orbitals
- Assuming that some atoms have electronic orbitals rather than electron orbits
- Simply learning the new information as an alternative to the existing learning, so that alternative conceptions of the same target are available

In the latter case, if the learner does not appreciate the nature of these ideas as models, then they will have multiple understandings of the same concept of atomic structure. However, if they understand that these alternatives refer to different models of atomic structure – which have proved useful to scientists at different times, for different purposes, and each of which has a range of application, whilst as a model only being a partial representation of atomic structure – then it is possible to have meta-understanding that integrates these apparently inconsistent ideas within a coherent overall understanding. That requires sufficient epistemological insight into the nature of science to consider orbits and orbitals to have the ontological status of model components rather than descriptions of physical entities that exist in the world.

An Example of Relating Science to Belief

Similarly, a student from a faith background that involved a belief in a creator God as set out in Abrahamic traditions, who learnt that scientists consider that the entire universe was formed in a 'big bang', might have difficulty in bringing these ideas into coherence *if* they were encouraged to take a very literal reading of scripture, but might have no difficulty in matching the very different accounts if their religious tradition saw scripture as representing a deep truth poetically, in figurative terms.

So, in terms of the nomenclature used in Chap. 6, the learner who believed science had 'proved' the universe was created in a big bang singularity in school science, but themselves believed the universe was created by a number of discrete acts over a period of 6 days when in Church, would be considered to hold multiple personal understandings. In principle, there is a contradiction in their system of knowledge, although this might well be tolerated if the two understandings were compartmentalised as 'science knowledge' and 'religious belief' and considered to apply to different domains (cf. Solomon, 1983). Gould (2001) suggested that religion and science should be considered as nonoverlapping realms (magisteria) with different concerns; however, some religious communities do read scripture as meant to offer a historical, scientific account of the formation of the world and its biota, from which perspective there are clear inconsistencies with scientific accounts (Long, 2011). Many school age learners do experience religious and scientific accounts as competing (Taber, Billingsley, Riga, & Newdick, 2011), and Billingsley (2004) found from interviews with undergraduates reflecting on their schooling, that a common response was to deliberately compartmentalise knowledge as applicable in different classroom contexts.

However, a student who accepted the big bang account offered in science, and believed in a Creator, but viewed scripture as a poetic narrative offering an allegorical account of creation, does not have the same contradiction in their knowledge systems. That same student may, however, understand how some other people view scriptural accounts as literal and technical reports of how the universe was created, without committing to such a position, and so demonstrate meta-understanding.

As suggested in Chap. 6, personal knowledge is always a 'work in progress', and the possibility of inconsistent knowledge is a necessary corollary to the flexibility of our knowledge systems. Learning mechanisms that only allowed consistent additions to existing knowledge structures would not allow recovery from flawed initial learning in a topic.

Discussions of Integration and Coherence in Students' Science Knowledge

The question of the nature and characteristics of students' science knowledge has been an active part of scholarship in science education. Whether students' knowledge deserved to be considered theory-like or not, and whether student conceptions could be considered as part of coherent knowledge structures rather than relatively isolated 'islands' of knowledge with limited connections have been vigorously debated. I have discussed this issue previously (Taber, 2009b) and argued that the research base available strongly suggests that learners' knowledge varies across such dimensions. So the same person may have theory-like, highly organised, integrated and coherent knowledge that is applied consistently in one topic, but may seem to hold inconsistent and fragmentary knowledge in another topic.

We would expect variations within a class of students and changes with age and experience of learning about different areas. The question should therefore not be 'whether' student knowledge is theory-like or not, but under what conditions it becomes more theory-like. Such a position certainly does 'save the phenomenon', as it encompasses a range of knowledge characteristics as admissible, and so this position can be consistent with the results of different studies which report very different knowledge characteristics. However, this is not some unprincipled compromise position, but rather is highly consistent with what would be expected when we model the learner as supported by a cognitive system along the lines set out in Part II of this book. From that perspective, we should *expect* the cognitive system to include a range of types of knowledge elements (e.g. as represented in Fig. 11.1) with different characteristics.

The model developed here includes implicit knowledge that is applied precon-sciously and is inflexible, and explicit propositional knowledge that is open to introspection, and can be reflected upon and even modified during application. So when Claxton (1993, p. 46) characterised young people's knowledge in terms of what he call *mini-theories* that are 'piecemeal…fragmentary and local', this was a fair comment on some of the knowledge of young people. This is reflected in what Claxton refers to as 'wild inconsistencies' because the 'range of convenience' of a particular 'mini-theory' 'may be rather circumscribed' (p. 47).

However, research has also shown some conceptions that students hold may have very broad ranges of application: for example, ideas about motion dying away naturally (Gilbert & Zylbersztajn, 1985) or a wide range of chemical phenomena being explained in terms of the need of atoms to fill electron shells (Taber, 1998a). Research suggests some 'young people' are perfectly capable of constructing complex knowledge structures, which may be applied consistently, even if this is not always obvious because the set of conceptions they are fitting together (the conceptual structure they are constructing) do not parse the world along the same distinctions as the scientific models.

Even where learners do seem to offer alternative inconsistent ways of thinking about a scientific topic, it needs careful investigative work to determine whether this actually reflects fragmentary knowledge, rather than discrete components of what might be considered some overarching system (Pope & Denicolo, 1986). An uncritical application of the mini-theory perspective might characterise a research physicist as holding fragmentary, piecemeal knowledge for explaining the photoelectric effect and electron diffraction in terms of apparently inconsistent properties of electrons!

Consider the example of a learner discussed earlier in the book, Tajinder, who during his college course offered three alternative types of explanations for why

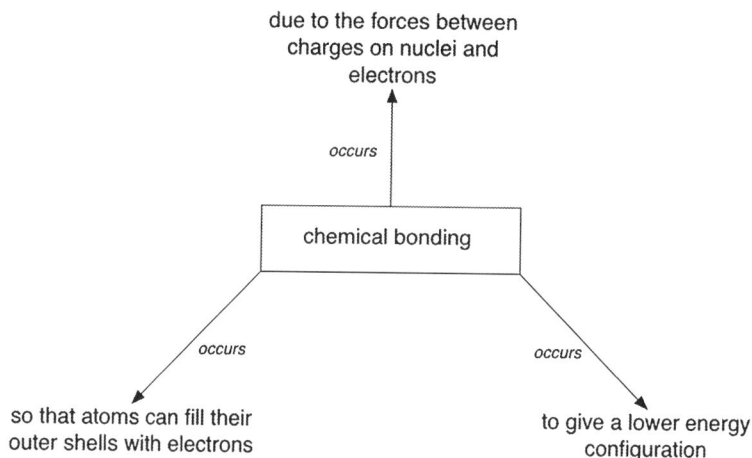

Fig. 12.4 Alternative conceptions of chemical bonding or alternative explanatory principles subsumed within a student's personal conception

chemical bonding occurs (see Chap. 5). These three alternatives are represented in Fig. 12.4, and we might ask if these amount to alternative conceptions of chemical bonding.

If these three types of explanation had been elicited from different students, for example, as responses to a question (such as 'why do chemical bonds form?') in a 'paper and pencil' instrument, then a researcher might well consider them as discrete alternative conceptions. Alternative is used here in the literal sense: they are alternatives to each other. If Tajinder had been responding to such an instrument, it is likely he would have offered one of these explanations and so perhaps been assigned as holding that one particular conception. However, for much of his college chemistry course, Tajinder would offer each of these explanations at different points during the same interview.

That could suggest that Tajinder held a number of conceptions linked to the concept label of 'chemical bonding', which were applied in different contexts. That is, a student may construct mental models of the world that demarcate phenomena differently from scientists and so fail to appreciate where science subsumes a range of cases into one category. Had this been the case here, it would have been expected to find that Tajinder had nonoverlapping ranges of application for the different conceptions. However, that would not reflect the way Tajinder used these three 'explanatory principles', as he would sometimes switch between explanations during an interview whilst discussing a single context. For example, in discussing bonding in molecular oxygen in one interview, he shifted from an account based on how 'to become stable [an oxygen atom] wants an octet state, well it wants eight electrons in its outermost shell to become stable', to describing 'to become more stable, or at a lower energy, it can gain two electrons, to move down in the energy state, therefore becoming more stable', and then to how 'that plus six charge can attract electrons from another species to pull into there, or just to gain an attraction

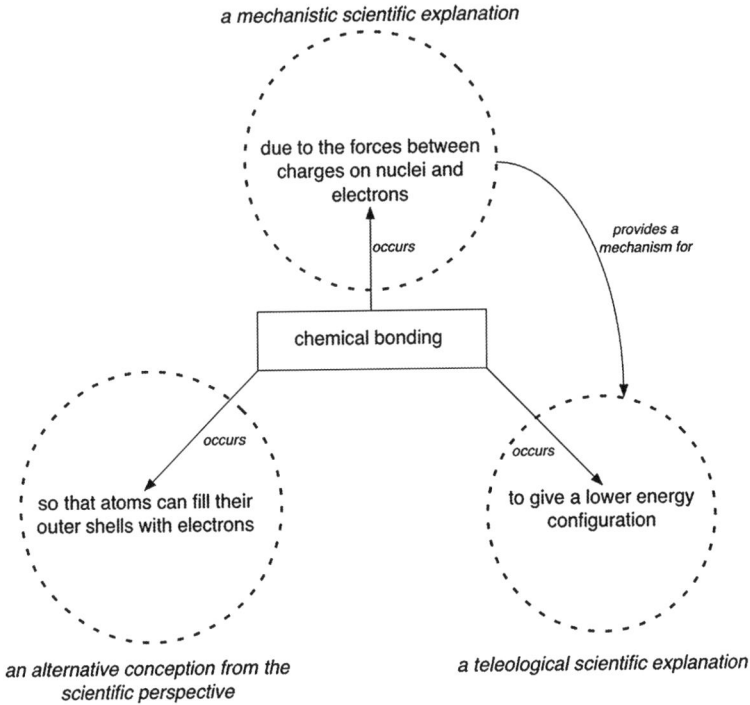

Fig. 12.5 Alternative explanations related to curriculum target knowledge

for it' (see Taber, 2000b, p. 410, where more extensive extracts from the interview are presented). Tajinder was aware of his shifts in explanation, which he justified by reporting that he did not think any 'one is totally correct, I think you can take bits out of each of them to make a best answer' (p. 411).

From the perspective of curriculum science, two of Tajinder's explanatory principles would be acceptable as part of an explanation of bonding phenomena and can be considered as different aspects of a scientifically acceptable understanding: although explaining phenomena as occurring to minimise energy would be considered teleological if presented on its own as a sufficient explanation (see Fig. 12.5).

Tajinder, however, did not seem to consider that there were logical links between his three explanatory principles, but saw them more as a repertoire of explanatory tools from which he would select when asked to provide an explanation. Tajinder's manifold conception of chemical bonding was neither a single integrated explanatory scheme, nor a set of alternative conceptions of bonding with different ranges of application, but more a confederation of explanatory principles understood as independent, yet potentially relevant to the same broader range of contexts where he might be asked about aspects of chemical bonding. This example supports Pope and Denicolo's (1986) warning about the risk of ignoring the complexity of individual thinking when seeking to classify learners' thinking using a limited number of discrete categories (see Chap. 6).

Integration Across Science Topics

That example considers only one individual student's conceptualisation of one topic area (chemical bonding), and studies where student thinking can be explored in some detail across a wide range of topics would be very informative for investigating conceptual integration in students' science knowledge. However, such studies are rare, as most research into student thinking has a particular conceptual focus, allowing in-depth questioning. Studies which elicit knowledge across a range of topics either need extensive access to a learner or need to compromise some aspects of the in-depth interviewing approach often needed to explore student thinking in detail.

One such study (Taber, 2008a) was based upon an interview schedule set up to elicit student knowledge across a range of topics in physical science from interviewees studying both chemistry and physics at college ('sixth form') level, but using a sequence of questions informed by a consideration of links between topics (see Fig. 12.3). Although some follow-up questioning was used, a pre-structured question sequence was followed that had been designed to allow opportunities to see if links between concept areas had been established.

Part of the motivation for the study was the finding from previous work that even when students studied both physics and chemistry, they would often fail to appreciate how the models of chemical structures and processes they studied in chemistry were meant to draw upon basic physical principles they studied in physics. This has been discussed in various examples referred to earlier in the book: for example, thinking that chemical bonding is due to atoms seeking full electron shells, rather than explained in terms of interactions between the charged components of molecules, atom and ions. In the reported case study,

> Alice used key ideas of force, energy, and particles widely in her explanations, and certainly did use ideas about forces and energy in some of her explanations in chemical contexts [although] her explanations of chemical reactions seemed to be a mixture of ideas based upon electrical interactions and alternative ideas based upon the significance of full shells. (Taber, 2008a, pp. 1928–1929)

One of the questions in the interview asked why a balloon rubbed on a jumper is able to remain attached to a wall. Alice recognised this as being an electrostatic effect, with 'some sort of interaction if you like with the electrons and things, and you have a positive and negative charge which allows, a glue effect if you like, attraction between two areas, one of positive and one of negative'. Alice explained that the balloon was charged, as 'when you're rubbing the balloon you're transferring electrons either onto it or away from it'. The balloon would stick to the wall 'because you've got opposite charges, you've got the, say, negatively charged balloon, and then your positively charged wall'.

Now although the hypothetical balloon had been (hypothetically) charged by friction, Alice acknowledged that the wall 'hasn't had anything done to it as such' to give it an opposite charge. Alice suggested that 'maybe in comparison to your very negatively charged balloon, it's still likely to attract'. That is, the wall had a 'relative' charge because, although neutral, it was relatively positive by comparison with the negative balloon. This explanation was not correct, as the canonical explanation relates to the induced polarisation of neutral molecules in the wall: that is, the electrical field due to the charged balloon reconfigures the charge distribution in the wall sufficiently for there to be a net interaction between balloon and wall.

Alice's alternative explanation, whilst technically incorrect, did reflect the way electrical potential is understood, in that a point at 0 V potential is relatively positive compared to a point at a negative potential, and the potential difference may drive a current. Indeed it is quite common in electrical contexts that either the positive or negative side of a system is earthed, that is, at 0 V. However, of more interest in the context of the interview schedule was how Alice's suggested explanation here compared to her explanations on two other questions.

One of these contexts involved nuclear stability. Alice did not think that the nucleus of a sodium atom could fall apart 'spontaneously', although she could not offer any idea for what holds the nucleus together. She was asked whether her earlier idea about the balloon and wall could help, since if the neutrons were neutral, then they were more negative than the positive protons. Alice rejected this idea. So her understanding of electrostatic phenomena in one context did not seem to be applied in a somewhat different context.

The other relevant context concerned what held solids composed of discrete molecules together. Alice thought that a molecular solid would have 'intermolecular forces holding things together' that might be 'van der Waals' forces' (although she was talking about NaCl which she considered molecular, cf. Table 6.2). These occurred where:

> you've got if you like an electron cloud between, surrounding … each molecule, and as these clouds don't stay in one fixed place, there's always going to be erm sort of momentary areas of dipole. And that's where you get your positive and negatives attracting each other again.

So Alice had learnt about intermolecular bonding based upon induced dipoles that give rise to forces between net neutrally charged objects. However, when asked about how the balloon could stick to a neutral wall, she did not think to apply this idea to the novel context.

This is clearly only one case, and indeed as pointed out earlier in the book, responses given on one occasion can only be taken to reliably report on aspects of student thinking *at that time* and do not necessarily tell us what the same student might have suggested on a different occasion or even had the question sequence had been different on that occasion. Alice knew about induced polarity of neutral molecules, remembered that it was used to explain some types of intermolecular bonding, but during *this* interview did not make a link between that idea and what might happen when a balloon charged by friction sticks to a wall. We might say this schema was not activated in the context of the balloon 'trick'.

It is also possible to conjecture that this might in part reflect a lack of integration between the kind of explanations applied at the level of submicroscopic models and level of everyday phenomena (i.e. the jagged boundary shown in Fig. 12.3).

Alice did not apply her explanation of induced charge separation that she used in discussing molecules to the macroscopic phenomenon of the balloon attached to a wall, just as she did not apply the notion of a neutral object having 'relative' charge which she used to explain the balloon's behaviour to the unexplained stability of a clump of protons and neutrons making up a stable nucleus.

Possibly this suggests something about the way conceptual knowledge was structured in Alice's cognitive system with knowledge of the molecular world somewhat compartmentalised away from macroscopic phenomena: although one would wish to collect more examples, on more than one occasion, before drawing any strong conclusions. As is often the case, a researcher is able to interpret data from research to suggest potential features of a learner's knowledge, but these must remain tentative unless based on a very robust evidence base, given the various caveats that need to be applied in this kind of work. What does seem quite clear, however, is that there is much potential for research to explore aspects of student conceptual integration in science across topics and contexts.

Domain-Based Learning

When Solomon (1983) explored school children's thinking about energy, she proposed that they were thinking in two distinct domains: that their understanding from the lifeworld was quite distinct in nature from their learning of formal scientific models in schools, such that these ideas did not become integrated but rather operated as two independent systems of knowledge between which they have to make transitions to link everyday and formal school thinking. Such a view certainly has implications for how alternative conceptions might best be treated in science teaching, if they are actually represented in cognitive structure quite separately from school knowledge (Claxton, 1986).

One particular issue which could be of particular importance here concerns the (related) questions of (1) the extent to which the human cognitive system should be considered as modular and (2) whether there are domains within which cognition (learning, problem-solving, etc.) occur (Hirschfeld & Gelman, 1994b). In the earlier chapters of the book, the cognitive system was represented as comprised of various components that took on different roles in cognition (e.g. Fig. 4.8). It seems sensible to consider the cognitive system in this way, to the extent that the nervous system seems to naturally have identifiable areas that are specialised to perform specific functions. What is more open to debate is the extent to which this specialisation might operate within different levels of the system, through modules that are self-contained or 'encapsulated' in their operation – that is, modules which process specific input to give particular output, based upon some kind of inherent but perhaps very limited, knowledge base, but without having access to the broader context of knowledge represented more widely in the cognitive system (Hirschfeld & Gelman, 1994a).

Such units certainly seem to operate at some levels in the system, that is, at those levels primarily concerned with perception. So key features of the interpretation of visual input and of auditory input, for example, are quite distinct and seem highly specialised. Young children learn to recognise and distinguish the specific sounds heard in their local language, that is to reorganise the field of vocalisations heard into a particular system of sounds (Kuhl, 2004) and to do this completely automatically without input from 'higher' levels of the system, such as those which interpret the meaning of utterances and which could offer in principle context for supporting the earlier stage of processing. Similarly, the identification of edges and movements in the visual field is carried out by specific neural components without reference to the knowledge base available 'higher' in the system, and we have seen this can lead to optical illusions where we 'know' that what we 'see' cannot be correct (see Chap. 3).

The notion of p-prims, considered as a type of intuitive knowledge component in Chap. 11, assumes processing units of this kind. That *parts* of the cognitive system operate in this way is not contended, but there are different views on the extent to which such specialism may operate, or have implications, at the later ('higher') stages of processing. In particular there is much debate about the extent to which there should be considered to be domains of cognition or knowledge represented in the cognitive system.

Domains of Knowledge

Domains are of course acknowledged in public knowledge systems. This is true both in terms of recognising that certain ideas 'belong to', or are part of the theoretical apparatus of, different academic areas – electromagnetic induction falls under physics; acidity falls under chemistry – and also in terms of wider areas of public knowledge such as cookery, chess and gardening knowledge. However, such recognition could itself be simply something learnt from culture: we learn the publically recognised spheres of activity and so learn to recognise the contexts of experiences – as when, for example, watching television and considering we are watching a 'gardening programme' – and categorise and characterise these experiences within and according to specific expectations from these different areas.

This certainly seems to happen: indeed, school children often seem to find particular difficulty in applying knowledge learnt in the context of one school subject in a different curriculum area, and indeed the 'problem' of transfer of learning is considered a major issue in education (Hammer et al., 2005; Lobato, 2006). Yet it is also sometimes claimed that our cognition is to some extent organised into domains in more fundamental ways: that is, ways that are inherent, rather than a result of individual experience.

That is, it is often suggested that because for much of our evolutionary history, humans and their primate ancestors shared certain key areas of experience where efficient problem-solving might well be a major selective advantage, natural

Table 12.1 Strengths and limitations of modularisation of cognition

Advantages of increased modularisation	Disadvantages of increased modularisation
Speed: quick responses to signals in the environment	Limited: knowledge represented in the module is minimal, and processing occurs without access to more extensive knowledge represented elsewhere in the system
Automation: decision-making does not rely upon referring to executive	Inflexibility: there is no conscious reflection upon, and control of, processing

selection has led to domains of cognition. From this perspective, just as it is an advantage for humans to have genetic instructions to develop specific neural apparatus for interpreting the visual field in terms of edges, object movement, etc., particular ways of processing certain areas of experience may have given sufficient advantage for the genes associated with them to have been selected. Such ways of processing those areas of experience may have become 'hard-wired', and so automatic and fast, and in effect innate (Sperber, 1994). Such efficiency has costs in terms of processing that is inflexible, and not readily overruled, and presumably the trade-off between strengths and disadvantages is a key factor in determining whether such features are selected (see Table 12.1).

Viable Domains in Cognition

Studies of expertise suggest that a wide range of areas of knowledge may act as domains in terms of an individual's learning: so, for example, chess masters may show extraordinary memory for chess positions, whilst showing quite normal memory performance outside their area of expertise. This is understood to largely be a matter of extensive experience and practice within the particular domain (Gardner, 1998). However, it is less clear what domains should be considered to be present inherently in human cognition rather than just as a response to commonalities in experience. Chomsky (1999) has made strong arguments for aspects of language processing to be innate. Three other areas that have been commonly mooted as the basis of cognitive domains are mechanics, natural history and social relations (Sperber, 1994). In each of these areas there is some evidence for innate aspects of development, for example, leading to folk physics, folk biology and acquiring theory of mind. This idea is explored further in the next section (see Chap. 14).

Part IV
Modelling Development
and Learning

Chapter 13
Introduction to Part IV: Development and Learning

This part of the book will build upon the earlier chapters to consider how we can model development and learning in science education. Part II explored the idea of the learner as a cognitive system. It built up a model of how the individual can take in and process information from the environment and then represent the output of cognition in the public space. It was argued that an important feature of the human learner as a cognitive system is the intimate relationship between *the apparatus* of cognition and cognitive *processing*: that is, that the processing of new data through the existing cognitive apparatus has the potential to modify the apparatus itself.

The first chapter in this present part (Chap. 14) will consider cognitive development: the way our cognitive processes become more sophisticated so that the cognitive apparatus of the adult is qualitatively different from that of the neonate. Part III looked in some detail at what we might understand as knowledge in the context of a human learner. In particular, it considered important distinctions such as that between implicit and explicit conceptual knowledge and the notion of how a person's conceptual knowledge can be understood to be organised into a 'conceptual structure'.

The second chapter in this part (Chap. 15) considers how conceptual knowledge changes as a result of cognitive processing. At a gross level, these two chapters are about different things, as one is about how the cognitive apparatus develops and the other is about how we can change the knowledge represented in the cognitive system. However, that is clearly not an absolute distinction given the intimate relationship discussed above.

So, for example, when perceptions are interpreted through what have become automatic processing components such as p-prims (see Chap. 11) developed by abstraction from common general patterns, then we might think of these processing components as part of the cognitive apparatus. However, they are also representing implicit knowledge that has been acquired (constructed) by the system. They are both part of the machinery of knowledge acquisition and also previously established

K.S. Taber, *Modelling Learners and Learning in Science Education: Developing Representations of Concepts, Conceptual Structure and Conceptual Change to Inform Teaching and Research*, DOI 10.1007/978-94-007-7648-7_13,

Table 13.1 A model of the first-order distinction between development and learning

	Cognitive development	Conceptual learning
Is primarily about	Changes in kind of thinking available	Changes in knowledge represented
Depends upon	Largely under genetic control, but supported by normal experiences common to environments where humans develop	Limited by cognitive development, but dependent upon specific resources in the environment
Path	General nature is common for all human development	Highly contingent, leading to somewhat idiosyncratic outcomes

elements of acquired knowledge. As described earlier, a person's memory is not a discrete store accessed by his or her processing apparatus but rather an integrated part of that apparatus. In effect our brains work in a way that prioritises supporting the ability to provide an interpretation of current experience informed by our previous experiences rather than having access to a high-fidelity record of those past experiences.

Despite this complication, it is useful to separate out cognitive development from conceptual learning, at least as a first-order simplification. To some extent there is a link here with nature/nurture issues, as can be seen from Table 13.1.

Discussions of the relative importance of genetic and environmental factors can become emotionally charged as, for example, when considering such matters as the role genetic factors might play in explaining criminal behaviour or the significance of sex for aptitude for science. Arguments setting out whether nature or nurture is more important may sometimes seem to miss the point that there is no absolute way to measure similarity in either genetic make-up or environment conditions. The extent to which genetics or environment is more significant in determining whether individuals can be successful on verbal intelligence items, for example, will look rather different when the 'individuals' are all 'normal' human beings, rather than a mixture of humans, chimpanzees and orangutans. If that seems a contrived example because 'of course' we are only interested in humans, then we need to bear in mind that what makes an individual a human rather than an orangutan is their genetics, and that all people have sets of genes that have a great deal in common.

To offer extreme examples, if a new born baby was ejected into space without an environmental suit, then environmental rather than genetic factors would dominate the course of the hypothetical baby's (tragically short) development and learning compared with other learners in more typical environments; just as an orange tree's classroom learning would be extremely limited by its genetics, regardless of the quality of the 'learning environment' and teaching in the class.

Bearing in mind, then, the proviso that all processes of development and learning are the result of interactions between genes and environment, Table 13.1 offers a model of a first-order distinction between two types of change that human cognition undergoes. These types of processes are commonly referred to as cognitive development and conceptual development, but it can sometimes be useful to actually distinguish between them by using the term development for one and learning for the other.

Thinking of Development as Under Genetic Control

A key distinction then between development and learning is that development can be understood to be largely a 'normal' process that humans will undergo so that in general terms we all follow much the same path of development. So neonates tend to have very similar cognitive abilities, and by adulthood these have generally developed in much the same way in nearly all of us, even if the extent of development may be greater in some than others (see Chap. 14). There is a danger of tautology here in suggesting that all normal individuals follow the same developmental path: whilst excluding from the category 'normal' anyone who does not. Some severely retarded individuals never fully develop normal adult thinking abilities, something that is understood to be due to genetic deficiencies or some form of 'damage' to the cognitive apparatus - such as, for example, might be due to insufficient oxygen reaching the brain during a problematic birth.

We might think of development being 'under genetic control' in the sense that all 'normal' (sic) humans have the genetic resources to facilitate a particular general developmental path subject to typical environmental conditions – where 'oxygen starvation' would represent an atypical environmental condition for human development. So this is certainly not to say that the environment does not play a major role in development, but rather that the necessary features of the environment required to support 'normal' development tend to be common enough not to be a limiting factor, and that, in particular, there is considerable redundancy in the precise stimuli and experiences which are able to provide such support. The developing child needs experience of objects to push and pull and squeeze and so forth to support normal development, but a small selection of the wide range of particular objects potentially available would suffice to do the job.

Thinking of Learning as Learning Environmentally Contingent

By contrast, we can think of learning as being primarily contingent upon specific learning opportunities. Whereas acquiring certain cognitive abilities is supported by a wide range of environments, learning Newton's laws of motion or orbital models of atomic structure, or the theory of natural selection, is much less likely to happen to occur 'by chance' and is indeed only likely to occur in particular cultural contexts where environments are especially engineered to support this specific learning. And even then, as we have seen, acquired understanding will not necessarily match intended understanding.

Again, this is not to ignore genetic factors. Newtonian mechanics, atomic structure and evolutionary theory will only be learnt by those who have developed suitable cognitive apparatus through the genetically driven processes of development discussed above. Our orange tree would not make progress here and indeed nor would our orangutan. So in both the cases of cognitive development and conceptual learning, genetics and environment are essential, but to a first approximation we can identity situations where we can largely take one or the other as given.

Chapter 14
Models of Cognitive Development

Cognitive development is not about the specific conceptions we have acquired, but rather about the nature of our ability to process information. The assumption, and one which is rather well supported by the evidence, is that the difference in intellectual capacity between a child and an adult is not simply that the adult has over time accumulated more knowledge of the world through experience, but rather that there is something qualitatively different in how an adult can build up new knowledge and tackle problems compared to the child. However, the development of a wider base of knowledge and experience in a particular domain *is* necessary to facilitate the most developed styles of thinking in that domain.

According to currently influential models of cognitive development, this is not the only reason why cognitive abilities cannot be completely divorced from levels of knowledge: it is thought that development is itself in some ways facilitated by experience/learning. This is certainly the case when considering the best known theory of cognitive development, as proposed by Piaget.

Piaget's Stage Theory of Cognitive Development

It is difficult to overestimate the importance to this area of scholarship of the work of Jean Piaget. Piaget set up a research programme of 'genetic epistemology', a choice of label that reflected 'a need to study the origins of knowledge' (Piaget, 1972, p. 15). Piaget's background was as a biologist, and this informed a view of the developing child as an organism which needed to adapt itself to the environment. Piaget considered this quite different from just a simple 'stimulus-response' interaction and rather saw the individual as *constructing* solutions to the challenges faced.

Such challenges were experienced as 'tension or disequilibrium' and motivated an active response that would bring about 'a new form of equilibration' (p. 54). A 'new' form, because in responding to environmental situations, the individual is actually changed. In Piaget's scheme there can be 'assimilation' of new input from

K.S. Taber, *Modelling Learners and Learning in Science Education: Developing Representations of Concepts, Conceptual Structure and Conceptual Change to Inform Teaching and Research*, DOI 10.1007/978-94-007-7648-7_14,
© Springer Science+Business Media Dordrecht 2013

the environment, but this has to be 'accommodated' to ensure good fit, not just added by some form of accretion. So Piaget was a constructivist, who considered that the individual was building up an internal model of the world that was subject to ongoing modification developed in response to new information.

Piaget was active in developing his research programme over several decades, and clearly his own ideas changed during this time, but central to much of his work was the notion that cognitive development took place through a series of discrete and invariant stages (i.e. a 'stage theory').

His scheme had various graduations, but the four main stages are labelled:

- Sensorimotor
- Pre-operational
- Concrete operational
- Formal operational

According to Piaget, the newborn baby is in the sensorimotor stage of development, which lasts for something like 2 years. As the name suggests, in this stage of development, children experience their world in terms of their senses and their actions in the world through use of their motor capabilities, that is, pushing, grabbing and sucking.

A model of the human cognitive system was presented in part II that suggested that people have components or modules within the nervous system supporting perception, motor action, etc. (e.g. see Fig. 4.7). It was also noted that as the individual is able to sense the environment in which they act, there is scope for feedback on the effect of actions (e.g. see the introduction to part "Modelling Mental Processes in the Science Learner"). We can consider the child in the sensorimotor stage to have the general components of the cognitive system discussed earlier in the book, so the *overall* model of cognition, when viewed at this level of generality, is no different here than it would be in a teenager or adult Nobel laureate (see Fig. 14.1). What is different, however, in this model is the actual processing possible within these components.

A key point is that through sensing the environment, the child can begin to build up a model of the world as it is experienced, and that by action on the world informed by that model, the child gets feedback on how well the model fits experience and can so modify the model. This perspective *on learning* through experience reflects the pragmatist philosophy of Dewey (Biesta & Burbules, 2003) and acts as a key source for constructivist notions of learning (Glasersfeld, 1989) – including the metaphor of the child as acting in the world as a kind of naive scientist (Driver, 1983) – which have been so influential in science education (Taber, 2009b).

However, for Piaget, a key aspect of his constructivist thinking was that it applied to cognitive *development* as well as learning. So, for Piaget, it was not simply that we build up more nuanced and more extensively tested conceptual models, but that the very apparatus of cognition was built up through action in and on the world. So the experience of acting on the world and building up models using the neonate's cognitive apparatus facilitated the development of new structures which provided more sophisticated cognitive apparatus that would then support a more powerful form of cognition.

Fig. 14.1 Using feedback from the environment during the sensorimotor stage of development

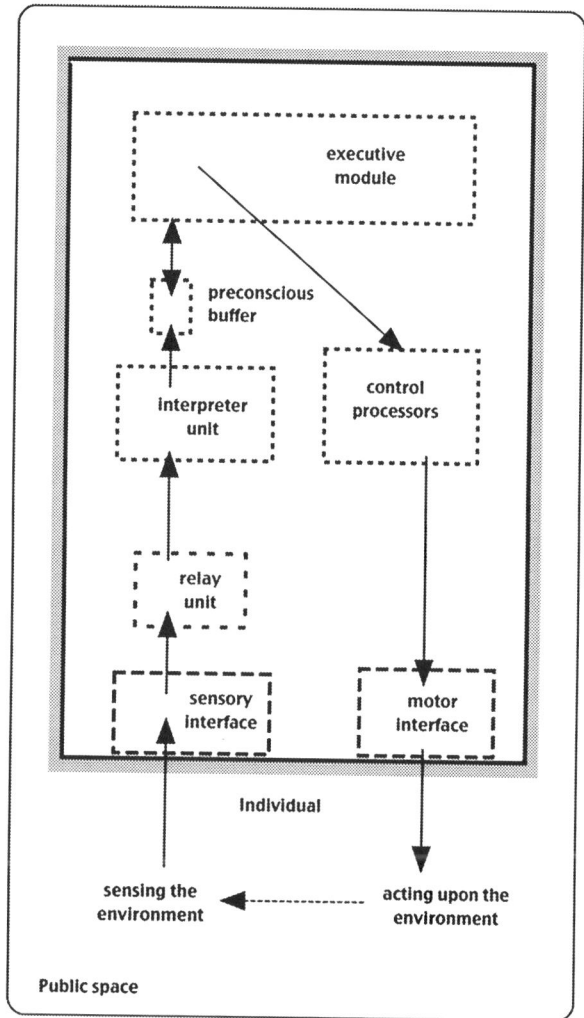

Fig. 14.1 Using feedback from the environment during the sensorimotor stage of development

In terms of Fig. 14.1, although the baby has the basic modules for cognition, it is – through its actions in and experience of the world – modifying the forms of processing that are possible within those component parts of the system, so that the nature of the processing possible becomes more sophisticated. So early in life, the initial 'blooming, buzzing confusion' of experience (to borrow a much-copied phrase from William James) resolves into recognisable objects in the visual field, and hearing discriminations are tuned to the phonemes used in the human language spoken in the child's environment.

According to Piaget, the child's experience in the sensorimotor phase supports the development of new structures that allow transition into a qualitatively different phase, known as the pre-operational phase that might typically last from age 2 to 7 years. During the sensorimotor stage, the child is said to be egocentric, meaning that she or he is unable to consider another's perspective but is tied to considering the world from his or her own position. This can be taken quite literally, so that the child cannot recognise, for example, that another person may not be able to see something that they can see from their position. When shown an incongruous object being hidden in a container, the child will expect that others not present when the object was hidden will know it is there. This was very significant as the Piagetian perspective is that 'an individual who was incapable of appreciating other points of view or of separating self from reality was thereby incapable of construing the world objectively, incapable of genuine deduction, effective communication, or truly moral judgement' (Sugarman, 1987, p. 5).

This egocentrism gradually weakens during the pre-operational stage, but the thinking of children at this stage is said to be 'magical' and they are considered not to be capable of logical thinking. Piaget put great emphasis on the inability of youngsters in this phase to conserve in such tasks as comparing sets of tokens that had different spacing so that one set appeared 'larger' in terms of its extent or understanding that volume is conserved when liquids are poured between vessels. (This is one area where Piaget's work has been critiqued quite severely, and where some details of his reported limits on what youngsters of this age can achieve have been challenged, e.g. Donaldson, 1978). Piaget saw conservation as important, as 'to appreciate the full class of conditions under which an object remains identical with itself is, in his view, to have a concept of object' (Sugarman, 1987, p. 149).

The child's language undergoes considerable development during the pre-operational stage. The child's experience in the world during this stage provides the basis for developing new structures that allow more sophisticated thinking, and the child moves into the concrete operational stage (e.g. from about ages 7 to 11).

Egocentrism no longer restricts the child once he or she has acquired concrete operations, as children in this stage are able to appreciate viewpoints other than their own. According to Piaget, the stage of concrete operations allows logical thinking, and the child at this stage can demonstrate that they conserve, for example, appreciate that increasing the spacing in a line of counters does not increase the number of counters. However, the child is limited to undertaking mental operations relating to actual ('concrete') objects and cannot yet operate on the hypothetical and abstract. Again, however, Piaget's notion was that experiences during the concrete operations stage support the further development of the cognitive apparatus to support the next stage, formal operations.

Formal operations were said to typically develop from about 11 years of age during the secondary school years – although other studies suggest this may be optimistic for many populations – and to allow abstract reasoning. For Piaget (1970/1972, p. 47), the 'chief characteristic' of formal operational thinking was the 'capacity to deal with hypotheses instead of simply with objects', as the learner at the stage of formal operations could carry out operations on operations themselves. According to Piaget (p. 47)

> It is this power of forming operations of operations which enables knowledge to transcend reality, and which by means of a combinatorial system makes available to it an infinite range of possibilities, while operations cease to be restricted, as are concrete operations, to step-by-step constructions.

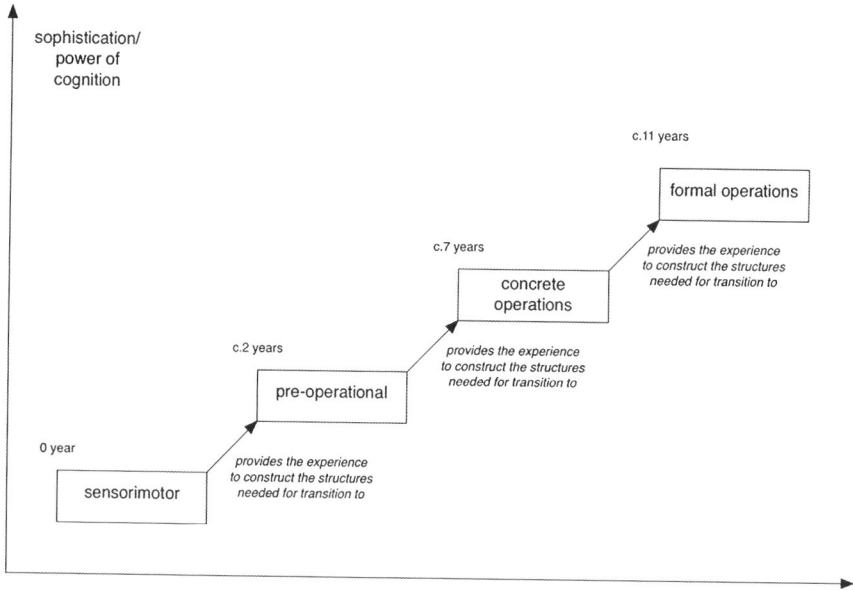

sophistication/
power of
cognition

c.11 years

formal operations

c.7 years

*provides the experience
to construct the structures
needed for transition to*

concrete
operations

c.2 years

*provides the experience
to construct the structures
needed for transition to*

pre-operational

0 year

*provides the experience
to construct the structures
needed for transition to*

sensorimotor

Fig. 14.2 Piaget's model of cognitive development occurring in stages

A Constructivist Model of Cognitive Development

A key feature of Piaget's stage theory was that he saw cognitive development in terms of the individual developing their own cognitive apparatus through their interaction with the environment: that each stage of development facilitated new types of thinking which enabled more sophisticated modelling of the world and so prepared the way for the next stage (see Fig. 14.2).

Relevance of Piaget's Theory

Piaget's work has been widely criticised from various perspectives (Donaldson, 1978; Sutherland, 1992). Details of his scheme have been criticised – for example, in terms of limitations of the tasks he employed to give insight into what youngsters might be capable of, his focus on the individual working alone and the unrepresentative nature of his sample as a basis for making pronouncements about the ages at which children reach different stages given that he was proposing a theory that claimed to be about human development regardless of national/cultural context. Despite this, Piaget's programme was widely extended by others (Elkind & Flavell, 1969; Modgil, 1974), and his ideas have been very influential in science education (Bliss, 1995).

Piaget's theory of development has very clear implications for science learning, as it suggests that learners are inherently limited in the types of thinking they can display, according to their 'stage' of development. So, for example, Shayer and Adey, using a more finely graduated version of Piaget's scheme than the 'first order' model discussed here, published a critique of the school-leaving examinations curriculum for science in England arguing that many of the concepts that 14–16-year-olds were asked to study would seem to require cognitive development beyond that which students of that age would have reached (Shayer & Adey, 1981).

Piaget's Theory and Models of the Cognitive System

One area where Piaget's work has caused considerable discussion is the extent to which it implies that cognition is due to a set of capabilities that are independent of context. That is, whether the apparatus of cognition is a general purpose (content-free) processing apparatus that can be applied to any subject matter along the lines of the 'science-as-logic' notion of scientific thinking skills met in Chap. 7. Piaget's stage theory considered cognitive development to offer new structures of thought, which once attained could be applied in different areas, and so provided basic processing capabilities that were not tied to specific areas of experience or knowledge.

However, a problem for Piaget's theory in this regard was that when children were tested, they often demonstrated attaining a stage of development according to some tests, which was not initially reflected in performance in parallel tasks set in different contexts. Yet similar levels of performance across contexts would be expected if cognition was primarily resourced by a general-purpose problem-solving apparatus. This phenomenon was referred to in the Piagetian programme as horizontal décalage (Flavell, 1963) which meant there was a temporal lag before new cognitive abilities could applied across all contexts. This was 'horizontal', as opposed to vertical décalage which referred to the time lag between the appearance of different cognitive abilities on related tasks; that is, a child learns to *recognise* that an object when seen from different viewpoints is indeed the same object some time before it is able to mentipulate a representation of the object to *predict* how it will appear from other viewpoints.

An example of horizontal décalage was that when children were tested on conservation talks by being asked about a ball of clay that was then reshaped, children tended to be able to report that the amount of clay remained the same after reshaping as that in a reference ball, acknowledged to be the same before the reshaping, about a year earlier than recognising that the weight would be conserved under the same transformation (Morton & Munakata, 2002).

The issue of horizontal décalage suggests that it is not appropriate to see cognition purely in terms of a general purpose processing system that can operate on any data with similar efficiency. This does not *necessarily* mean that there is not general purpose processing *as part of* the cognitive system, as horizontal décalage could reflect differences relating to other stages of processing prior or subsequent to processing within such general apparatus. However, this does mean that Piaget's stage theory needs to be interpreted with care, as the implications of a child's Piagetian stage for the tasks they can successfully complete unsupported are not always straightforward.

Beyond Formal Operations

Piaget's programme of research was heavily focused on the development of logical, mathematico-scientific thinking. The individual who has fully acquired formal operational thinking should be capable of understanding 'fair' testing, and so designing and interpreting experiments with controls, as well as following the logic of mathematical proof. However, arguably, logical thinking that does not go beyond what can be demonstrated through accepted premises or straightforward interpretation of unambiguous evidence is not the most sophisticated type of form of, and so 'highest' level of, thinking that people use – both within and outside scientific work.

Dealing with incomplete, uncertain and contested situations can be intellectually more demanding than solving logically closed problems. This is clearly significant for science education. For example, it is just this kind of thinking which is needed when a science curriculum prescribes learning about the interactions between science and societal issues (Levinson, 2007; Sadler, 2011). Sadler, Klosterman and Topcu (2011, p. 48) refer to a construct they label 'socio-scientific reasoning' that involves:

- Recognising the inherent complexity and multifaceted nature of socio-scientific issues
- Analysing issues from multiple perspectives
- Appreciating the need for ongoing inquiry related to such issues (n.b., current knowledge is not final)
- Employing scepticism in the review of information presented by parties with vested interests

Some commentators have suggested that Piaget's model needs to be extended to include 'a fifth stage' of post-formal operations (Arlin, 1975; Commons, Richards, & Armon, 1984; Kramer, 1983).

Sternberg (2009a), a leading intelligence theorist, has, for example, mooted the idea that wisdom, as 'a very special case of practical intelligence, one that requires balancing of multiple and often competing interests' (p. 363), could be viewed 'in terms of post-formal-operational thinking, thereby viewing wisdom as extending beyond the Piagetian stages of intelligence…Wisdom thus might be a stage of thought beyond Piagetian formal operations' (p. 357). Given the ages proposed by Piaget as associated with attainment of the main stages of his theory, it is perhaps not surprising that research suggests these ('post-formal') intellectual skills are not likely to be fully developed by many school-age learners.

Perry's Model of Intellectual and Ethical Development

A particularly significant longitudinal study was carried out under the direction of William Perry (1970). Perry interviewed undergraduate students at the elite Harvard and Radcliffe colleges. He produced a scheme describing positions learners could take during 'intellectual and ethical' development, which has been widely recognised as significant for those undertaking university teaching, including in the sciences (Finster, 1991). Perry's scheme has the nature of a stage theory, similar to Piaget's scheme, in the sense that

> The sequence of structures we observe in our data qualifies as a 'developmental' pattern in the special sense originally derived from biology in that it consists of an orderly progress in which more complex forms are created by the differentiation and reintegration of earlier simple forms (Perry, 1970, p. 48).

Perry's scheme involved three parts, each of which involved three 'Positions' (as in Table 14.1). Unlike in Piaget's theory, Perry saw possibilities for some variation in an individual's route through his scheme – at least that individuals might avoid progressing through the normal stages through processes he labelled 'retreat' and 'escape'. One commentator explained that 'he conceptualizes "backsliding" as a normal part of development' (Mary Belen reported in Ashton-Jones & Thomas, 1990, p. 287).

Perry discussed how his scheme

> begins with those simplistic forms in which a person construes his [or her] world in unqualified polar terms of absolute right-wrong, good-bad; it ends with those complex forms through which he [or she] undertakes to affirm his [or her] own commitments in a world of contingent knowledge and relative values. The intervening forms and transitions in the scheme outline the major steps through which the person, as evidenced in our students' reports, appears to extend his [or her] power to make meaning in success in confrontations with diversity (Perry, 1970, p. 3).

Perry characterised the core changes in each of the three main parts (Perry, 1970, pp. 64–65):

- During the first part, 'a person modifies an absolutistic right-wrong outlook to make room, in some minimal way, for that simple pluralism we have called Multiplicity'.

Table 14.1 The positions taken by learners in terms of the nature of knowledge, according to Perry

Part	Position
The modifying of dualism	1. Basic duality
	2. Multiplicity pre-legitimate
	3. Multiplicity subordinate
The realising of relativism	4. Multiplicity correlate or relativism subordinate
	5. Relativism correlate, competing or diffuse
	6. Commitment foreseen
The evolving of commitments	7. Initial commitment
	8. Orientation in implications of commitment
	9. Developing commitment (s)

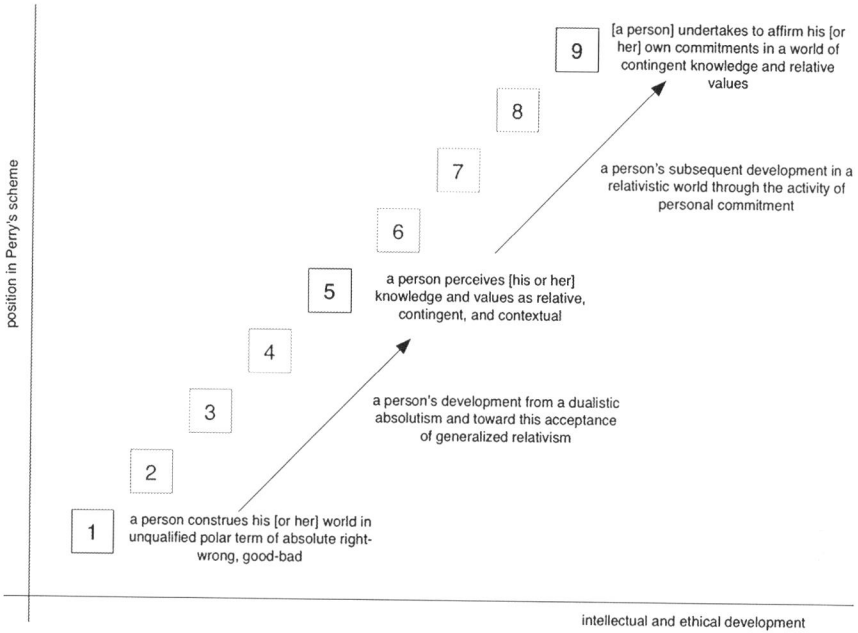

Fig. 14.3 Main themes in Perry's scheme of intellectual and ethical development

- During the second part, 'a person accords the diversity of human outlook its full problematic stature, next transmutes the simple pluralism of Multiplicity into contextual Relativism, and then comes to foresee the necessity of personal Commitment in a relativistic world'.
- The three positions in the final part 'trace the development of Commitments in the person's actual experience'.

Perry also described how his scheme could be considered to be centred on position 5, as this was *preceded by* coming to accept relativism, and then *followed by* responding to that relativism by developing personal commitments (as shown in Fig. 14.3).

What is especially noteworthy about Perry's research is that in working with some of the most gifted and/or educationally advantaged young men and women, Perry found many students struggled to deal with teaching that asked them to (a) consider multiple perspectives where there was no clear absolute 'correct' viewpoint; and (b) move past a simple relativistic notion that given there is no clear 'right' perspective, everyone is equally entitled to their own opinion, and all opinions are therefore of the same merit. The latter is necessary to be able to acknowledge the importance of context and one's own stance, whilst still offering some form of argued evaluation of the merits of different perspectives – something that is the standard fare of areas as diverse as political debate and art/literary criticism.

Perry's Model of Development and Models of Cognitive Processing

Perry's work, whilst clearly having some similarities with Piaget's, cannot easily be seen as simply extending the Piagetian model. Piaget's focus on logico-mathematical thinking assumed distinct stages in cognitive development that were invariant and resulted in transformation of the structures of thinking that were irreversible. That is, an adult who has passed through normal development is no longer able to revert to experience what it is like to be a child at the sensorimotor stage of development. Of course we can try and imagine what it must have been like, but we cannot actually experience the world that way – bypassing the higher structures of mind – as the apparatus which provided that childhood experience has itself been transformed through developmental processes. Indeed, in trying to imagine what it must have been like, we are calling upon cognitive resources that would not have been available to us when we were actually experiencing the world that way. This is why few of us report having clear memories from this early stage of our lives.

Perry, however, refers to 'positions' along a developmental path: language which seems to reflect an element of choice and his notion of 'retreat' suggest that it is perfectly possible for us to adopt patterns of thinking which are less advanced than those we are capable of; in a sense the person makes a choice at some level not to engage in the most sophisticated forms of thinking available, at least with regard to certain aspects of our experience. Perry (1970, p. 204) suggested 'one may retrogress at any point or range of development in our scheme'. Arguably, the 'decision' to retreat, even if largely a result of preconscious thinking, itself reflects quite sophisticated processing. However, the key difference with the Piagetian theory remains as in Perry's scheme earlier positions remain available to the individual.

Other Studies

One criticism of Perry's work is that his sample was disproportionately male. He interviewed more male students (from Harvard) than females (from Radcliffe) and largely drew on the data from male students to illustrate his scheme. Perry acknowledged that his 'judges' who rated the interview transcripts against the scheme "engaged in a lively discussion of the differences between men and women" but concluded that any "differences were evident in the content and manner of the students' reports rather than in those structurings of experience relevant to the developmental scheme" (p. 17): in effect these were seen as differences of style within a common overall developmental pattern. Despite this claim, the gender imbalance in the sample interviewed by Perry as well as in other studies by other researchers led to a programme of work seeking to explore what became known as 'women's ways of knowing' (Hofer & Pintrich, 1997). Whilst this work revealed important patterns in women's thinking (Ashton-Jones & Thomas, 1990), these could be seen as occurring in conjunction with the general pattern of development found by Perry (Clinchy & Zimmerman, 1985).

Another well-known and highly influential scheme describing the development of ethical and moral reasoning was developed by Lawrence Kohlberg, who like Perry found that development extended well beyond adolescence (Crain, 1992). Kohlberg was strongly influenced by Piaget's work and built upon the work Piaget had undertaken on the moral development of the child (Piaget, 1932/1977).

Kohlberg's scheme, like Piaget's, involved a sequence of stages, which he saw 'as representing increasingly adequate conceptions of justice and as reflecting an expanding capacity for empathy, for taking the role of the other' (Kohlberg & Hersh, 1977, p. 56). Kohlberg suggested that development continued well beyond the age at which students completed compulsory schooling with 'new stages developing only in adulthood' (Kohlberg, 1973, p. 500).

Reviewing work related to development in thinking patterns, Deanna Kuhn (1999) has argued that critical thinking should be understood in a developmental framework, and she proposed four stages:

- Realist: where reality is directly knowable, and certain knowledge is acquired from an external source, and where assertions are copies of external reality
- Absolutist: where reality remains directly knowable, and certain knowledge is still acquired from an external source, but where assertions may be correct or incorrect, so people can have false beliefs
- Multiplist: where knowledge is generated by human minds (and so is uncertain), and individuals freely form their own opinions of the way things are
- Evaluative: where knowledge is generated by human minds (and so is uncertain) but where different assertions may be evaluated and compared by argument from evidence, using established criteria

The final stage of Kuhn's model clearly links to the modern (post-positivist) understanding of scientific knowledge and the key challenge to science education to teach scientific knowledge as capable of being simultaneously provisional and yet sufficiently robust as to be able to inform rational decision-making. However, Perry's work would suggest that many college students are still at the relativist 'multiplist' stage, where the loss of absolute knowledge is understood as necessitating intellectual anarchy.

Eastwood, Schlegel and Cook (2011) have reviewed development frameworks for college students and adults proposed by different researchers from a range of perspectives, with a view to considering how such schemes may inform the development of socio-scientific reasoning. They suggest there are sufficient 'similarities among their findings [to] reinforce and validate the existence of particular trends' (p. 93). Eastwood and colleagues suggest that generally such schemes show (p. 93):

- Early stages 'characterised by conceptions of knowledge as absolute and derived from authority, understanding of reality as directly observable, and difficulty recognising complexity or different perspectives'
- Middle stages 'characterized by perception of complexity, uncertainty and multiple perspectives, although reasoning may be inconsistent and decisions or commitment may be hindered by complexity'
- Highest levels where 'knowledge is seen as complex, uncertain, and a product of inquiry. Individuals apply consistent criteria to form evidence-based decisions and recognize and incorporate multiple perspectives in their reasoning'

All these various studies converge on the notion that thinking skills develop over time, with the caveat that many learners, especially school-age learners, may not be capable of readily demonstrating the types of thinking which may be necessary to fully engage with teachers' expectations in science, particularly in terms of epistemological sophistication.

Social Processes in Cognitive Development

The notion of development implies a sense of programmed change, and indeed the Piagetian stage model and other similar schemes are considered to report normal stages through which people have the potential to pass through due to their common genetic inheritance. However, development occurs in a suitable environment and can be delayed or even brought to a stop if there is not suitable experience. In Piaget's model, experience of acting in, and receiving feedback from, the environment is considered to be an essential part of the developmental process – providing the disequilibrium that was seen as the driver for internal changes. This has been commonly seen in terms of physical environment: experience with objects and materials leading to the acquisition of conservation rules, for example.

However, a sociocultural, or socio-historical, perspective emphasises the social and cultural environment that supports development. Vygotsky (1978), in particular, argued that the individual builds up internal structures of thought reflecting what is experienced in the social plane. The individual has to experience these ways of thinking modelled by others before they can be internalised and so become available to be used independently. From this perspective, the rate, and even nature of development, is culturally dependent to the extent that different cultures may offer different resources for supporting development (Luria, 1976).

Bruner (1964, p. 1) conceptualised cognitive development in terms of 'a series of technological advances in the use of mind' which were largely 'transmitted with varying efficiency and success by the culture'. Bruner, strongly influenced by Vygotsky, referred to 'the role of heuristics in the growth of perception and problem solving' that children 'picked up' from the culture and which acted as metaphorical 'crutches' to support growth (Bruner, 1967b, p. xi). Bruner saw a key role for the forms of representation available to the developing child (cf. Chap. 11): initially just enactive through physically acting in the world, then through imagery ('ikonic'), and then symbolically through forms of language (Bruner, 1967a).

Modelling the Development of the Cognitive Apparatus

Bruner (1960, p. 33) famously invoked 'the hypothesis that any subject can be taught effectively in some intellectually honest form to any child at any stage of development'. In the context of the ideas reviewed in this chapter, this needs to be seen as advice for the teacher to seek to match the level of teaching to the intellectual development of the learner, but without oversimplification that loses the essence of what is being taught.

The 'cognitive system' model of the learner presented earlier in the book offers limited support for the teacher or science education researcher in this context, as cognitive development does not concern gross changes in the overall structure of the system such as the addition of major new system components. Rather development of the system through interaction with its (physical and social) environment refines its operation. New structures are built into the system, but at a finer level of detail than the gross system components represented in the figures used earlier in the book (e.g. Fig. 4.11). So to allow for stages of cognitive development within the model of the cognitive system, it would be necessary to work with a much finer grain model, which identified more detailed features of processing and more specific processing units within the basic processing modules.

That will not be attempted here, as the current state of knowledge does not provide a firm basis for offering definitive models. Perhaps such fine grain models will prove difficult to devise at the system level of analysis, without blending somewhat into the neurological (physical, cf. Chap. 3) level. So, for example, Morton and Munakata (2002) have used a neural network approach to consider how developments in parts of the prefrontal cortex allow children to overcome 'perseveration' – where previously appropriate responses continue to be made when no longer appropriate after a task has been changed. Although a child seems to understand the new task (e.g. sorting cards by shape) they seem unable to switch from previous task behaviour (e.g. sorting cards by colour). This is considered to be a problem due to 'latent' memory traces formed in posterior cortex taking precedence over the representation of the current task in the prefrontal cortex. It is argued that development of the prefrontal cortex provides the ability to better coordinate the 'latent' and 'active' memories and so offers the flexibility to switch to the behaviour needed for the new task.

Domains of Cognition and Modularity of Mind

One topic of considerable interest is the question of the extent to which cognition may have domain-specific features reflecting modular aspects at the finer-grained level of cognitive architecture (see Chap. 12). In the schematic models of the cognitive system presented earlier in the book, different system components have been shown as blocks with the assumption that it is helpful to see cognition as based around apparatus made up of somewhat discrete components, such as working memory, for example, with information being routed between components during (cognitive) processing. However, it was also suggested that it was not always clear what comprised discrete components of the system: for example, research has suggested that it is not appropriate to consider long-term memory (LTM) as a function of system components discrete from the parts of the system involved in interpreting and making sense of experience. That is, it may not be appropriate to think of an interpreter unit with draws upon a separate LTM store when interpreting information from sensory units in the system. Rather, it is suggested that the functions of LTM and interpreting experience are more intimately connected (cf. Fig. 5.2), with memory being modified in the process of interpreting new experience.

An analogy here might be made with the way coding takes place in research studies. In some studies, such as those with a confirmatory purpose, looking to test some hypothesis, a formal coding system is developed to code data in advance and is used to interpret data and remains fixed through the analysis process. The fixed coding system may be seen as analogous to a memory of prior experience that is being used to make sense of current experience. By contrast, the iterative coding processes used in grounded theory (GT) methodology (i.e. the constant comparison method), for more open-ended research – discovery studies, where the existing state of knowledge does not provide a clear conceptual framework for making sense of data – may be seen as analogous to the way LTM acts in human cognition. In GT studies the coding system itself is constantly modified as new data is considered: the codes are the GT analyst's 'memory' of how previous data has been understood, but that memory is always in flux. Where the analogy breaks down is that in GT the analyst is required to constantly return to the previously coded data to reconsider it in terms of the modifications to the coding scheme. By contrast, in human cognition, the original (sensory) data is no longer available to the system, just the codings as they are *now* understood in terms of the shifting coding scheme. As pointed out in Chap. 5, human memory appears to be adapted to make the best available sense of current experience rather than to keep accurate records of the past – *LTM is primarily an organ of interpretation, not an organ of record.*

To What Extent Are Our Minds Modular?

The question of the extent to which mind might be 'modular' was explored by Fodor (1983). For Fodor (1985) the notion of a module implies a system which is 'informationally encapsulated' in the sense that it has access to its own 'proprietary database' to inform the processing of input, but not to knowledge in the wider system outside of that database. The existence of modules in that sense can explain, for example, why we 'see' optical illusions despite knowing that is precisely what they are (see Chap. 4) – our perceptual system reports seeing a face in the clouds because it has matched sensory input to its database, and so we become conscious of seeing the face, although we also 'know' we are not seeing a face at all based on information elsewhere in the cognitive system that the perceptual module cannot access.

In these terms, aspects of our preconscious processing, and the implicit knowledge such processing accesses, can be modular, but this clearly does not apply to our explicit knowledge. For we have executive control over explicit knowledge that allows us to access it – at least in principle, if not always readily – and mentipulate it in working memory. Much human creativity, including in science, depends upon the non-modular nature of our explicit knowledge, allowing us to compare and make links between intrinsically very different areas of our knowledge base.

Are There Inherent Domains of Knowledge?

Although our explicit knowledge is not encapsulated into bound modules, there have been claims that human cognition is structured through domains – in that our cognitive architecture recognises different areas of experience that to some extent can be considered as discrete. This can be understood to have some evolutionary basis in at least two possible ways.

One possibility is that there may have been some selective advantage in recognising different areas of experience and to some extent partitioning them in mind. There could be efficiency advantages, for example, in restricting a search for relevant information if the entire knowledge base does not need to be searched, and it may be easier to select between different ways to process information if different aspects of experience can be readily classified as falling under particular domains. We can consider that there may be a trade-off here between greater efficiency and reduced creativity.

Alternatively, there may be no or limited selection advantage to domain-based cognition, but it could be an artefact of evolutionary history: that is, domains in human cognition that we have now could be contingent upon encapsulated modules acting in preconscious thinking upon which conscious cognitive processes have built during evolution. Certainly a number of such domains have been proposed: there is commonly considered to be at least the three domains relating to common-sense mechanics, to folk biology and to theory of mind (folk psychology). These would each deal with different areas of experience – the mechanical properties of objects in the environment, the classification and properties of fauna and flora and social relationships within the tribe, family group or other community – and could each relate to the development of a largely discrete knowledge base.

Demetriou's Model of the Mind

One model of mind which has been informed by much research and includes both general and possible domain-specific features has been developed by Demetriou and colleagues and summarised by Demetriou and Mouyi (2011). This model considers there to be three 'functionally distinguished levels' (p. 70):

- A basic 'processing potentials level' that determines such matters as speed of processing and working memory capacity and constrains the operation of the other levels
- A hypercognitive system (with consciousness as an integral feature) concerned with self-monitoring, self-representation and self-regulation
- A number of specialised domains of thought concerned with aspects of the individual's environment

The term 'hyper'-cognition is considered more appropriate than the more usual metacognition (cf. Chap. 7) as it implies a level that overlays cognition.

For Demetriou and Mouyi, the domains 'specialize in the representation and processing of a particular type of information and relations' (p. 72) and so involve specialised mental operations and processes relating to the particular feature of experience concerned. Their mooted domains are:

- The categorical system (for identifying similarities and differences; classifying)
- The quantitative system (dealing with numerical processing)
- The causal system (identifying cause and effect)
- The spatial system
- The propositional system (considering logic)
- The social system

These domains are themselves considered to be comprised of three levels: (1) core processes; (2) mental operations, processing skills and principles; and (3) acquired knowledge and formulated beliefs.

The core processes are considered to be innate, or to develop early in life, having evolved through natural selection, and to be encapsulated so that no further learning at this level of processing is possible once they are formed. They support the development of the other two levels.

The system accumulates knowledge through an iterative process (cf. Chap. 5):

> Each system involves knowledge accumulating over the years as a result of the interactions between a particular system and its respective domain and beliefs formulated as a result of the exposure to and the experiences from the domain with which it is affiliated (Demetriou & Mouyi, 2011, p. 73)

The systems are considered to be both domain specific in terms of the areas of experience addressed and 'symbolically biased' in that they each use symbolic systems 'most conducive to the representation of the domain's own elements, properties and relations' (p. 72).

The systems are also considered to be 'procedurally specific in the sense that they involve mental operations and processes which reflect the peculiarities of the elements and the relations that characterize a specific part of reality' (p. 72). These operations develop from the interaction of the core processes with experience of the domain. They are considered to 'emerge as a process of differentiation and expansion of the core processes' (p. 73) when those core processes are considered inadequate to support understanding of experience and problem-solving within the domain. This intermediate level remains open to further change, with processes being modified in response to new knowledge to both more efficiently process new information from the domain and better handle the knowledge from the domain already represented in the system.

These changes are supported by the hypercognitive system, which allows reflection on, and monitoring of, aspects of cognition. Demetriou and Mouyi divide this system (p. 77) into what they call 'working hypercognition' – which is involved in directing attention, formulating plans, monitoring progress and feeding back based on discrepancies between current and desired situations and overall evaluation – and 'long-term hypercognition' which retains representations of past cognitive experiences (mediated by working hypercognition) as a knowledge base for informing future hypercognition. There is a hypercognitive process Demetriou and Mouyi label 'metarepresentation' (p. 91) which acts to classify, compare and model different mental experiences and which is considered to contribute to enhancing understanding and improve problem-solving.

Demetriou and colleagues' system therefore comprises both innate features and contingent features that are 'bootstrapped' on the innate features and the interaction of domain-specific and general cognitive operations – with a hypercognitive system having a major integrating role. Their research has found evidence of phases of development in aspect of hypercognition occurring typically over the age ranges 3–7, 8–12 and 13–18 years. This seems reasonably consistent with the Piagetian stages of pre-operational, concrete operational and formal operational cognition discussed earlier in the chapter.

Reflecting on the Demetriou Model

The Demetriou model is an example of an attempt to characterise the human cognitive system, supported by a programme of research (Demetriou & Mouyi, 2011). A key feature of the model, according to its developers, is that it provides a basis for taking a position on the long-standing debate in intelligence theory over whether intelligence is primary a general function of processing underpinning a person's cognition or rather is better understood as having quite discrete facets that develop and act largely independently (Gardner, 1993; Lawson, 1985; Sternberg, 1980, 2009b). According to Demetriou and Mouyi (p. 98), their studies support neither the case for a general factor of fluid intelligence which is primarily responsible for an individual's intelligence nor the notions of a set of largely discrete intelligences in different areas of experience, but rather something more complex with both common features that act across all aspects of cognition and more specialised domains that develop independently in the individual.

The Demetriou model is of particular interest, as it seems to link well with a wide range of other research and scholarship that could inform research in areas such as science education. It therefore offers an example of what a more detailed model of human cognition, suitable for informing educational work, might look like. It includes innate elements, reflecting much evidence that aspects of human cognition are in effect 'hardwired' (e.g. see Chap. 11). Yet it also fits with a broad constructivist notion that development involves iterative stages, where the current capabilities of the cognitive system provide a based for interpreting experience of the world in ways that facilitate cognitive development. Indeed, as noted above, the age ranges suggested for the cycles of hypercognitive development seem quite well matched to Piaget's stages of pre-operational, concrete operation and formal operational thought, as discussed earlier in the chapter.

Moreover, the model proposed by Demetriou and Mouyi includes both system components which operate in a largely automatic way (i.e. the core processes, could be candidate mechanisms to explain p-prims and intuitive rules; see Chap. 7) which are not open to conscious control or development and components which are accessible to consciousness and which can be developed through experience. Demetriou's model also includes a number of distinct domains of cognition, including a social domain that would seem to link with the development of 'theory of mind' (see Chap. 2).

Domains of Cognition

Demetriou and Mouyi's (2011) model does not however include specified domains relating to physics/mechanics or natural history/folk biology that have commonly been mooted and which would seem to be of particular potential significance to science education. The possibility of there being some inherent aspects of common alternative conceptions in science, in that these may be linked to some aspect of normal cognitive development, could be of importance in considering how teachers should plan for and respond to some of the alternative conceptions students commonly acquire. Certainly, these two areas in particular, folk physics and folk biology, offer common ways of thinking that are distinct from the canonical scientific concepts and which seem to occur in different cultures.

Folk Physics

Research has found that aspects of mechanics are counter-intuitive to most learners, and the same common alternative conceptions seem to be very widespread. In particular, a way of thinking reflecting the Aristotelian notion of impetus seems to form the basis for a very common alternative conceptual framework (Gilbert & Zylbersztajn, 1985). People tend to expect moving objects to soon exhaust whatever gives them motive power, quite in disregard of Newtonian principles. Arguably this could just derive from common experience: all humans soon discover that in practice an inanimate object does not usually start to move unless acted upon, and then it will soon stop moving, depending upon how much effort one puts into bringing about the movement. However, there is an also an argument that this pattern is so common that natural selection might have acted to select for the development of innate cognitive elements that in effect have 'knowledge' that this is how objects behave. Although this is not recognised as a discrete domain in Demetriou's model, folk physics could be considered from that perspective to have in part at least developed with support of the mooted 'causal system', which identifies links between cause and effect.

Another common feature of everyday thinking about physics, or folk physics, is the expectation that an object moving in a circular path will continue to do so; that is, that there is a sense in which circular motion is a form of 'natural' motion.

This is a little more difficult to explain in terms of everyday experience – an object swung in an arc clearly does not continue to orbit when we cease holding it – or the discus and hammer competitions at athletics events would certainly be very different. However, if this is not easily explained as due to common human experience in the world, it seems just as difficult to imagine how it was selected for to become part of an innate mechanics module that would support a somewhat distinct domain of cognition.

Indeed, there is some evidence that expectations of continued circular motion in the absence of an accelerating force are actually greater among school students than pre-school-age children (Kaiser, McCloskey, & Proffitt, 1986). However, that does not rule out this common conception as being due to the activation of an innate mechanism that was selected for some other reason. For example, there could be evolutionary value in a mechanism which spots patterns in movement and predicts future positions of a moving object based on those patterns.

Folk Biology

Another area of science where there is suggestion of innate mechanisms at work is that of natural history: the classification of the biota. Human societies traditionally had good reason to pay careful attention to the similarities and differences between living things in their environments when different species might be predators or prey, food plants or sources of toxins. It might therefore make sense that natural selection would operate to support a strong ability to classify organisms into different species. We certainly find that the recognition of 'natural kinds' of living things (Keil, 1992) seems to impede an acceptance that species have evolutionary origins and actually blend into each other over geological timescales (Taber, 2013g). Generally recognised natural kinds are often not actually species (although they can be, or even varieties, in some cases) but less distinct groupings from a biological perspective (Medin & Scott, 1999). Nonetheless, the general idea that individual living things belong in distinct ontological groupings, each reflecting some specific 'essence', seems very strongly ingrained in our thinking (Gelman, 2009; Keil).

However, the evidence is that such innate mechanisms may go beyond the existence of a kind of a discrete categorical system as recognised in the Demetriou model. Rather, it seems that humans the world over seem to classify living things in remarkably similar ways considering that the actual sample of biota met and considered significant varies with the local ecology as well as cultural traditions. Despite such differences in salience, there seem to be strong commonalities in how people from different parts of the world recognise natural kinds. Hirschfeld and Gelman (1994a, p. 19) report that 'the basic principles of classification of biological kinds are extremely stable over significant differences in learning environment and exposure'.

Development of Domains

There seems to be a lot more to be learnt about this aspect of cognition: in particular, the precise commonalities and extent of innate aspects of specialised processing. One question concerns the number of domains that may have some sort of innate basis. It is also important to understand how the existence of such innate tendencies might influence knowledge representation and the ability to integrate knowledge across domains and so seek greater coherence across the knowledge represented within any particular individual's mind. That is, key questions for research would include (1) to what extent is any person's knowledge of the world a matter of development and (2) how do features of development channel and constrain individual learning – so that the individual can form knowledge based on their own personal experiences in the world.

Chapter 15
Modelling Conceptual Learning

The 'learning paradox', as it has come to be called…poses a fundamental problem for constructivism: If learners construct their own knowledge, how is it possible for them to create a cognitive structure more complex than the one they already have?…The only creditable solutions are ones that posit some form of self-organization…At the level of the neural substrate, self-organization is pervasive and characterizes learning of all kinds… Explaining conceptual development, however, entails self-organization at the level of ideas – explaining how more complex ideas can emerge from interactions of simpler ideas and percepts. (Scardamalia & Bereiter, 2006, p. 103)

Theories of cognitive development discussed in the previous chapter focus on the way the apparatus of cognition develops (Leslie, 1984), rather than on how specific learning occurs. Although research on cognitive development has certainly been of interest to science educators (Bliss, 1995; Shayer & Adey, 1981), in recent decades there has been more interest in issues of specific learning and – in particular – conceptual change (Vosniadou, 2008b). In part this reflects an understandable division of labour, with developmental psychologists and other cognitive scientists primarily interested in development and mechanisms of learning and science teachers and science education researchers primarily interested in building up a body of knowledge that can inform science teaching. In that context, the work of Piaget and Perry (see the previous chapter) may seem to largely illuminate constraints on learning and so perhaps inform sensible choices of curriculum aims for different learner groups, rather than offer guidance on how to develop effective subject pedagogy in the science disciplines.

As I have discussed elsewhere (Taber, 2009b), from the late 1970s a research programme developed in science education commonly identified as 'constructivism' or the alternative conceptions movement (Gilbert & Swift, 1985), which focused on the contingent nature of learning, and in particular how new learning is shaped by current knowledge (Gilbert, Osborne, & Fensham, 1982). Piaget's theories were certainly constructivist, in the sense that he considered the operational stages of development to reflect structures of thought that were built upon and through earlier stages, and which provided the apparatus for developing new ways of thinking that could allow higher levels of thought to emerge (see the previous

K.S. Taber, *Modelling Learners and Learning in Science Education: Developing Representations of Concepts, Conceptual Structure and Conceptual Change to Inform Teaching and Research*, DOI 10.1007/978-94-007-7648-7_15,
© Springer Science+Business Media Dordrecht 2013

chapter). Such a model might seem to suggest that science teaching should be straightforward as long as the material to be taught was delayed until students reached the necessary level of operations. Given this, careful conceptual analysis could determine the sequencing of instruction that would facilitate attentive students to acquire canonical knowledge.

Yet it was well recognised, and has become increasingly well documented since, that carefully designed instruction given to apparently ready and suitably motivated learners often led to learning that was at odds with what was taught. Knowledge is not just information that can simply be transmitted as long as transmitter and receiver are functioning well and clear lines of communication have been established (see Chap. 9) – the student can see and hear the teacher and they speak the same native language.

Such a learning-as-information-transfer model is simplistic and does not reflect classroom experiences. Thus, earlier in this book (see Chap. 4), a model was set out of how we publically represent our knowledge in the public space using various symbolic systems (speech, writing, drawing, gesture, etc.) and others then have to not only sense those representations but interpret them (perceive them) in terms of their own sense-making resources. Thus, Ausubel's (1968, p. vi) dictum that 'the most important single factor influencing learning is what the learner already knows'.

Time and again research (e.g. as outlined in Taber, 2009b) has suggested that often:

- Learners' pre-instructional ideas can be stable despite being contradicted during instruction.
- Learners' acquired versions of taught concepts are distorted compared with what was intended, in senses that reflect aspects of their pre-existing thinking.

This is certainly NOT always the case (Gilbert et al., 1982), but it is very commonly so. From the perspective of Ausubel's theory of meaningful learning, this is not surprising:

> …the most important factor influencing learning is the quantity, clarity, and organisation of the learner's present knowledge…which consists of the facts, concepts, propositions, theories, and raw perceptual data that the learner has available to him [or her] at any point in time…The second important focus is the nature of the material to be learned. (Ausubel & Robinson, 1971, pp. 50–51)

Meaningful learning is a process whereby learners relate new information to existing conceptual structures, and so those pre-existing conceptual structures are inevitably critical for what will be learnt, as they determine the nature of new conceptual knowledge constructed.

Is There a Learning Paradox?

Earlier in this volume (Chap. 4), perception of objects and events in the world was considered, and it was suggested that after the stage of external stimuli triggering an initial sensory response, there is then a further process of 'processing' of the sensory signal before it is (sometimes) presented to consciousness: so that perceptions are usually considerable *interpretations* of raw sensory data. That is, we normally actually perceive objects and events rather than experience the 'blooming buzzing confusion' that William James (1890) suggested would comprise the newborn's experience of the world. This interpretation is an automatic part of the processing of information in the brain, before it reaches the stage at which the perception enters consciousness. A good deal of our experience is of this form.

However, this is not always the case. Sometimes we perceive objects or events in the world *without* recognising what they are. In these situations we may feel uneasy, or at least curious, and may actively seek more information – perhaps turn on the light, move closer, change our angle of view, clap our hands or wave our arms to see if there seems to be a response. Usually the perception resolves and we feel more at ease and sometimes foolish at not recognising something that now seems perfectly familiar. During such periods when we are unsure what we are sensing, we may consciously attempt to identify the object or event by a logical process of reviewing the information available (size, colour, etc.), although whether this plays a role in solving the 'problem' rather than simply helping us feel in control whilst the usual subconscious processes continue to search for an interpretation to present to consciousness is less clear.

On other occasions we experience phenomena that do seem to require us to actively (consciously) *make sense* of them. Here we are talking about something more than perception in the usual meaning. We may recognise the events and objects in our surroundings quite clearly – for example, who did what to whom and with what. However, we may seek a deeper form of understanding – perhaps to understand why something happened, what motivated certain behaviour, who was to blame for particular events, etc. This may require something more than perception in the sense of the interpretations that are presented to consciousness. In such situations we create a mental model to explain what we have perceived (see Chap. 11).

In the motto at the head of this chapter, Scardamalia and Bereiter (2006) raise the issue of the 'learning paradox', which – put simply – asks how we can teach ourselves things we do not already know. Under the traditional, folk pedagogy, notion of teaching as 'transmission' of knowledge, it was assumed there is a more advanced knower such as the teacher or the textbook author who can impart knowledge to the less advanced learner. Yet, if constructivists argue that each learner has to construct knowledge anew, then this creates the question of how it is possible to build up more advanced learning based only on existing less advanced knowledge.

There Is No Viable Alternative to Construction of Conceptual Knowledge

Hopefully, readers who have read this far into the book will appreciate two points about this alleged paradox. The first is that the paradox exists whether one is a constructivist or not, unless one accepts that conceptual knowledge exists in the world independently of minds in a form which allows us to acquire it. In this book it has been strongly argued that conceptual knowledge can only exist in minds.

The qualification 'conceptual' is important here. If we consider knowledge more generally as a kind of 'know-how' then there is plenty of knowledge around that does not rely on minds (Collins, 2010). Trees 'know-how' to grow taller than humans, and some species can commonly manage to outlive us without apparently ever forming any conceptions. Every cell has the 'know-how' to control a complex set of chemical processes; viruses have the 'know-how' to invade cells and make use of their resources; a zygote has the know-how to become a fully developed person (environmental conditions allowing). Yet the physical world itself has no conceptual knowledge of trees or metabolism or epigenesis in any helpful sense. A world without people would continue to have the 'know-how' for trees to grow and reproduce, but this is not conceptual knowledge. Conceptual knowledge is not out there waiting for humans to absorb it, but must be constructed through concept formation within a cognitive system.

If we accept that conceptual knowledge must be constructed then, the learning paradox exists regardless of whether we accept that once formed such knowledge can somehow be transferred or, better, reproduced, from person to person, or not. In this book, it has been argued that a person's conceptual knowledge cannot be reproduced in a strict sense, only represented, allowing those representations to be used as information sources for others to construct their own knowledge. But regardless of this, somewhere along the line, someone formed a conception of 'atom', of 'electromagnetic field', of 'chromosome', etc. when such concepts had never existed before. This is not just the case for scientific concepts; of course, the same applies to the concepts of 'inflation', 'prison', 'symphony', 'irony', 'ismism', etc.

So if we reject the existence of some platonic world, where concepts have an independent existence *but are able to be accessed by people*, we must acknowledge that all concepts are constructed, that is, invented. The common misconception that Newton discovered gravity a few centuries ago (Pugh, 2004) is absurd, but although the detail is certainly wrong, some human being, or possibly protohuman, somewhere first reflected on regularities in their environment and conceptualised them in terms we might recognise as gravity. Since then, millions of others have constructed their own versions of a gravity concept: partly through their direct experience of the world and partly mediated through representations produced by teachers or through media such as textbooks of how others conceptualise gravity – often including representations of their teachers' conceptualisations of how Newton conceptualised gravity.

Emergence Is a Widespread Phenomenon

The second point we might make about the learning paradox is that our experience of the world *is* that more complex structures do commonly emerge. Whether we consider the structure of galaxies, the earth, the ecosystem or individual living organisms, we find that the forces of nature bring about structures that did not previously exist. This is especially clear in the case of living things, where evolution has allowed the construction of incredible complexity. Incredible, literally, because it intuitively seems to most of us that the variety of living things, with their myriad special features, can only be the outcome of providence – that is guided by some higher intelligence capable of foresight in planning such complicated systems (Taber, 2013g).

Indeed, over 150 years after Darwin (1859/1968) published on the origin of species, some scholars continue to argue that the complexity of living organisms requires the involvement of an intelligent designer at least at some points in the process (Behe, 2007). Yet the theory of natural selection posits that, given enough time, such high degrees of complexity are possible through a combination of modest natural processes and an environment that in effect (i.e. without any sense of purpose or deliberation) selects those outcomes that better 'fit' in some way.

Emergence means that when a system is formed from several component parts that can interact, the system has new properties. Fully describing the system cannot be achieved by simply cataloguing the characteristics of the components as there are now interactions that were not present before, and so characterising the system requires including the relationships between components as the well as the components themselves.

There is of course nothing mystical about this. Arguably, if we want to fully characterise a single entity, which could become a system component, we should detail its behaviour in all potential contexts. A chemical analogy might be helpful here: to fully characterise an element (e.g. oxygen) we report its chemical *as well as* its physical properties. That is, we describe what happens to a sample of the element when it is warmed, cooled, pressured, subject to a potential difference, etc. and also what classes of substance it reacts with, under what conditions and which products are formed in each case.

If we react oxygen with hydrogen it demonstrates specific behaviour that is not due to the oxygen itself but rather is a restricted part of its potential behaviour when we select from all its possible potential behaviours by structuring the conditions under which we observe the oxygen. Any particular sample of oxygen cannot realise all of these potentials – once it has reacted with sodium it is not present in an elemental form to demonstrate its reaction with phosphorus. From *this* perspective, emergent behaviour is not something additional, but rather the narrowing of the vast potential field of interactions by the selection of a specific configuration.

This perspective can also be applied to conceptual development. We might consider, as an example, the concept of electromagnetism, which may be considered to be built from pre-existing concepts of electricity and magnetism. These previously distinct concepts came to be seen as related (as creative discovery, cf. Chap. 7), and, over time, as elements of an overarching concept of electromagnetism. This example is historical, but one that senior school and college students are often expected to recapitulate when studying physics.

The 'new' concept of electromagnetism is more than simply the concept of electricity 'plus' the concept of magnetism, as it also inherently involves the ways in which electricity is considered to be related to magnetism within a recognisable overall pattern. Yet these new relational features were always *potentially* present even if not actually formed before links were made. This is purely an argument about what can be conceptualised and is not referring to the nature of reality: the potential for phlogiston to be conceptualised existed both before it was accepted as a useful scientific notion and after it had been discredited. Learners demonstrate that all kinds of conceptualisations are potentially possible through the range of conceptions they develop about natural phenomena – whether it be orbital motion not requiring any force, atoms seeking to fill electron shells or trees that induce pregnancy if one takes advantage of their shade.

That is, inherent in the concept of electricity was the potential for it to be linked to the concept of magnetism in certain ways, and vice versa. So the formation of the new subsuming concept of electromagnetism brings 'new' features to light – and indeed facilitates a conceptual link with the concept of light! The recognition of these new features justifies the re-conceptualisation of electricity and magnetism under the subsuming concept of electromagnetism – there is no learning paradox here providing the cognitive apparatus supports the ability to (a) form concepts, (b) seek relationships between concepts and so (c) reconceptualise these existing concepts in terms of new identified/mooted propositional links.

Of course this example is meant purely for illustration. The logic of earlier chapters in the book suggests the example cannot refer to 'the' concepts of electricity, magnetism and electromagnetism, but rather particular instances of knowledge representations within particular minds, that is, the concept*ions* of individuals. That is, there are a great many potential ways one could find to relate the concepts of electricity and magnetism, not all of which would match the canonical science perspective which itself is informed and constrained by the interpretation of empirical evidence. Particular learners will have reasons for conjecturing certain relations to be possible and of potential importance – often informed by teaching, reading and their own experience of relevant practical work – but as we have seen that does not necessarily mean their knowledge is a 'true' account of the world.

Two provisos should be highlighted here. It is not implied that the pre-existing concepts of electricity and magnetism are unchanged by this process. They are certainly changed in acquiring new links, but also the formulation of the overarching concept, the new system of concepts, may lead to inconsistencies or absurdities that can motivate changes in the 'original' conceptions, that is, the features the distinct subsumed conceptions were assigned before the re-conceptualisation. This point is picked up below.

The second point to be addressed is a possible objection that my statement 'new relational features were always potentially present even if not actually formed before links were made' could be seen as acknowledging that the concept of electromagnetism does indeed already exist in some Platonic sense prior to its discovery by our hypothetical learner here or indeed any 'knower'. As so often, that depends how we define and understand our terms.

Given the existence of (e.g. human) minds able to develop conceptions, we can imagine a kind of conceptual phase space of the different conceptualisations that could exist. Within that conceptual space must, by definition, occur the concept of electromagnetism as developed by my hypothetical learner, and indeed that space must include ALL the conceptualisations of electromagnetism (as well of course anything else) that have ever formed in minds or ever will. Moreover, a vast, if not infinite, set of other conceptualisations that could be produced by relating electricity and magnetism also potentially exist in this space even if they will never be conceived in any mind. If that is what we mean by concepts having independent existence, then it is a trivial sense – and indeed relies for its existence on someone having a mind to conceptualise the conceptual phase space itself. This is rather different from the notion of ideas having independent ('World 3', see Chap. 4) existence outside of ('World 2') minds.

The Task of Modelling Learning

This rather philosophical exploration of the learning paradox leads to my discounting it as a serious problem for constructivist models of learning in science (or any other disciplines) providing that:

1. Learners have suitable apparatus to facilitate the concept construction process.
2. Learners have suitable apparatus to evaluate and select between potential conceptual constructions.

In modelling conceptual learning, then, we need to consider the mechanisms by which concept construction, modification and evaluation can take place. The cognitive system modelled earlier in the book (especially Part II) inherently offers that apparatus, so that we can agree with George Kelly (1963, p. 75) who argued that 'learning is not a special class of psychological processes. It is not something that happens to a person on occasions; it is what makes him [or her] a person in the first place'.

Concepts and Conceptions (Revisited)

Chapter 11 offered an analysis of the main types of knowledge component that might be considered present in a learner's cognitive structure. In that chapter it was suggested that the approach to understanding the terms concept and conception recommended by Gilbert and Watts (1983) would be followed there. The term 'concept' was reserved for 'formal meanings as part of public knowledge systems', whereas 'conception' was used to refer to the personal understandings of individuals.

Ezcurdia (1998) suggests that people may be said to have acquired the same concept (e.g. metal, an example used below) but would each have their own particular conceptions (e.g. of metal). So from this usage we could refer to someone acquiring a *concept* of metal – but their *conception* may not match the concept of metal held by others or some notion of the canonical concept. The terms concept and conception are therefore doing useful work in making the distinction between what is common and what is distinct: different people have the 'same' concept but their conceptions are different.

Similarly we could say that a person's *concept* (e.g. of metal) may change over time because at different times they have different *conceptions*. Again we can use the term concept to refer to what is considered to have continuity and conception to refer to the particular – here at a moment in time. It is the 'same' concept in the sense in which the reader is the same person they were as toddler or the way a mature tree in the garden is the same plant as the sapling planted there many years before. This sameness is a matter of identity rather than being identical: in the way a ship is the same ship after a major refit or the way a political party or university department is considered the same party or department even after all the personnel have changed over time.

So the following parts follow the common usage of referring to individuals forming concepts (rather than conceptions) and undergoing conceptual (not conceptional) change. However, in keeping with Chap. 11 the term conception will be used when either specifying the particulars of a concept that distinguishes it from another person's conception of the 'same' concept or the sequences of different conceptions involved when an individual undergoes conceptual change (e.g. see Figs. 15.4, 15.5, 15.6, 15.7 and 15.8).

Concept Formation: Developing Spontaneous Concepts

Concept formation seems to be a key attribute of the human cognitive system. We have innate tendencies to recognise certain regularities in our environment (see Chap. 4). The recognition of these patterns supports categorisations that have been in effect been tested for utility over human evolution. So, for example, we recognise 'natural kinds' of living things and moreover do so spontaneously. Someone has to tell us that a particular type of living thing is called a 'cat' or a 'horse' in our local language community, but we are born with mechanisms for identifying such types.

Many other concepts may be formed spontaneously without needing to be innate. The ability to recognise repeated patterns and so develop what are in effect expectations that allow us to categorise experience is an intrinsic feature of the operation of the cognitive system (see Chap. 4). Spontaneous concepts are not based on reflection upon experience, but the automatic working of the cognitive system in interpreting information from the environment. That is, such concepts are formed spontaneously because of the nature of our perceptual systems and the inherent pattern recognition mechanisms built into human cognition.

It is important to be clear that this neither means that we all spontaneously develop precisely the same conception of horse, nor that the conceptions we do develop are necessarily 'correct' in some technical sense such that any particular conception will match a canonical scientific concept. But we do readily develop working concepts in this way. As we have substantial common genetic inheritance as humans, and as we often experience similar environments, we would expect there to be strong similarities in many of the spontaneous concepts we develop.

This is the basis of implicit knowledge, such as that represented in p-prims (as discussed in Chap. 11), which is often used to guide our behaviour without conscious control of attention. The advantages of speed and automation that implicit knowledge provides allow us to act on such knowledge without pausing to reflect upon its nature or origins.

Introspection on Spontaneous Concepts

The implicit nature of this knowledge does not imply that we must completely lack self-knowledge in this regard. As was discussed in the last chapter, a key feature of human development is the acquisition of the ability to reflect upon our own thinking: metacognition (see Chap. 7) – perhaps due to something like Demetriou's hypercognitive system (Demetriou & Mouyi, 2011, see Chap. 14) model. Although some aspects of the processing in human cognitive systems occur away from conscious awareness and control, we can still reflect upon *the outcomes* of such thinking (e.g. I know I'm seeing it as a face, but actually it's just a coincidental pattern in the clouds, cf. Fig. 4.6).

There has been a debate about the nature of our concepts (Gilbert & Watts, 1983), for example, whether they are based on membership criteria (if it is a large animal, with four legs, a mane, a particular head shape, etc.) or prototypes (if it looks like this mental image I have of a typical horse), etc. Such an argument may be unhelpful if the apparatus for forming spontaneous concepts is based on neural networks which become tuned to perceived regularities in the environment, as these types of cognitive components would inherently seek a match through an inbuilt feedback mechanism and not in any directly verbalisable manner (see Chap. 5). So we can *reflect upon* how we know when we have seen a horse and might refer to the number of legs, the tail, the mane, the size of the beast, etc., but we do not actually know how we make the judgement using preconscious processing – we can only offer conjectures for how we know.

Forming 'Reflective' Concepts

Scardamalia and Bereiter (2006, p. 104) discuss how 'a dynamic systems explanation of conceptual growth posits (along with other kinds of interactions) ideas interacting with ideas to generate new ideas'. So, as an example, I may, through a creative process (Chap. 7), coin a new concept I label as 'supermarines'. My conception of supermarines is vessels that may be used for transport on the surface of the sea, and

I will include as examples boats, ships, rafts, dinghy, catamarans, canoes, yachts, seaplanes (flying boats), etc.

Karmiloff-Smith (1994, 1996) has argued that a key feature of the cognitive system is an ability she calls 'representational redescription', which allows the cognitive system to form new types of representations of the information already represented: 'to exploit internally the information that it has already stored, by redescribing its representations or, more precisely, by iteratively rerepresenting in different representational formats what its internal representations represent' (1994, p. 699). In general, Karmiloff-Smith argues, this process allows knowledge to be represented at increasingly explicit levels. So, an implicit knowledge element such a p-prim might become re-represented at a level at which *the new representation* can be consciously accessed and reflected upon, perhaps in an iconic form, or perhaps as a verbalisable concept.

That is, in terms of our typology of knowledge elements discussed in Chap. 11, elements of conceptual knowledge which fall under the 'implicit' branch of Fig. 11.1 are not, and cannot themselves be, promoted to an explicit status but can *become represented* elsewhere within the system by a new representation formed within a category on the 'explicit' branch – which allows us to consciously access and operate with (not the original implicit knowledge element itself but) the explicit representation of the implicit knowledge element.

Acquiring Academic Concepts

A key distinction made by Ausubel (1968, 2000) was between 'rote' learning, which is learning material so it could be recalled verbatim without understanding, and 'meaningful' learning where the material to be learnt was subsumed within existing conceptual structure (see Chap. 5). Although it will be suggested that this should be considered a matter of degree rather than a dichotomous classification, it is a commonly used distinction and one of practical importance in teaching and learning.

Learning by Rote?

Learning that is *purely* 'by rote' may seem to offer the strongest example of the learning paradox, as it seems to suggest learning of material for which there is no relevant existing structure within the cognitive system to provide interpretation or linkage. However, when discussing rote learning, we are normally considering learning of verbal material, and so learning which is supported by the language 'modules' that we know are part of the normally developing cognitive apparatus of all humans (see Chap. 6). So if we learn some lines of poetry without understanding their meaning, they are still likely to include some familiar words and to follow a familiar grammatical form. Even 'nonsense' poetry follows grammatical rules and uses the phonemes of the local language.

Rote learning used to be very common in formal instruction and indeed still is in some national traditions (Boyle, 2006; Eickelman, 1978). In extreme form, this might mean reciting phrases such as 'the square of the hypotenuse in a right-angled triangle is equal to the sum of the squares of the other two side' or 'the rate of change of momentum of a body is directly proportional to the applied force and occurs in the direction of the applied force' as if being able to produce the statements accurately at will is itself a worthwhile learning goal. Learning by rote is nowadays generally considered poor educational practice, and in science teaching we seek meaningful learning as far as possible: that is, learning that makes sense to the learner though being related to an existing conceptual structure (Ausubel, 2000).

Rote learning clearly works in one sense, in that people *can* learn material by rote and for some material this may be useful or even necessary. An example might be when acting in a play – although, even there, performance based simply on memorising a text is unlikely to be of high quality. Generally, rote learning by itself is not very useful if it does not lead to understanding, but rote learning may sometimes be *part of* the process leading to understanding (Tavakol & Dennick, 2010) or may at least allow the learner to enter into discourse with others about what they have learnt.

The potential for rote learning seems consistent with the types of mechanisms responsible for implicit knowledge elements, in the sense that repetition provides repeated experience of the same pattern which at the physical level, see Table 3.4, can presumably lead to sequenced firing of the same neural components, leading in turn to strengthening synaptic connections; that is, the tuning of neural circuits that will more readily be activated.

Conceptual Growth: Subsuming Learning Under Existing Conceptual Structures

We can learn that Paris is the capital city of France, that sodium has atomic number eleven, that Equus refers to horse and a great deal of other apparently arbitrary information. This information is arbitrary when *to the learner* there is no obvious rationale for why Paris is called Paris rather than something else. Such items would need to be learnt by rote when they cannot be understood in terms of existing conceptual structure, although this is always a matter of degree.

Someone who did not speak English and learnt to recite that 'Paris is the capital city of France' without knowing what the terms France, Paris or capital city referred to would have learnt by rote to a much greater extent than an English speaker who knows of France and already had a concept of capital city. The challenge of the rote learning for this latter person is largely remembering the name of Paris, whereas for the non-English speaker the task is to learn an incomprehensible string of sounds. That is, for the person who has acquired the concept of capital cities, the new knowledge element, the meaning associated with the name Paris, fits within an existing schema (see Chap. 11) in which capital cities have a particular relationship with countries and where a number of existing examples are likely known, for example:

Country	Capital
England	London
USA	Washington, DC
France	[slot for the name of the capital city of France]

A lot of learning involves these kinds of processes – learning new examples or properties that can be added to our existing 'conceptual map' of the world. Mnemonics used as memory aids work in a similar way, by making links with material already present in conceptual structure. So, a learner who knows that gold and iron and copper are metals might add additional examples such as sodium and uranium (this example is developed further below). Indeed, this type of learning can be represented by showing additional propositions added to a concept map (see Chap. 12).

Learning 'Academic' Concepts

Vygotsky (1934/1994) referred to 'scientific' or 'academic' concepts, which are only acquired through formal instruction – and so are 'the purest type of nonspontaneous concepts' (p. 365) – as opposed to what Piaget had called 'spontaneous' concepts that the individual can acquire through their direct action in the physical world. Those spontaneous concepts (i.e. conceptions) derive from everyday experience, and although they do not necessarily remain tacit, this origin is significant:

> The child becomes conscious of his spontaneous concepts relatively late; the ability to define them in words, to operate with them at will, appears long after he has acquired the concepts. He has the concept (i.e., knows the object to which the concept refers), but is not conscious of his own act of thought. The development of a scientific concept, on the other hand, usually begins with its verbal definition and its use in nonspontaneous operations – with working on the concept itself. It starts life in the child's mind at the level that his spontaneous concepts reach only later. (Vygotsky, 1934/1986, p. 192)

Figure 15.1 represents the difference between spontaneous and academic conceptions using the form of representation adopted earlier in the book. In the figure, the student is shown directly perceiving an object, a plant, and being taught about the concept of 'primary producers'.

Academic Concept Formation

Spontaneous concepts are likely to derive from the cognitive system's inherent pattern recognition ability, when experiences that seem similar lead to the formation of a knowledge element for that pattern of experience. For example, certain types of

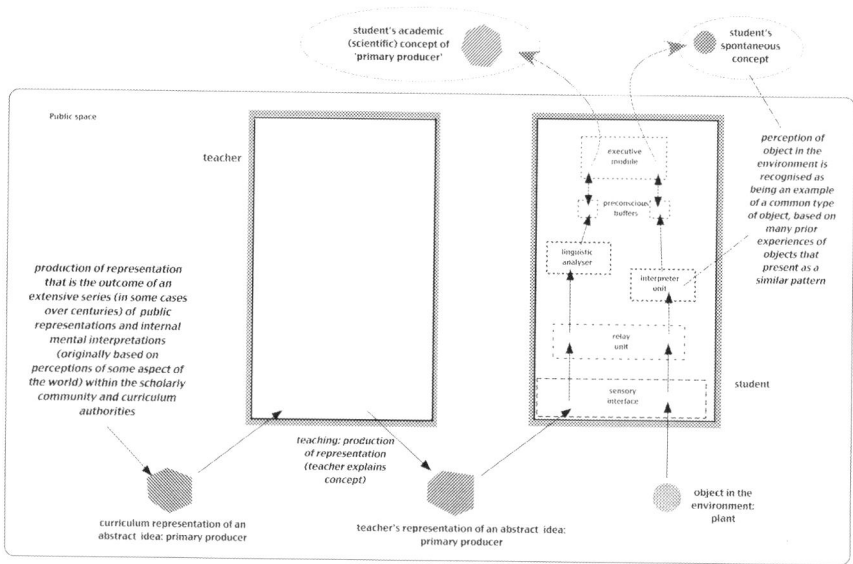

Fig. 15.1 The origins of academic and spontaneous concepts

objects in the environment, which are green, anchored in the soil, have laminal structures, etc. come to be seen as a class of things.

Academic concepts are usually presented through language, which is analysed in specific areas of the brain where specialised interpretative apparatus has evolved to handle this type of input (see Chap. 6). This might seem to 'short-circuit' the learning process by allowing the individual to develop concepts without extensive personal experience of the referents (Karmiloff-Smith, 1996). However, as suggested above, such concept learning is only meaningful when it can be interpreted as making sense in terms of existing conceptual structures – which ultimately means it depends upon direct experience of the world (Lakoff & Johnson, 1980b). Thus, verbal learning can occur providing both the apparatus and some relatable conceptual substrate are available to the learner. From this perspective what is learnt has the meaning imposed by being related to an existing conceptual structure and will not necessarily acquire the meaning intended by a teacher. This is a premise of the constructivist perspective on learning (Taber, 2009b).

Concept Modification

Earlier in the book (see Chap. 11) the knowledge components that were tacit were compared with those, such as a learner's conceptions, which they can directly access and reflect on. Once conceptual knowledge has been formed, there is the potential for it to be modified within the conceptual system. Caravita and Halldén offer a view of learning, where:

> [learning is not seen] as an event of mere replacement of old ideas by new ones, but as a process which occurs in a system where conceptions of specific phenomena are only one of the components. Organization, refinement and differentiation among contexts are other important and observable aspects which continuously enlarge the power of the system to perceive and interpret reality. (Caravita & Halldén, 1994, p. 90)

Whereas tacit knowledge components such as p-prims are encapsulated, so that once established in the system, they remain stable and unchanged, conceptual knowledge has the potential to be related, compared and interlinked in various ways. Various types of modifications are possible.

Piaget (1970/1972) saw a process whereby experience provided new material to be incorporated into conceptual structure (assimilation), sometimes leading to inconsistencies in the system (disequilibrium), which could be fixed by modifying existing knowledge (accommodation) to bring the system back to coherence (equilibrium).

Disequilibrium only occurs when we notice something that does not fit with existing ideas, whereas the nature of the perceptual system is such that most commonly we manage to *interpret* new information in ways that are consistent with our existing conceptual structure (see Chap. 4). Therefore, only when a new learning experience cannot be made sense of in terms of current knowledge are we likely to experience the 'cognitive dissonance' (Chapanis & Chapanis, 1964; Cooper, 2007) produced by something that confounds our expectations and therefore cannot be perceived in terms of existing knowledge.

We might envisage one type of conceptual development as simple growth in the range of application of a conception as more examples that can be subsumed are discovered. So a student who has a conception of metal that includes knowing that iron and copper are metals may go on to learn that manganese and zinc are additional examples of metals, without substantially changing their existing conception of what a metal is.

We can also envisage conceptual development that brings about more fundamental changes in the nature of the learner's conception – such as when the everyday notion of what it is to be a metal is related to new learning about the canonical chemical concept of metal. Sometimes there may be potential for 'changing one's mind' such that existing conceptions are found to be inadequate, requiring changes in aspects of existing understanding. A conception of metals, for example, incorporating lifeworld ideas that metals are magnetic, metals are hard and metals are solids may be challenged by new learning (this example is developed below).

It would seem that characteristics of progression in learning might be understood as:

- Increasing integration of conceptual knowledge by identifying links between conceptions
- Increasing coherence of knowledge by identifying apparent inconsistencies and seeking to interpret them

Interpretation may involve resolving the apparent inconsistency (as, e.g. recognising it as due to alternative models of the same target) or recognising an apparent flaw in personal knowledge that requires attention.

It is known that the human cognitive system has inbuilt mechanisms for seeking greater integration of knowledge that do not rely on conscious interrogation of a person's knowledge base (see the discussion of memory consolidation in Chap. 5); however, metacognition – conscious interrogation of and reflection on one's own knowledge (see Chap. 7) – also plays an important role in identifying apparent inconsistencies between different (explicit) knowledge components.

Vygotsky's Notion of Concept Development

Vygotsky's model of conceptual development involved interaction, and a kind of convergence or hybridisation, between spontaneous and academic concepts. Vygotsky suggested that in effect spontaneous concepts allow academic concepts to be meaningful and academic concepts provided the framework for making spontaneous concepts explicit. Vygotsky (1934/1986, p. 148) described this process using a spatial metaphor involving 'two different paths in the development of two different forms of reasoning'. Vygotsky talks of academic or scientific concepts being formed higher in the system than spontaneous concepts, as 'a scientific concept… starts its life in the child's mind at the level that his spontaneous concepts reach only later' (p. 192). This is possible because such concepts are 'mediated' (p. 194).

The two types of concepts interact and converge: the academic concepts moving 'downward to a more elementary and concrete level' (p. 193) and the spontaneous concepts moving upwards. That is, they 'develop in reverse directions: starting far apart, they move to meet each other' (p. 192):

> In the case of scientific thinking, the primary role is played by initial verbal definition, which being applied systematically, gradually comes down to concrete phenomena. The development of spontaneous concepts knows no systematicity and goes from the phenomena upwards towards generalizations. (Vygotsky, 1934/1986, p. 148)

This spatial metaphor, of vertical movement towards convergence, is represented in Fig. 15.2. This metaphor, focusing on shifts along a concrete-abstract dimension, however, oversimplifies the process Vygotsky describes. For Vygotsky, the shifts that occur in these initially quite distinct types of conception are facilitated though being related to concepts from the other category, through a kind of mutual development:

> In working its slow way upward, an everyday concept clears a path for the scientific concept and its downward development. It creates a series of structures necessary for the evolution of a concept's more primitive, elementary aspects, which give it body and vitality. Scientific concepts, in turn, supply structures for the upward development of the child's spontaneous concepts toward consciousness and deliberate use. Scientific concepts grew downward through spontaneous concepts; spontaneous concerts grow upward through scientific concepts. (Vygotsky, 1934/1986, p. 194)

acquired through
definition and
directed application

```
                    ┌─────────────────┐   leads      social
                    │    academic     │    to      interaction
                    │    concepts     │ ◄──────
                    └─────────────────┘
```

become
concrete
through links

become
abstracted
through links

```
   everyday        leads    ┌─────────────────┐
   experience       to      │  spontaneous    │   initially
                  ──────►    │   concepts      │   tacit
                            └─────────────────┘
```

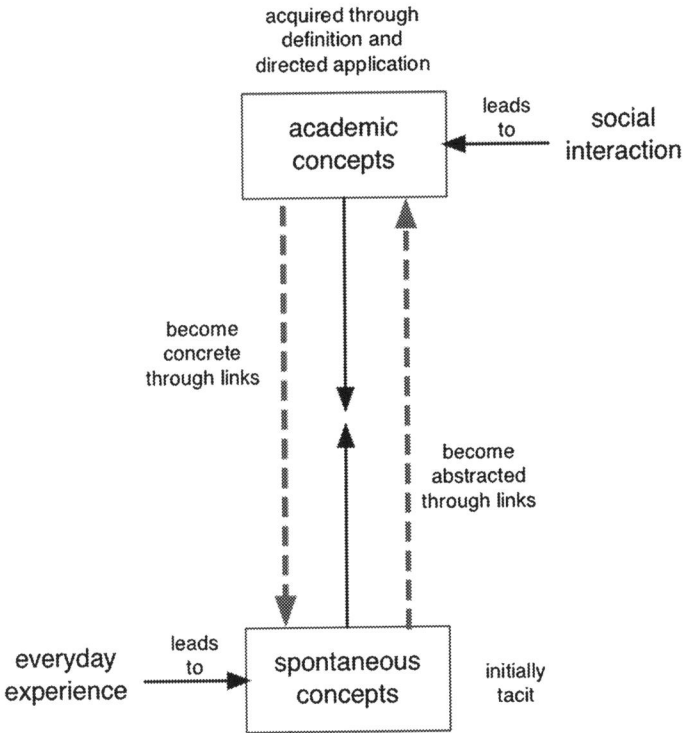

Fig. 15.2 A representation of Vygotsky's spatial metaphor for conceptual development

Vygotsky was writing the best part of a century ago, and parts of his description may now seem outdated. He notes that spontaneous concepts appear before the learner is aware of them or is able to define them or to consciously apply them at will (p. 148). That is, this type of knowledge is initially implicit. Vygotsky suggests that scientific concepts facilitate the shift to explicit knowledge. This, however, need not be the case: the presence of something like the hypercognitive system (see Chap. 14 and the discussion of metacognition in Chap. 7) allows us to *become aware* of at least some of our initially implicit knowledge (e.g. personal constructs; see Chap. 11) and so make them explicit. This need not require the mediation of academic (taught) concepts but can occur through the process of representational redescription discussed above.

Even if the formation of what might be termed 'reflective' concepts, that is, concepts open to conscious reflection, from initially tacit spontaneous concepts may not require the mediation of taught academic concepts as Vygotsky suggests, once the reflective concepts have themselves been formed, there is a question of their relationship with academic concepts acquired through language and social processes (e.g. teaching).

As noted earlier (see Chap. 12), some commentators have suggested that concepts derived from everyday experience largely form a discrete system represented separately in conceptual structure from taught concepts (Claxton, 1993; Solomon, 1992). From this perspective, the task of relating these two systems is seen as challenging for the learner. Vygotsky's research, however, led him to conclude that 'the development of spontaneous and academic concepts turns out as processes which are tightly bound up with one another and which constantly influence one another' (1934/1994, p. 365).

Rather than forming somewhat isolated categories of thought, Vygotsky argued that the two 'types of concept are not encapsulated in the child's consciousness, are not separated from one another by an impermeable barrier' but rather that 'the dividing line between these two types of concepts turns out to be highly fluid, passing from one side to the other side an infinite number of times in the actual course of development' (p. 365). According to Vygotsky, our spontaneous and academic concepts 'do not flow along two isolated channels, but are in the process of continual, unceasing interaction' (p. 356).

This does not seem consistent with Solomon's (1983, 1992) notion of lifeworld knowledge being a separate domain to school learning of concepts. However, if Solomon's notion of there being distinct domains of knowledge is understood in terms of the topography of learners' cognitive structures – there being separate systems for representing lifeworld and school science concepts in different locations – then it runs into difficulties as her distinction does not seem to fit well with the actual distinctions between different types of knowledge elements elicited from learners (as discussed in some detail in Taber, 2009b, pp. 241–251). Solomon's ideas may be better understood in terms of students having to learn to participate in different discourse practices in science classes (Gunckel, Mohan, Covitt, & Anderson, 2012), rather than being about the representation of conceptual knowledge itself. This point is developed later in the chapter.

Vygotsky's model then refers to a high level of interaction between spontaneous and academic concepts and the development of each of these types of concepts towards a more hybrid state: spontaneous concepts deriving from concrete experience acquiring abstract nature and academic concepts acquiring concrete referents.

Melded Concepts

Vygotsky's description of this process maintains the labels of spontaneous and academic for the different types of concept that are interacting. Yet the implication of his account is that this distinction cannot be fully retained. Rather, by relating the academic concepts mediated through social processes to the spontaneous concepts developed through direct experience of the natural world, new conceptual structures form that subsume both. Through such a process people develop concepts that are hybrid forms: melded concepts.

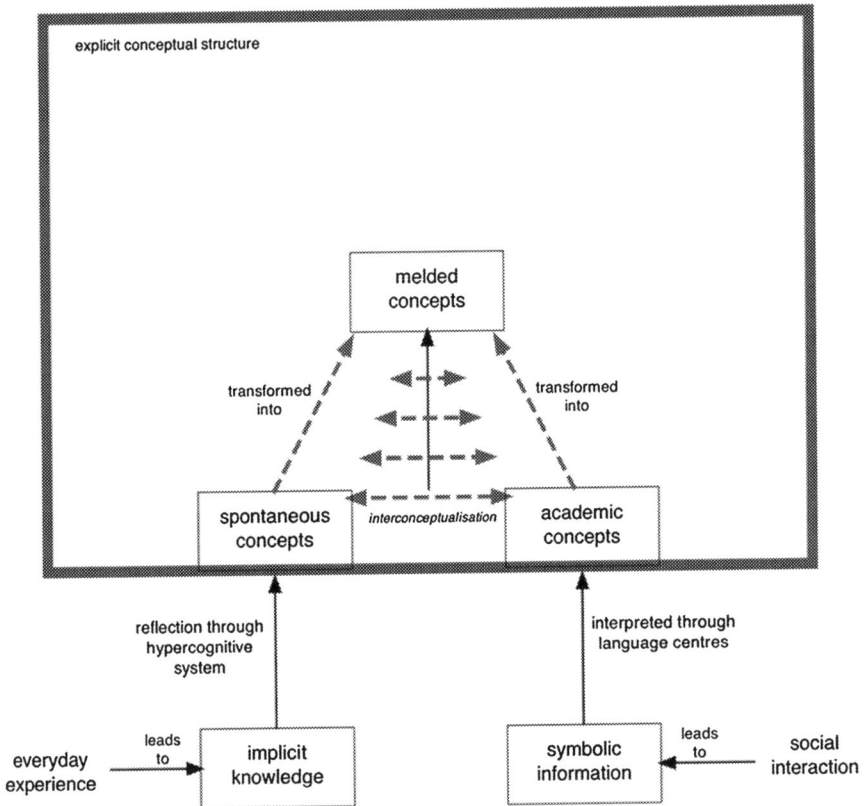

Fig. 15.3 A modification of Vygotsky's scheme: the development of melded concepts

This suggests that the distinction between spontaneous and academic concepts relates primarily to *the origins* of concepts, but due to the dynamic nature of memory this distinction may afterwards be broken down by the interactions Vygotsky describes. In some cases there might be considerable integration of what begins as a purely spontaneous concept with what has been learnt through verbal instruction. Through what might be termed 'interconceptualisation', what were originally discrete spontaneous and academic concepts evolve into a melded concept that draws upon both an experiential base in direct experience of the natural world *and* culturally mediated learning relying on communication through a form of language. Certainly it would seem this is often the ideal we look for in teaching science.

We might therefore reconceptualise Vygotsky's description as something more like Fig. 15.3, where both spontaneous concepts and academic learning are processed initially through perceptual apparatus, before becoming represented as explicit knowledge in conceptual structure, allowing the potential for linking, and possibly some level of integration.

Here, in Fig. 15.3, an alternative spatial metaphor is employed, where rather than spontaneous and academic concepts moving up and down (respectfully) to meet each other, both are originally represented near the periphery of conceptual structure, but through being linked and used to interpret each other, come to occupy a more central position. This representation borrows from the metaphor of surface and deep learning (Chin & Brown, 2000). That is, conceptual structure is here conceptualised spatially not in relation to physical location within the cortex, but in terms of connectedness, with more connected material seen as more central.

This process highlights the value of language and social mediation in human learning. Without such mediation, the individual learner would only form concepts based on interpretation of direct experience. Reflective concepts could certainly form, and be modified by new experience in the sense Piaget describes. Moreover, the inbuilt tendencies of the cognitive system to relate the contents of conceptual structure, notice inconsistency (disequilibrium) and modify the system of concepts towards greater coherence would act on spontaneous concepts – but would always be limited to the data provided by perceptions of personal experience of the world.

Vygotsky points out how cultural mediation through language allows us to also develop what are initially spontaneous concepts not only through reflection upon our own experiences but also through our interpretations of the public representations of the reflections of others. Sometimes, of course, this process involves many stages of iteration such that we can consider there to be, at least in principle, canonical versions of concepts (see the discussion of public knowledge in Chap. 10).

As Vygotsky recognised, academic concepts can only be meaningful through being related to existing concepts that ultimately are grounded in spontaneous concepts formed through personal experience and this inevitably means that academic concepts may be acquired in idiosyncratic ways. Moreover, the possibility of forming melded concepts opens up the conceptual system of any learner to a potentially vast sources of 'secondary data' based upon the public representations of the knowledge of other members of the community. In a global society with books, radio, television, the Internet, etc., this in effect means that every learner can be part of a network of billions of people able to represent their personal knowledge in the public space where it can be perceived and interpreted by others.

A Hypothetical Example of Concept Development

It certainly seems that conceptual development involves a number of distinct types of changes to conceptual structure. This can be illustrated by using the example of a student's learning about the 'metal' concept. The example here is hypothetical, designed to highlight some of the different aspects of concept development, but reflects the kinds of changes reported in studies.

Figure 15.4 represents a hypothetical student's concept of metals before formal instruction in the topic in middle or secondary school science. Most students will

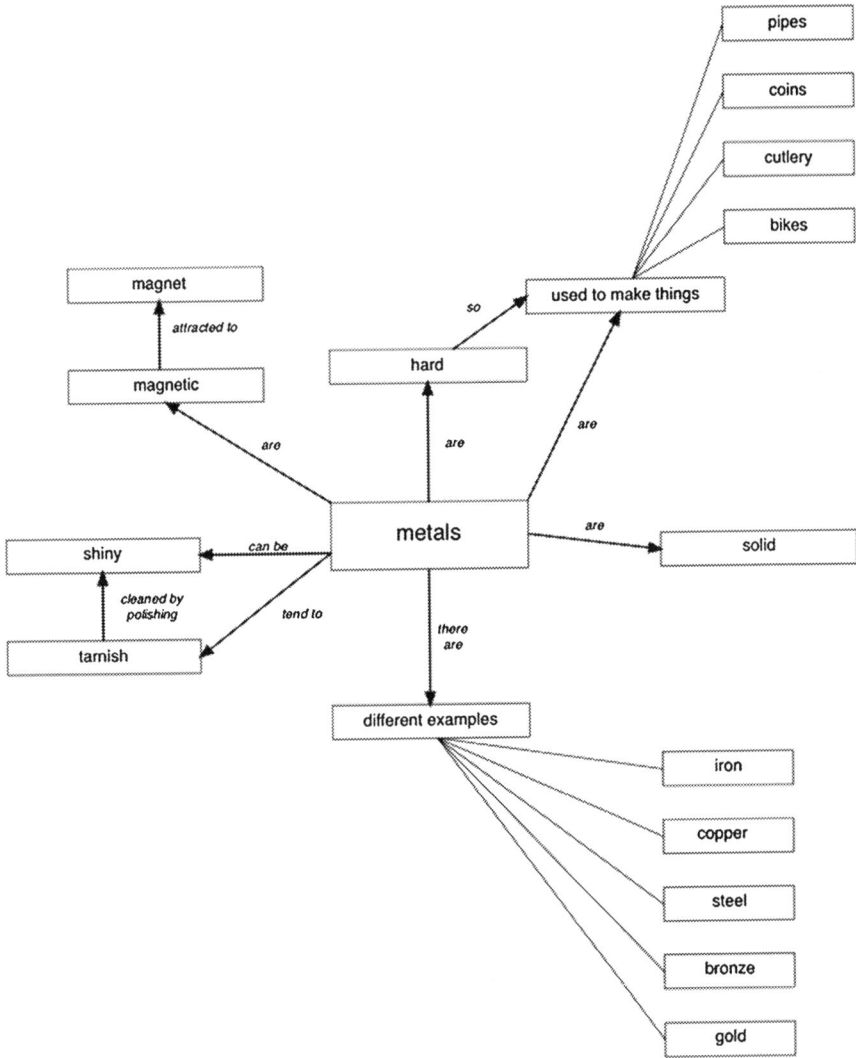

Fig. 15.4 Conception 1. Representation of a hypothetical pre-instruction conception of 'metal'

have acquired a concept of metal from a combination of their spontaneous experience of materials, and the tendency to find patterns in and categorise experience, and the way the term metal is used in lifeworld contexts. So an initially spontaneous concept will have acquired explicit representation in verbalisable form through exposure to many references to metal and to common examples of metals.

We would expect a young learner to typically consider metal to be a category of material, which includes some common types (iron, copper, etc.) and which has some common properties which are related to common uses of metals that are regularly experienced (such as coinage and knives). As a lifeworld concept, the notion of metal overlaps with, but is not entirely consistent with, the scientific concept.

So metals in everyday discourse are materials with certain useful properties that make them suitable for being formed into materials. So, for example, metals are hard and strong solids, allowing us to use them to make bridges that span rivers. It is also quite possible that our hypothetical learner will have acquired the common alternative conception that metals are (i.e. generally) magnetic (Hickey & Schibeci, 1999) in the sense that they 'stick' to a magnet.

When our learner meets the concept of metal formally in science class, they are likely to make sense of teaching about metals in terms of the pre-existing conception of metals. When the teacher refers to metals, this will be recognised as a reference to the types of materials the student already understands metals to be. Meaningful learning involves making sense of teaching in terms of existing conceptual structure, and references to metals will be interpreted through existing understanding of that concept.

Some modifications to existing conceptual structure can be seen as little more than additions to the existing conception, so Fig. 15.5 reflects that, for example, the learner may do some school practical work to show the electrical conductivity of metals – something commonly included in lower secondary courses. Probably, only a few examples will be investigated, but this is likely to be enough to acquire the generalisation that all metals conduct electricity, just as previous experience with magnets might have suggested that metals were generally magnetic.

Where new examples of metals are encountered either physically in the school laboratory (e.g. perhaps zinc) or mentioned by the teacher (e.g. perhaps manganese) these can be readily subsumed under the existing conception of metal – especially when they seem from the way they are discussed to fit the prototype of solid, hard materials useful for forming into structures. New properties of metals may be encountered, so, for example, our learner may be taught that metals have a property of being 'sonorous', which may be linked to new applications such as being formed into bells.

However, not all new learning can be fitted into existing conceptual structures so readily (see Fig. 15.6). The learner may be taught that actually most metals are not magnetic, and that only three common metals have this property: iron, nickel and cobalt. Indeed, if our learner sticks with the physical sciences long enough, this can change again when the magnetism concept expands to represent various forms of magnetism – paramagnetism, diamagnetism, antiferromagnetism – and it will transpire that the everyday notion of magnetism only refers to one type: ferromagnetism. At that point the magnetism concept would be some way removed from the simple notion of what a magnet can pick up.

Our hypothetical learner may also be taught that metals are considered a major category of the chemical elements. However, this may be accompanied by the start of a shift in the concept or perhaps a sense that the metal concept is ambiguous and has several foci; see below. So some of the examples of metals that were already familiar, such as steel and bronze, are not elements, but rather mixtures of elements and so perhaps not actually chemically metals, but something else: alloys. Our learner

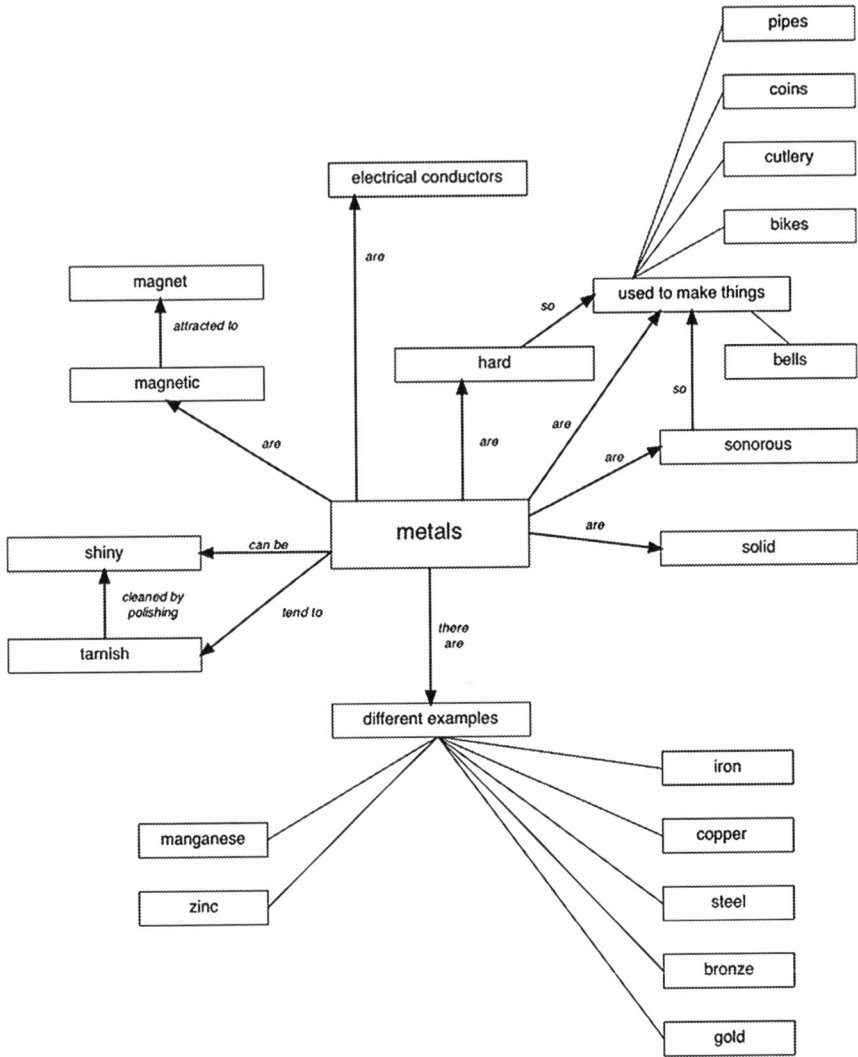

Fig. 15.5 Conception 2. Some new information may be assimilated by being subsumed into the existing structure – new examples, new properties

will also be taught about new examples of elements considered metals, which do not fit the stereotypical metallic properties – so sodium is a soft metal that reacts vigorously with water, and mercury is considered a metal whilst being a liquid at room temperature. This rather different, chemical, notion of the metal can undergo further development as study continues, as is suggested in Fig. 15.7.

So the primary properties of a metal *from a chemical perspective* relate not to its physical characteristics but its chemical behaviour: that is, to the nature of the

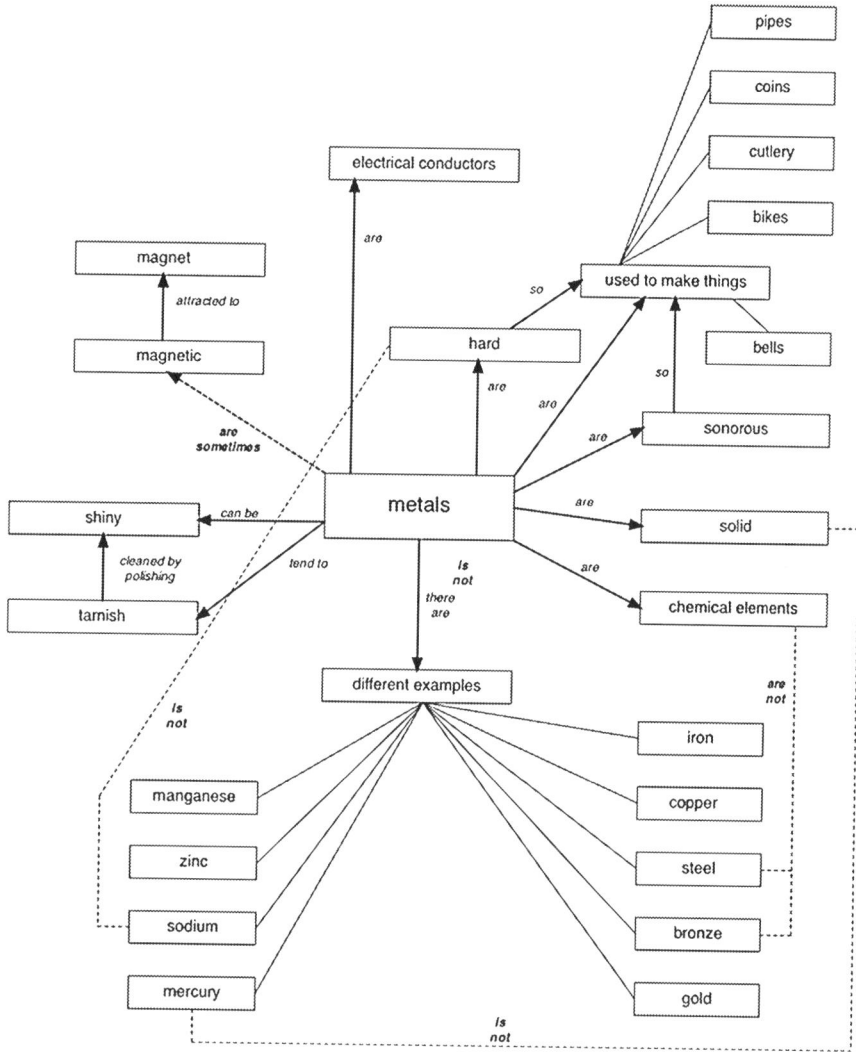

Fig. 15.6 Conception 3. Some new information assimilated into the structure may lead to inconsistencies with existing aspects of conceptual knowledge

reactions it undergoes. So the existing (lifeworld) notion that metals commonly tarnish will be related to a new idea that metallic elements may commonly be oxidised – and so linked to a more general chemical concept of oxidation. This will be accompanied by a shift in focus from how this affects the appearance of the metal to the nature of the product: an oxide that is basic or amphoteric, so potentially linking to developing concepts of acidity and alkalinity. Similarly, metals will produce salts when reacted with acids and will be classed as electropositive elements.

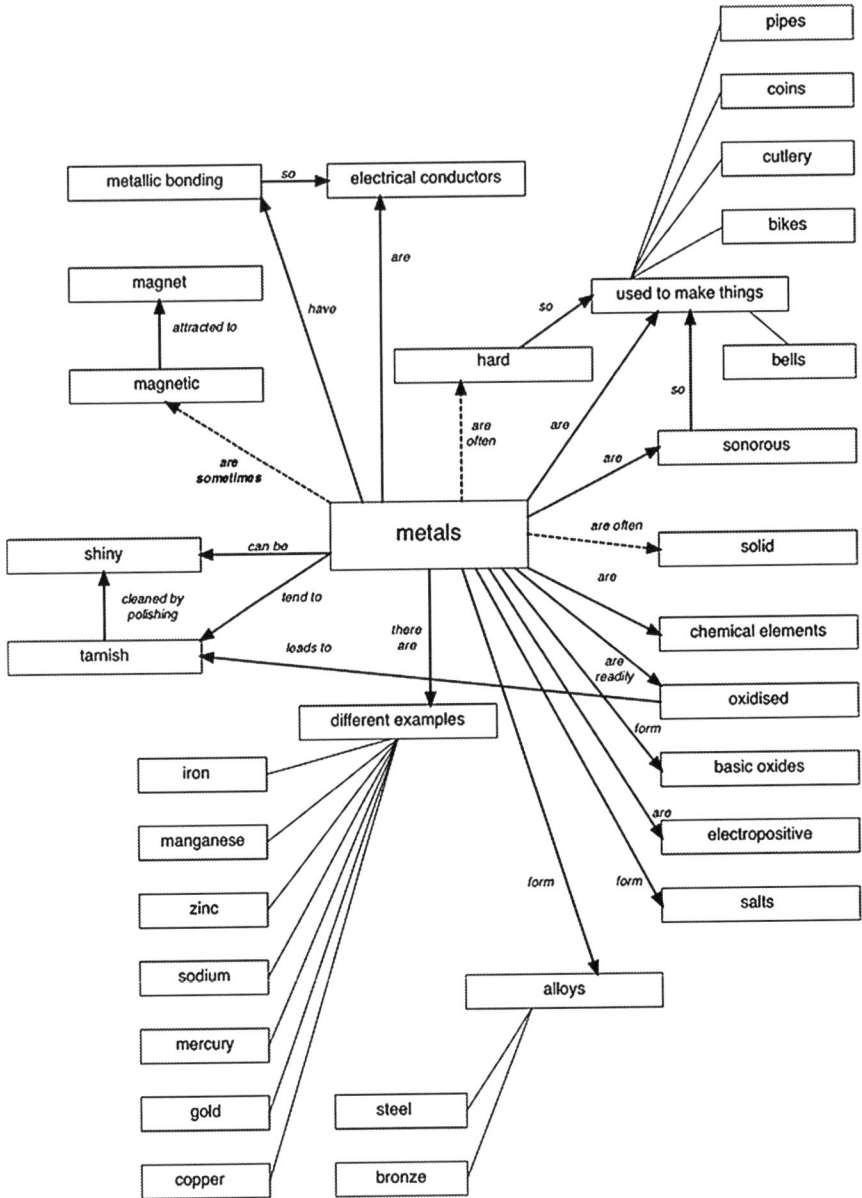

Fig. 15.7 Conception 4. Some new information assimilated into the structure may be accommodated by modifications of previous understanding and may offer potential for new linkages

This latter property may be explained in terms of submicroscopic models of atomic structure, which may also be used to characterise the crystalline structure of metals and the form of bonding found in metals. At this point our hypothetical learner's

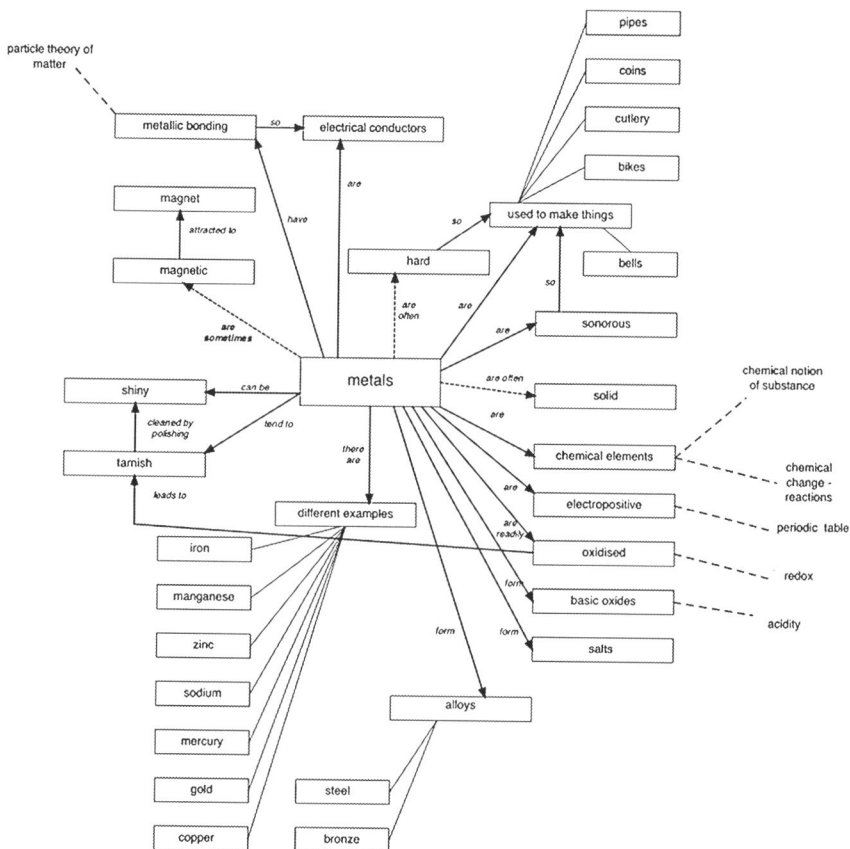

Fig. 15.8 Conception 5. Ongoing evolution of the concept within conceptual structure may lead to a shift in the nature of the concept (e.g. metal as everyday category to metal as a chemical category) and offer potential for extensive new linkage with other parts of conceptual structure

concept of a metal will have shifted quite considerably and will have become firmly embedded within a network of chemical concepts (see Fig. 15.8).

The Challenge of the Separate Domains Model to Conceptual Development

This form of representation (Figs. 15.4, 15.5, 15.6, 15.7 and 15.8), albeit here demonstrating a hypothetical case, seems to suggest that there *is a single conception* of metal that is evolving. Yet arguably this oversimplifies the nature of conceptual change. Caravita and Halldén (1994, p. 90) argue that 'organization, refinement and differentiation among contexts are other important and observable aspects which continuously enlarge the power of the system to perceive and interpret reality'. Commentators such as Solomon (1983) and Claxton (1993) have argued that systems of lifeworld concepts exist *alongside* formally taught concepts (see Chap. 12), with successful learners discriminating from context which set of concepts may be appropriate for particular discourse. Yet, if (as Vygotsky suggested) spontaneous concepts evolve with the learning of academic concepts, then arguably the initial spontaneous concept is no longer present as it has been modified through the construction of a hybridised, more developed, melded concept (cf. Fig. 15.3).

Solomon (1992), after detailed work looking at how children used the energy concept both in and outside formal school contexts, suggested that people maintain two separate systems of concepts. If that also applied across other topic areas such as metals, then they would be expected to retain a lifeworld notion of metal in one domain of conceptual structure, whilst building an alternative school chemistry concept of metal in a separate domain of academic concepts.

Energy is an abstract concept that is often understood by young people in quite different ways from the formal physics concept, because of the way the term is commonly used in social discourse – and the formal physics concept is somewhat counter-intuitive in that energy does not refer to anything directly observable but is used more like a formal accounting device (Feynman, Leighton, & Sands, 1963). In effect, Solomon's model suggests that there are synonymous energy terms and students are expected to use the context of any reference to know which energy concept is being referred to. So the learner is meant to appreciate that in the lifeworld context, a person can *raise* their energy levels by some moderate exercise, whereas in the physics classroom the same activity would be understood as a process of transfer of conserved energy *from* the chemical stores associated with blood sugar and oxygen through processes of working and heating the environment. Confusion might be best avoided here if these two ways of thinking and talking about energy are not considered to refer to 'the same thing'.

So whereas the simplistic conceptual change notion might suggest that school science should challenge an existing alternative concept of energy and seek to replace it by a more scientific conception, in the example of 'energy' there might be a good case for arguing that the role of school science is actually to help students form a distinct new (canonical scientific) energy concept, to be maintained in parallel with the existing lifeworld concept. The latter would be technically inadequate but arguably remains more useful in everyday social discourse.

Claxton (1986) has argued that given the difficulties in getting learners to shift from their pre-formal alternative conceptions of scientific ideas to new conceptions reflecting the scientific concepts, it is ineffective to try to start from their existing ideas and expect them to substantially modify these, and it might be better to look to build new concepts completely independently. In the case of energy, this seems a sensible suggestion: if we expect students to build the formal concept of energy from their existing lifeworld notion, then the task will be challenging.

In this case Vygotsky's notion of the spontaneous and academic concepts coming together is inevitably going to be problematic when 'lifeworld energy' can be gained by eating sweets and running around, and can be readily used up, and 'school physics energy' is never created nor destroyed. The logic of Solomon's research is that effective

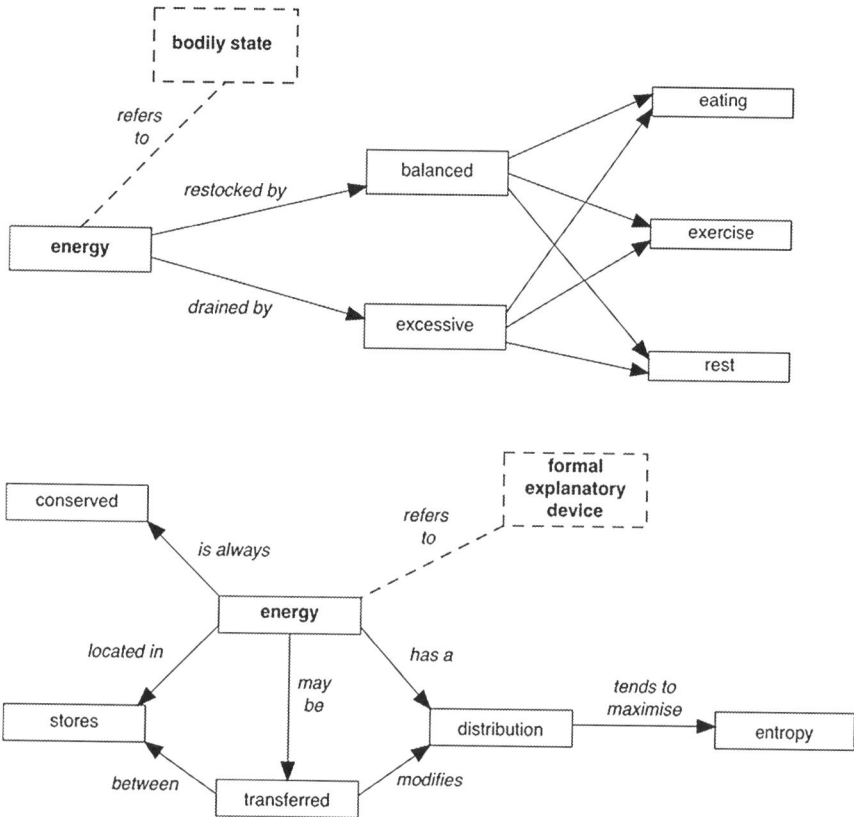

Fig. 15.9 The school science concept of energy (*bottom* scheme) is quite different in nature to the most common way the idea of energy is used in everyday discourse (*top* scheme)

students do create a new school science energy concept that they keep separated from their lifeworld concept and know when to use each in science classrooms and examinations versus in everyday discourse. By contrast, less successful students mistakenly attempt to make sense of school science tasks with their existing quite different lifeworld concept of energy. Arguably here, there is a good case for even avoiding the confusion of terms and basing teaching around free energy, given a suitable new label, rather than energy itself. As Fig. 15.9 suggests, the nature of these two concepts is quite different, as they have very different central concerns and core properties.

Multiple Conceptions or Manifold Conceptions

There is a significant challenge here for the research programme. In the case of energy, it seems the scientific concept is quite unlike the everyday notion, and Solomon's (1992) suggestion that conceptual development here is best understood as the formation of a new canonical energy concept in parallel to the existing lifeworld energy concept would seem credible and to avoid the issue of how people operate with a single energy concept which has quite different properties, and associated rules, depending upon context.

However, unlike in the case of energy, the lifeworld concept of a metal is not completely distinct from the formal chemical concept of metal – and indeed is quite close to how the term is commonly used in engineering contexts. There are many examples of metals – iron, copper, zinc, aluminium, etc. – that fit 'both' the lifeworld and the chemical use of the term. It seems much more credible here that the chemical concept of metal is not constructed separately from the everyday usage, but rather that the learner builds upon and modifies the spontaneous concept whilst learning the scientific nature of the concept.

This distinction would make sense in terms of the work of Chi and her colleagues (Chi, 1992, 2008; Chi & Slotta, 1993; Chi, Slotta, & de Leeuw, 1994). Chi has looked at student conceptual learning in terms of the way people build up their understanding of the ontology of the world. In particular, Chi has argued that such an ontology has distinct major trees of concepts and that a range of common learning difficulties in physics relate to students misidentifying scientific concepts that fundamentally refer to processes (e.g. heat) as material substances (Reiner, Slotta, Chi, & Resnick, 2000). So, for example, learners commonly think of heat more like the historical caloric (Cajori, 1922) than as a process of energy transfer due to temperature differences.

According to Chi, although conceptual change can bring about modest changes in the understanding of the nature of entities through modifications of a learner's ontological trees, it is not viable to switch a concept completely from one tree to another. So, for example, Chi would not think a student with a substance notion of heat can modify that to a process-based notion, but rather the learner would have to form the scientific concept of heat quite separately from any existing substance based notion. This ties in with Solomon's description of what happens in learning about energy in school and Claxton's prescriptions for avoiding attempts to build scientific concepts from students' own ideas where they are at odds with target knowledge.

From this perspective, teaching for conceptual development involves providing learners with alternatives to their existing concepts and supporting them to learn to access and apply the new school-learned concepts, rather than their prior conceptions, which remain unchanged. This will leave learners with multiple conceptions of energy, and heat, etc., each with different ranges of application. We might represent conceptual development in such as case as in Fig. 15.10 as an addition of a new concept.

If these two concepts are genuinely distinct, but just synonymous, then the context of a reference would be expected to activate one or other concept: just as references like 'Napoleon has been bringing home dead birds again' and 'Napoleon was an effective military leader' are likely to be recognised as referring to a family pet and a historical leader who share the same label. However, the example of developing the chemical concept of metal seems quite different, with the 'same' concept is incrementally modified over time. This would seem to be better represented by a scheme like that shown in Fig. 15.11.

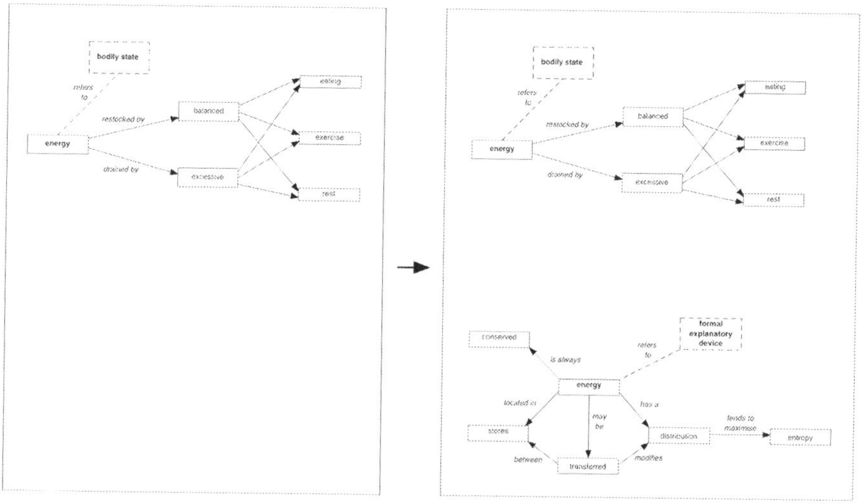

Fig. 15.10 Conceptual development of the scientific energy concept alongside a spontaneous concept (cf. Fig. 15.9)

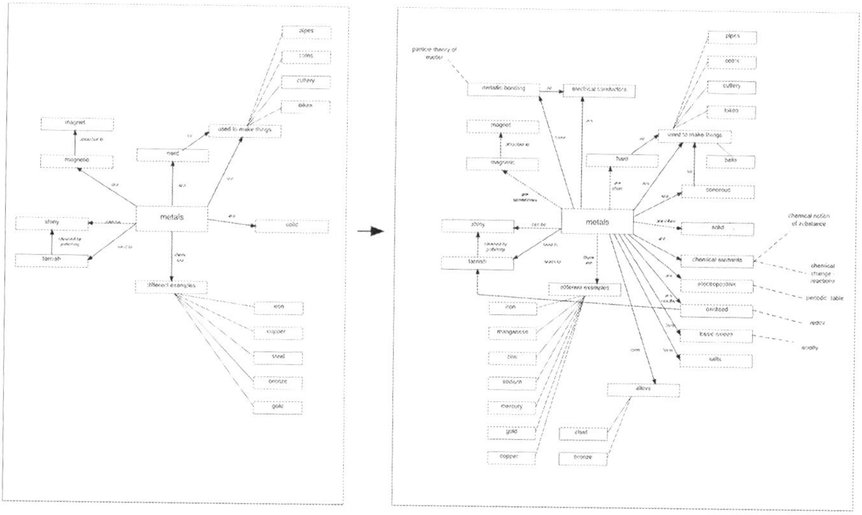

Fig. 15.11 Conceptual development of the scientific conceptions of metal (cf. Figs. 15.4–15.8)

In this case, our hypothetical learner will be able to work with a multifaceted metal concept, such that the term 'metal' can be understood differently in different contexts. So our learner will come to appreciate that, in the chemistry laboratory, sodium, potassium and mercury will be included in references to metals, but not say bronze or steel, whereas in the craft workshop the situation is reversed. Whilst it is simple enough to draw such figures, they do not explain how context provides the cues for different facets of the 'same' concept to be foregrounded in different circumstances. This presumably reflects extensive levels of interconnection between the different nodes of a figure such as Fig. 15.7, with various connection strengths, such that different patterns of activation across the concept become possible (such as in Fig. 15.12). This is an interesting area, but one which has not been explored within science education yet.

One promising idea that has been taken up for workers adopting a social con-structivist perspective on teaching and learning concerns the notion of discourse practices. Although Solomon referred to different domains, an alternative way of understanding her findings is to think in terms instead of the different discourses that learners partake in. So it has been suggested that we all initially learn a particu-lar discourse, with its norms and rules, in the home as infants. Later we enter other contexts, where new discourses become appropriate. From this context, learning progressions in school science involve switching from describing phenomena in terms of the 'home' discourse, to the more technical discourse of school science (Gunckel et al., 2012). The kind of cultural border crossing posited as a metaphor for learners entering the science classroom (Aikenhead, 1996) becomes a crossing into a different discourse community.

Multifaceted Conceptions in Science and Science Learning

The two examples of concept areas considered above, of energy and of metals, are quite different then in two important respects. The two energy concepts refer to dif-ferent kinds of things, and have very different properties, and so it seems feasible that they may be quite distinct in cognitive structure as there is little basis for form-ing any coherent account drawing on both notions. The scientific metal concept refers to a material substance, just as the lifeworld notion does, and there is considerable overlap in how these two facets of the metal concept can be understood, including a range of common examples. In that case a melded hybrid concept incorporating both spontaneous and formally taught aspects seems more feasible.

The example met in Chap. 12 of a student having several different ways of think-ing about chemical bonding (see Fig. 12.4) would at first sight seem to be an inter-mediate case: drawing upon both scientifically valid and alternative notions of the target concept. Yet, in practice it is much closer to a melded concept as the different narratives the student drew upon were, to his thinking at least, all based upon under-standing of formally taught chemistry. These explanations were all offered in a dis-course context of a student answering questions posed by a researcher who was also his teacher in the physical location of the college he attended. There is no lifeworld notion of chemical bonding, as the idea only has currency in academic settings, and although the idea that bonding forms so that atoms can fill their electron shells has no scientific validity, it was an interpretation of school learning and not an idea met outside of school science.

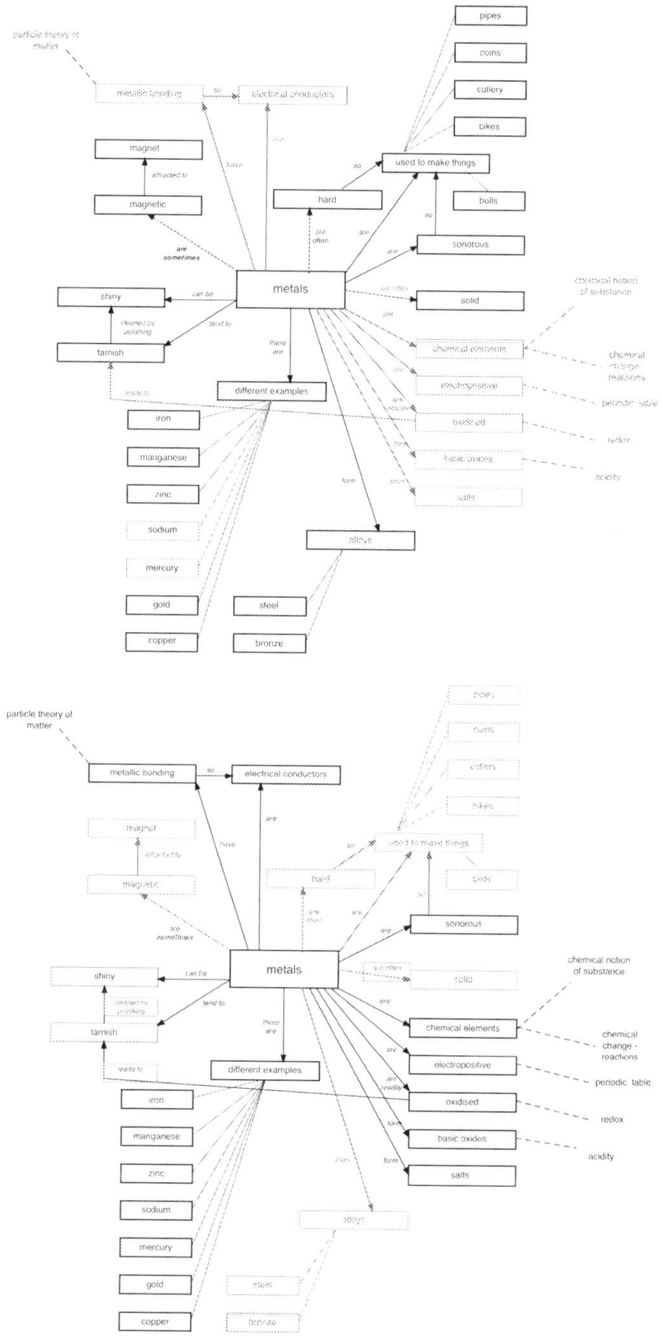

Fig. 15.12 Different patterns of activation of complex, multifaceted, conceptions (such as 'metal') seem to be triggered in different contexts

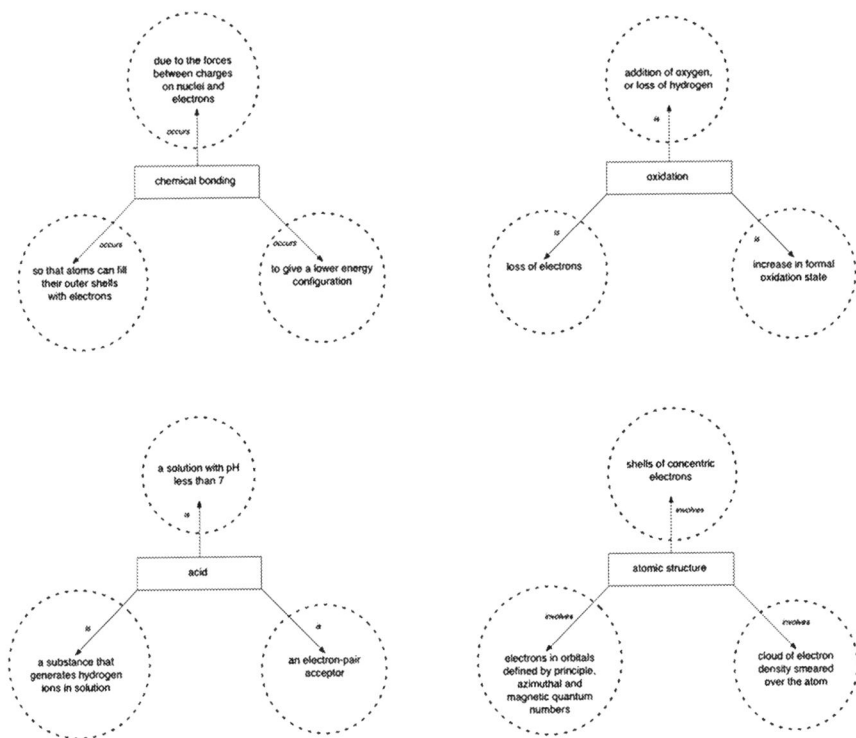

Fig. 15.13 Students' conceptions of chemical ideas may often be manifold

Whilst at first sight it might seem strange that a scientific concept taught in school could have such quite different manifestations, this is actually not so unlike a number of other concepts met in school and college chemistry. So the student of chemistry will meet sequences of definitions and models relating to such areas as acidity, oxidation and atomic structure. As with the student's multifaceted notion of chemical bonding discussed in Chap. 12, students are likely to develop conceptions in these other areas of chemistry which include inconsistent, alternative ways of understanding the concept due to the range of different models used in chemistry in these areas (see Fig. 15.13).

In three of the four cases the alternatives are all sanctioned within the curriculum and so are alternative conceptions of a scientific concept that are all in a sense canonical in they could be the 'right' answer in the context of certain questions that might be posed in the classroom. If we expect students to accept and distinguish between alternative scientific models for some scientific concepts met in their study, we should not be surprised if they develop a promiscuous conceptualisation of other concepts, especially when they consider they are adopting ideas presented in instruction.

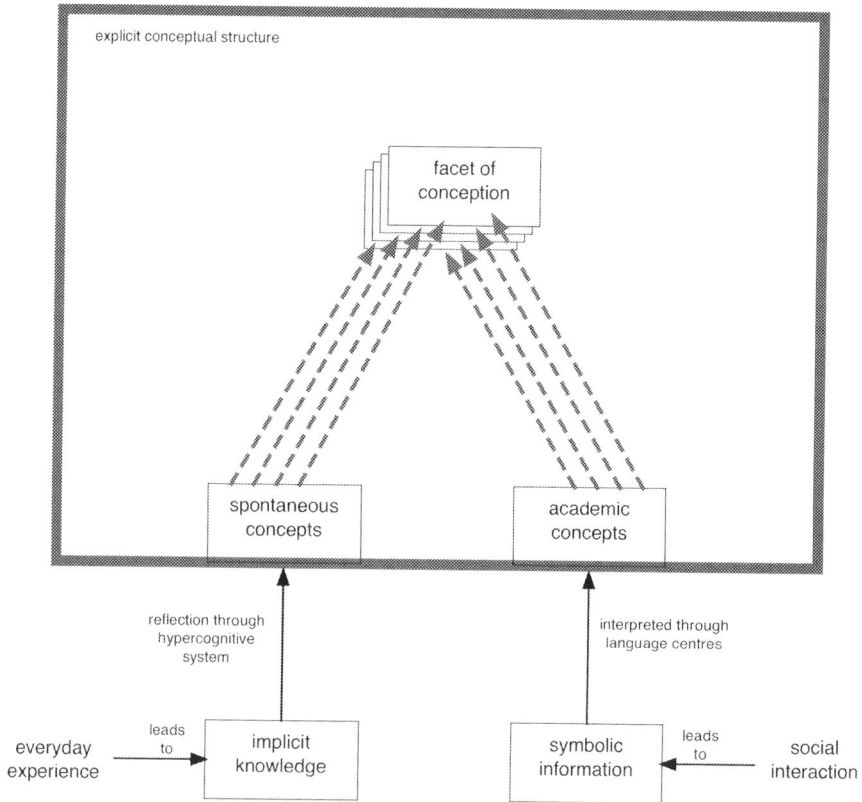

Fig. 15.14 Development of manifold conceptions

Petri and Niedderer (1998) explored one student's learning about atomic structure in a German physics class, as he, 'Carl', met different models of the atom. They concluded that

> Carl's statements in interviews, questionnaires and written tasks near the end of the instruction can be explained if we assume that the final state of Carl's cognitive system is an association of co-existing conceptions. To clarify, an association is when several conceptions co-exist and are connected to form different layers of the cognitive system, with a metacognitive layer on top. (Petri & Niedderer, 1998, p. 1083)

So Petri and Niedderer consider these different facets of a manifold conception to exist as distinct but connected 'layers' of the cognitive system. In Fig. 15.3 I have represented conceptual structure as a two-dimensional conceptual 'space', but Petri and Niedderer suggest an additional dimension is needed, allowing different facets of a complex concept to overlay each other.

This is reflected in Fig. 15.14, which suggests that each facet, or 'layer', of a manifold conception will draw upon both aspects of implicit knowledge and learning form others (e.g. teaching). Generally, different aspects of tacit knowledge may

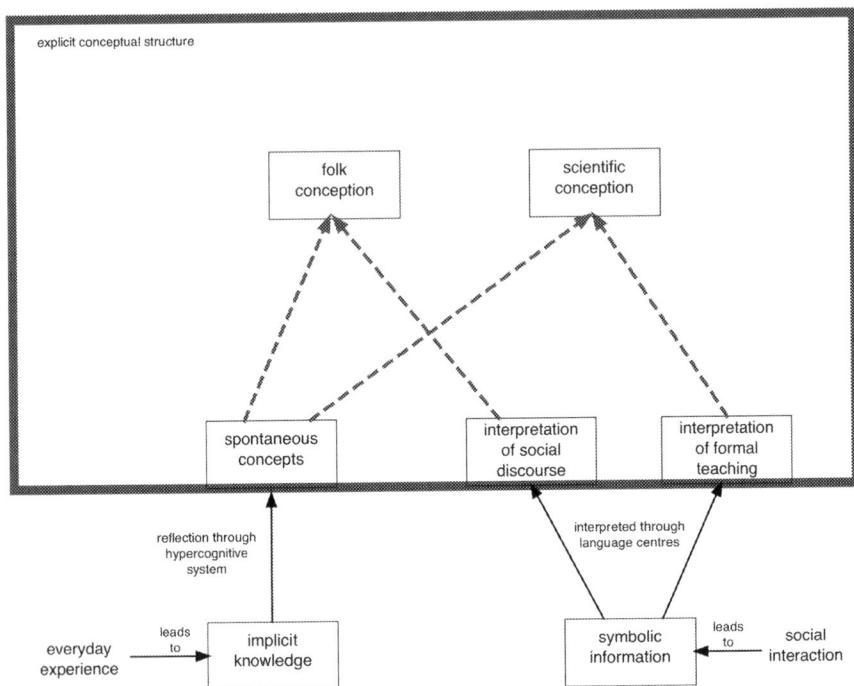

Fig. 15.15 Formation of discrete alternative conceptions

be drawn upon to provide the experiential basis of different facets. So, for example, it might be conjectured that the three facets of the acid concept suggested in Fig. 15.13 could possibly draw upon implicit knowledge (i.e. something like p-prims; see Chap. 11) related to extent (pH < 7), ejecting (generation of hydrogen ions) and engulfing or taking in (accepting electron pairs). This example is speculative, intended to illustrate the kinds of general patterns likely to be abstracted from experience at the level of phenomenological primitives. Establishing such links would be a matter for empirical research.

By comparison, the formation of distinct alternative conceptions, as suggested in the case of energy, whilst also drawing upon both linguistic information and implicit knowledge elements (e.g. relating perhaps to *balance* in one case and *conservation* in the other), would, it is suggested by Solomon's (1983) work, be represented quite separately in conceptual structure, as indicated in Fig. 15.15.

Revolutionary and Evolutionary Conceptual Change

It would seem we have three quite distinct models of how learners' conceptions of scientific concepts can develop. Vygotsky's argument that academic concepts draw upon spontaneous concepts allows the possibility of several largely independent conceptions developing in response to quite different discourses in different contexts and therefore being activated and applied in different contexts. So, arguably, for many learners, a conception of the folk world energy concept and a conception of the school science energy concept may be constructed largely independently.

However, much focus in science education (Taber, 2009b) and beyond (Vosniadou, 2008b) has explored how initially limited or 'alternative' student conceptions in science topics may be shifted towards canonical conceptualisations through teaching. Commonly a distinction has been made between evolutionary conceptual change and revolutionary conceptual change – although varying terminology has been used and some authors limit use of the term conceptual change for more abrupt, 'revolutionary', shifts. The simple models considered above can be related to this distinction.

So the hypothetical example of developing thinking about metals, discussed above, offers an example of an evolutionary conceptual change. Over a period of time, the learner's conception of metal shifts considerably (Fig. 15.11) but without any major discontinuities. New examples and properties are added without changing the basic type of thing that a metal is – a class of materials. Even though some previously accepted features have to be modified (as metals do not have to be solid, hard or magnetic), these modifications do not require a fundamental shift (as metal is still one type of stuff), and so we seem to be operating with the 'same' concept changing over time (whereas in the energy case we have added a whole new conception).

A more problematic case in many ways is how it is possible for learners to have revolutionary shifts in their conceptions: where they come to adopt a fundamentally different conception for the same target concept. Work on how to encourage such changes in learners has been the focus of much research and discussion in science education (Caravita & Halldén, 1994; Posner, Strike, Hewson, & Gertzog, 1982; Schwedes & Schmidt, 1992; Smith, 1991).

The ability to shift one's thinking between quite different conceptions of a topic has played a major role in the history of science and indeed was the basis of Kuhn's (1996) highly influential work on the 'structure' of scientific revolutions. However, inherent in that work was the assumption that even among professional scientists, such revolutionary 'changes of mind' were rare, with their spontaneous occurrence being limited to a few individuals. During so-called normal science most scientists adopt the canonical ideas of the field, supported by their induction into the disciplinary matrix through the discourses of the field. Science teachers are trying to encourage their students to adopt these canonical ideas, not make revolutionary breakthroughs in science by conceptualising the field in a more productive way than the rest of the scientific community. However, in some topics this may require a revolutionary shift from the students' current thinking.

Kuhn compared revolutionary insights to a paradigm shift, where a new pattern is recognised among familiar elements. In a sense that is what researchers recommending a knowledge-in-pieces approach (Hammer, 2000; Smith, diSessa, & Roschelle, 1993) to supporting conceptual change in students are looking to facilitate: that the teacher helps the learner construct new ways of understanding scientific concepts by building upon the most appropriate conceptual resources among the available implicit knowledge elements.

Thagard (1992) looked at this process of revolutionary change in scientific ideas from the perspective of explanatory coherence and suggested that such conceptual changes involved the construction of the new way of understanding the topic, in effect 'in the background', until a point was reached when the new way of thinking comes to make more sense, and fit more of the data, than the existing way of thinking. At that point the individual comes to consider the new way of thinking more fruitful and sets about persuading the field.

Arguably, representations such as Fig. 15.14 may be useful here, following Petri and Niedderer's (1998) metaphor of different layers of conceptual structure. Just as an individual might build up an 'association' of alternative ways of understanding concepts such as acidity and oxidation, and then *select* between them, so might they build up alternative conceptions of a target concept and over time *shift* between them. The suggestion is that in a revolutionary change, there is some kind of tipping point (Gilbert & Watts, 1983) where the balance of perceived strengths and limitations of two distinct conceptualisations switches to the new understanding being developed having more coherence, and this then becomes the preferred way of thinking about that target topic.

So whereas in the case of concepts which are understood in different ways (oxidation, acidity) we would expect the learner to retain the use of these different 'layers' as the different models are retained within science for different purposes, we can envisage how Lavoisier constructed his new understanding of chemical change as a new 'layer' 'over' the traditional phlogiston-based conceptualisation (Thagard, 1992), and over time came to consider the new conception (e.g. combustion is reaction with oxygen) as more fruitful than the traditional (e.g. combustion is release of phlogiston) conception. Using the visual metaphor of the representations in this chapter, making that comparison required Lavoisier to build his two conceptions as overlapping layers within conceptual structure allowing them to be directly compared (as in Fig. 15.14), rather than as discrete conceptions in different domains of conceptual structure (as in Fig. 15.9). However, it is important to keep in mind that figures such as those presented in this chapter only offer a schematic representation, a kind of spatial metaphor, as layers in the representational conceptual space do not relate directly to any obvious structural feature of the neurological substrate. The notion of layers may have much more to do with connectivity – how representations are associated through synaptic connections – than physical location in the brain.

Whether an individual retains manifold conceptions, or – in effect – shifts to a new conception leaving the earlier way of understanding in the background, but seldom activated, will presumably depend upon the extent to which the new way of thinking is found to make sense of all information and observations perceived as relevant to that target concept. During my study of the student discussed in Chap. 5, who developed his thinking about the nature of chemical bonding, there was the construction of new 'layers' within the 'association' (in Petri and Niedderer's terms) and a gradual shift in the extent to which the different layers were applied in discussing chemical phenomena. However, during that study, the original alternative conception never fell into disuse, although it ceased to dominate the learner's thinking across the range of application of the bonding concept (Taber, 2001b). It seems reasonable to consider this an incomplete learning pathway towards a revolutionary type of conceptual change: a revolution that was not completed during the two years of the learner's chemistry course.

When Is Revolutionary Change Required?

Given that revolutionary change is seen as so difficult to achieve, it must be questioned whether science teachers can reasonably be expected to encourage this type of change among their students. We have seen that Claxton (1986) has argued that often teachers would be better advised to avoid challenging existing conceptions and rather to seek to construct new conceptions to operate in parallel with their lifeworld understandings. This would make sense in those cases where Chi's (1992) work suggests the target knowledge is ontologically incompatible with the student' existing conceptions.

Watts and Pope (1982) suggested that it might to be useful to think about the learner's developing understanding of science topics as though the learner was working within a Lakatosian research programme (RP), and in the same year an influential paper about conceptual change made use of the idea implicitly (Posner et al., 1982, see below). For Lakatos (1970), a RP has a hard core of commitments, around which auxiliary theory is constructed. New evidence may lead to modification in the 'protective belt' of auxiliary theory, without challenging hard-core convictions. As long as new evidence can be accommodated within the programme, that is, without contradicting hard-core assumptions, then the learning trajectory will be shaped within that programme. If the core assumptions become seen as non-viable, then a new RP needs to be initiated built around new starting points (a different hard core); however, there is usually scope for reinterpreting new information within an RP using the malleable nature of the protective belt to insulate the hard core itself from the consequences of anomalies or counterexamples.

This seems a potentially productive way of thinking about student learning in science, although the perspective has seldom been taken up by science educators, that offers a useful perspective for making sense of some of the contrary claims in the literature as to whether students' ideas should be characterised as stable or readily modified (Taber, 2009b). Such an approach explains why learners are so resistant to change *some of* their ways of thinking. The kinds of ontological commitments that Chi (1992) suggested were so important would be strong contenders for hard-core assumptions of a student's personal RP. Perhaps the term personal *learning* programme, PLP, would be better in the context of individual learning. So a student who studies the physics of heating from within a PLP which has a hard commitment that heat is a kind of fluid is likely to make progress in learning – if not necessarily always quite the progress the teacher intends – as long as it is possible to interpret instruction in terms of heat as a fluid. For example, thermal conductivity could be understood in terms of how readily the heat fluid can pass through materials; temperature can be understood as a measure of the amount or concentration of the fluid in a particular place; and convection, conduction and radiation can be seen as means by which the fluid can get from one place to another. More detailed work is needed to explore how useful the PLP perspective might be in understanding student learning trajectories and in developing teaching to modify such trajectories.

In one study that has applied this perspective, Daniel Tan and I have suggested this may be a useful way of thinking about why research undertaken in Singapore found that graduates entering teaching as chemistry specialists showed similar levels of alternative conceptions (in the topic that was the focus of the study, ionisation energies) to the students they would be teaching, despite their opportunity to study the subject in depth in higher education (Taber & Tan, 2011). Presumably many of these new teachers had attended lectures, and read textbooks, and partaken in laboratory classes and tutorial and seminar sessions – but had managed to interpret a great deal of detailed information about their subject in line with the (alternative, non-canonical) hard-core assumptions about key chemical principles they brought from their own schooling.

The Notion of Conceptual Ecology

A basic premise of constructivist ideas in teaching is that the current state of a learners' cognitive system will influence the learning that takes place in the future. One aspect of the current state of the system is the actual available cognitive apparatus available to process new information – which might be said to depend upon the individual's level of cognitive development (see Chap. 13). However, just as important is the state of current knowledge as this provides the context in which new information can be interpreted and made sense of. I am using the term 'knowledge' here as suggested earlier in the book (in Chap. 9) to refer to what the learner believes to be the case or simply considers as a viable possibility: that is, the range of notions under current consideration as possibly reflecting some aspect of how the world is.

In essence the only alternative senses that can be made of teaching are those within the range of possible understandings of 'how the world is' available to the learner. Moreover, most commonly, we understand something we are told according to one out of those ways we have available to make sense of it, so normally the cognitive system will channel 'input' to activate some particular existing feature of conceptual structure that best seems to match the incoming information (cf. Chap. 4). In terms of the models being considered above, the context around what we hear or see will tend to cue activation of a particular conception: the folk conception of energy or the scientific conception, (Fig. 15.15), or a particular facet ('layer') of a manifold conception (Fig. 15.14). The context may be about 'who' and 'where' (children chatting in the playground versus the teacher presenting material formally in class) or more nuanced indicators. The teacher who asks how the students know whether something is an acid expects a different answer, based on a different facet of the acid concept, when she is dipping indicator paper into a solution in a flask, to when she is drawing 'curly arrows' on a symbolic representation of a reaction mechanism.

Posner et al. (1982) drew on the notion of 'conceptual ecology' (p. 214) and proposed conditions that need to be satisfied before major conceptual change (accommodation, in their account) could occur. They suggested four such conditions (p. 214):

1. There must be dissatisfaction with existing conceptions.
2. A new conception must be intelligible.
3. A new conception must appear initially plausible.
4. A new concept should suggest the possibility of a fruitful research programme.

The latter point relates to Lakatos' ideas: Lakatos (1970) suggested that scientists should continue to work within a research programme even when its flaws were apparent, until there was an alternative which looked more promising.

Limitations of the Conceptual Ecology Metaphor

Posner, Strike, Hewson and Gertzog's analysis was criticised by Pintrich, Marx and Boyle (1993) who argued that the strictly 'rational' basis for learning assumed by the Posner and colleagues model did not take into account the realities of the learning context in schools. Pintrich and colleagues argued that this approach ignored the way 'individual students' motivational beliefs may influence the process of conceptual change' and how 'individual learning in classrooms is not isolated but greatly influenced by peer and teacher' (p. 172). Pintrich and colleagues suggested that the operation of the conditions identified by Posner and colleagues was constrained or enabled by various extra-conceptual issues, and they nominated 'a range of theoretical entities in the field of motivational research that are possible candidates for incorporation in conceptual change theory and research'. In Fig. 15.16 the Pintrich et al. (1993) account is represented spatially as a set of terms of concentric circles.

Pintrich and colleagues argued that 'motivational constructs such as goal orientation, values, efficacy beliefs, and control beliefs that can serve as mediators of this process of conceptual change', and that students' 'intentions, goals, purposes, and beliefs' would 'influence the direction of thinking as the students attempt to adapt to the different constraints and demands placed on them by the tasks and activities they confront in classrooms' (p. 192). Pintrich and colleagues also suggested that a student's level of interest and the expectations implied by teaching styles and approaches would influence whether students would expect to process new information in any depth in a class and what they attend to in class. They suggested that the institutional and bureaucratic imperatives in schools may not always provide the

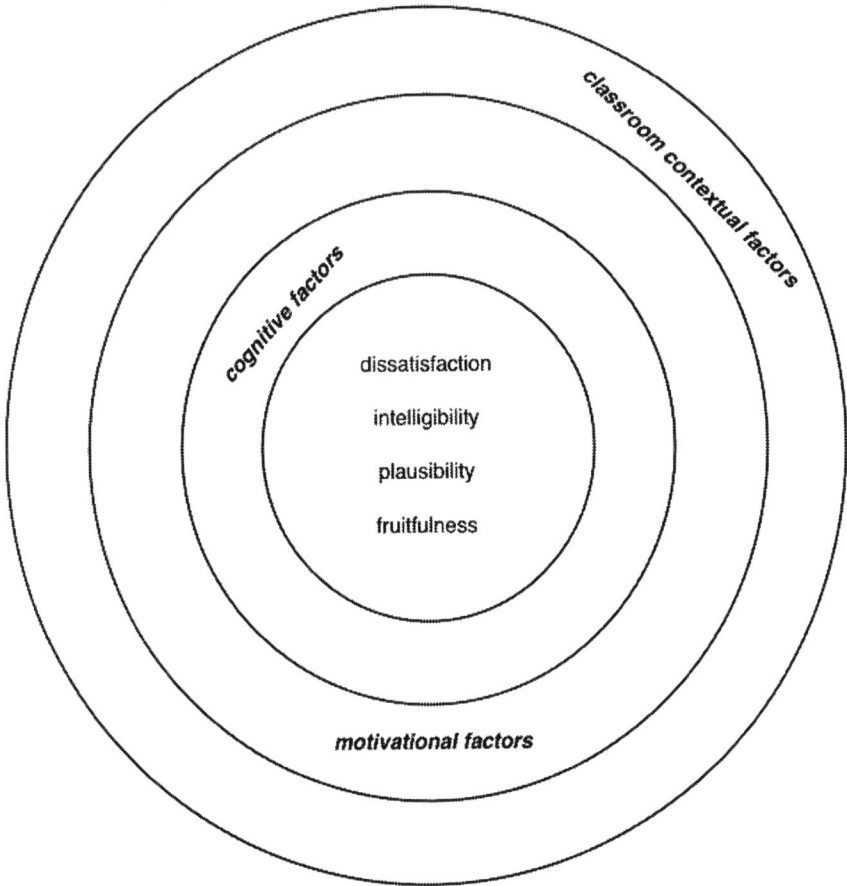

Fig. 15.16 Influences on conceptual change according to Pintrich et al. (1993)

environment for effective learning so that 'even if some students approach school learning as intentional learners with a goal of developing integrated and sophisticated understanding of a field of study, they might not believe that the goals of the schooling enterprise are to foster such understanding' (p. 193).

In effect, Pintrich and colleagues suggested that the factors identified by Posner and colleagues operate only to the extent that (a) the learner (i) has developed the necessary cognitive ability to apply them and (ii) has the motivational orientations for deep learning and (b) the institutional context offers norms and expectations that support this approach to learning (see Fig. 15.16). So whilst Pintrich and colleagues' argument is seen to offer criticism of the Posner and colleagues model, it does not negate that model, but highlights how it is incomplete.

Posner and colleagues' four conditions clearly refer to the way information is interpreted within a cognitive system, and as has been suggested earlier in this book

(Chap. 4) much information from the environment is not attended to and is filtered out before it reaches consciousness. It is not unknown for students in science classes to be thinking about 'something else', whilst the teacher is carefully setting out the arguments for accepting the scientific way of thinking about the topic of the lesson. It should also be borne in mind that although Posner and colleagues set out conceptual change as rational choice, the discussion earlier in this book suggests that need not mean explicit choice based on conscious reflection. It seems much of the cognitive processing that is involved in reaching such changes of mind takes place out of conscious awareness, although Pintrich and colleague would be right to point out that this is usually after periods of explicit consideration and exploration of the evidence that is being 'weighed'.

Figure 15.16 suggests that various factors filter, 'colour' and channel the information that will be 'weighed up' in such situations. As well as the norms and expectations of the classroom that might influence how study and learning is understood in that lesson, Pintrich, Marx and Boyle posit that such motivational factors as mastery goals, epistemic beliefs, personal interest, utility value, importance, self-efficacy and control beliefs will influence the level of student engagement with the material being presented. Where the student is engaged, Pintrich and colleagues list a range of cognitive processes/skills that will influence processing of available information, the 'data' in the system: selective attention, activation of prior knowledge, deeper processing (elaboration, organisation), problem-solving and finding, metacognitive evaluation and control and volitional control and regulation (p. 175). In effect, if the classroom conditions support deep engagement with learning, and if the learner is interested enough to be motivated to give full attention, and if the learner has developed the cognitive skills to be an effective learner, then the conditions for conceptual change may operate.

Pintrich and colleagues criticised the notion of the conceptual ecology because 'this metaphor is limited as a depiction of ontological change in learners in as much as learners are purposeful while ecosystems are not' (p. 192); however, arguably a learner's purposes are *emergent properties of their cognitive system*, partially responding to the learner-as-organism's inherent 'goals' which themselves are outcome of natural selection. As with much of the difference between Posner and colleagues and Pintrich and colleagues this seems to be a matter of how and where one focuses.

As an analogy, naturalists discussing the ecology of a particular habitat somewhere on earth would not normally feel the need to spell out much of the planetary-level context for what they are reporting. A hypothetical exobiologist from elsewhere in the galaxy might find it rather odd that such an account does not consider the rather significant factors of the radiation profile of our sun or the levels of oxygen in the earth's atmosphere. The exobiologist, perhaps having field experience of the ecosystems in many different planetary contexts, might feel these are rather major factors that have very important consequences for the biota in our focal habitat, and so consider it strange that they are not attended to. Yet, the earthbound naturalist who limits her reading to Earth-based journals might tend to assume that these factors can be considered as taken-for-granted background conditions.

Components of a Conceptual Ecology

In an earlier work I included a figure (Taber, 2009b, Figure 7.1) somewhat similar to Fig. 15.16 here, where I suggested that the individual conceptual ecology of a particular learner should be seen as nested within a series of other levels of context: a social environment roughly at the level of the classroom contextual factors in the Pintrich et al. (1993) model, within a cultural environment (the field of shared beliefs, values, norms, etc. in the society), within a natural environment (the biological constraints that shape the physiology and anatomy within which our cognition develops) and within a physical environment (which sets out the limits of what is possible in the universe).

Whether the cognitive and motivational levels of the Pintrich, Marx and Boyle model suggest there are missing components to that earlier representation depends upon what one includes within the scope of conceptual ecology. Posner and colleagues offered their own list of what might be important in influencing conceptual change (pp. 214–215):

- Anomalies
- Analogies and metaphors
- Epistemological commitments, including

    - Explanatory ideals
    - General views about the character of knowledge

- Metaphysical beliefs and concepts

    - Metaphysical beliefs about science
    - Metaphysical concepts of science

- Other knowledge

    - Knowledge in other fields
    - Competing concepts

Arguably the notion of a conceptual ecology can also encompass many of the factors Pintrich and colleagues considered at the cognitive and motivation level. The cognitive apparatus available to process the conceptual contents of a cognitive system is clearly highly relevant to how those concepts may be understood, related, compared, evaluated, etc. (see Chap. 13). Student interests and expectations about the nature of studying and learning, and what is expected and needed to function in the classroom, would all seem to be readily encompassed within conceptual ecology. For example, a belief that *learning in science class is about memorisation of material presented by the teacher* would be encompassed within the broad notion of knowledge used here. Similarly, a student view (whether explicit or not) that *studying science is unimportant and that minimal engagement in science classes saves valuable resources for thinking about more important things* is a judgement made in relation to that individual's overall conceptual structure. This will reflect prior learning about what is important and so should be valued, as acquired through prior experience and influenced by family, teachers, media, peers, etc.

So conceptual ecology is not just about what the learner thinks they know about the topic area under consideration, but also includes, inter alia, what they believe about their own learning abilities (generally or in that subject), what they believe about effective learning and what they believe about the importance of prioritising study in that subject over other competing demands on their attention. Pintrich and colleagues include notions about the nature of science as components of conceptual ecologies, and this might include features related to what might be termed 'scientific values' and 'scientific ways of thinking' (cf. Chap. 7). One particular aspect of an individual's way of making sense of the world that has attracted considerable attention in science education is what is known as worldview.

Worldviews, Scientific Attitudes and Religious Beliefs

Working as a scientist would seem to presuppose certain common values and assumptions (Kuhn, 1996): for example, the acceptance of some form of post-positivist position on the possibility of obtaining useful knowledge through systematic enquiry. Some would suggest there is a scientific 'worldview' that goes beyond this. Cobern describes a worldview as being 'about metaphysical levels antecedent to specific views that a person holds about natural phenomena' and providing 'the set of fundamental non rational presuppositions on which … conceptions of reality are grounded' (Cobern, 1994, p. 6). The position adopted here (developed in more detail in Taber, 2013e) is that scientists do not necessarily share the same worldview, but that scientists do share certain core commitments which would form *part of* their worldviews.

Arguably, (a) the consistency of the external world; (b) the presence of law-like regularities in that world; and (c) the possibility of obtaining viable knowledge of that world; can all be considered 'presuppositions' of science – principles that may then *seem* ratified by the findings of science itself. However, if this set of assumptions were to be viewed from a position that does not adopt such presuppositions, they might be considered tautological – as we interpret our observations in the light of these very presuppositions. Indeed, for most people with a scientific background, it is rather difficult to see how one could take a stance that does not include these particular assumptions about the world. A potential criticism of the scientific perspective is that its claims to knowledge rely upon metaphysical commitments (such as a–c), which are not in themselves open to genuine meaningful testing. Most readers of this book might wonder how it could be possible to live a structured, meaningful life in the world without taking for granted something like (a–c). These are assumptions that may well seem necessary and sensible, yet they are still a priori commitments. They inform our interpretation of experience rather than deriving from it.

Worldview has been described as a set of 'assumptions held by individuals and cultures about the physical and social universe… [including] the purpose or meaning of life' (Koltko-Rivera, 2006, pp. 309–310) or 'the principles and beliefs – including the epistemological and ontological underpinnings of those beliefs – which people have acquired to make sense of the world around them'

Table 15.1 Presuppositions of scientific and scientistic worldviews

Presuppositions	Worldview
The universe exhibits regularities reflecting some underlying stability (laws)	Scientific
Systematic enquiry into the world can bring knowledge that is in some sense valid (not necessarily 'absolute')	Scientific
All that exists is the physical world which can be probed by science	Scientistic
Science is the only approach which can provide genuine knowledge	Scientistic
Science can ultimately provide knowledge of all aspects of the world	Scientistic

(Kawagley, Norris-Tull, & Norris-Tull, 1998, p. 134). Certainly for some, there is considered to be a scientific worldview that goes beyond assumptions about the regularity and knowableness of the universe and encompasses more scientis*tic* assumptions such as:

- The physical universe is all there is (i.e. there is no *super*natural realm).
- The only type of worthwhile (or 'real' or 'true') human knowledge is that accessed by science.

And perhaps even that:

- All there is to know can one day be uncovered by science.

Clearly, holding such presuppositions can be very consistent with undertaking scientific work, but it is also clear that *not* holding these views need not undermine working in science (Taber, 2013e) – whereas, for example, not believing that the universe had some overall regularity to it would make scientific research, systematic enquiry into the natural world, rather pointless. These basic ideas, potential presuppositions of scientific work, are listed in Table 15.1. Whilst the first two principles would seem to be fundamental to all scientific work, the extent to which individual scientists adopt the final three will be much more variable.

A distinction that is sometimes made is between methodological materialism (which is about the assumptions necessary to do science) and metaphysical materialism (which is a broader assumption about the nature of the world). Methodological materialism does not allow supernatural causes and explanations to be introduced as part of science and is widely adopted by scientists. Metaphysical materialism goes beyond this and excludes the possibility of the supernatural completely – and is only adopted by some scientists. From this latter perspective, God or a spiritual realm is not only irrelevant to scientific explanation and argument but is necessarily rejected as a possibility (Taber, 2013f).

Some scientists exclude the possibility of there being any kind of God or other supernatural being and *consider this* to be part of their scientific approach to the world, whilst other scientists see no contradiction at all between undertaking scientific work and retaining a faith in a creator God who acts as a kind of 'ultimate' cause beyond the reach of science (Cray, Dawkins, & Collins, 2006). Indeed, many of the early pioneers of modern science were theists and did not see any need to exclude references to God from their scientific work given the cultural context in which they worked. Some other scientists are committed atheists but do not consider that their scientific colleagues need to share that commitment. Yet other scientists would consider the current evidence available about such matters as inconclusive and so would adopt the position that T. H. Huxley proposed as most suitable for scientists: agnosticism (Gilley & Loades, 1981; Lightman, 2002).

This would suggest that scientists do not all share a single scientific worldview but rather that there are different worldviews that have been found to be consistent with the necessary commitments for scientific work, even though *some* scientists argue that metaphysical materialism *should* be adopted as the basis of a scientific worldview (Taber, 2013e).

Worldview commitments provide a basic framework for making sense of the world which is not open to challenge – very much the 'hard core' of an individual's PLP (personal learning programme) in the terms discussed above. As everything is interpreted from the starting point of worldview, what is understood will not seem to contradict worldview commitments – and what others may suggest that seems contrary to worldview commitments may seem absurd. This is seen in debates about science and religion, with some scientists perfectly able to accommodate scientific work within a religious worldview, and indeed actually viewing all scientific knowledge within a theistic interpretation such that what science uncovers is how God maintains His creation through natural laws and mechanisms, whilst others seem incredulous at this, suggesting that any belief in the supernatural is inconsistent with a scientific attitude.

There have also been cases of devout scholars who whilst considering themselves scientists were able to dismiss widely accepted scientific ideas about the evolution of the universe and life on earth claiming that there was no real evidence for such ideas (Morris, 2000). Often this derives from worldview commitments to religious scripture as (a) the Word of God that (b) must be understood as offering a technically accurate account of the origins and history of the natural world, rather than offering theological truth sometimes presented through allusion, metaphor, myth, etc. This approach would not only be rejected by metaphysical materialists but also most theistic scientists who accept consensus scientific ideas about origins and believe that religious scriptures must be interpreted in the light of modern science, rather than scientific evidence needing to be fitted with a literal interpretation of scriptures.

However, in some educational contexts, many students do hold worldviews that are inconsistent with current scientific thinking about the origins of the universe and of life. For example, this has been a major issue in many parts of the USA (Long, 2011). In some Islamic countries, for example, Jordan (Dagher, 2009) and Oman (Ambusaidi & Al-Shuaili, 2009), science education is based on national curricula that explicitly reflect a theistic worldview. In National contexts such as these, students are actively taught that science reveals the wonders of God's creation.

Student Worldviews Inconsistent with Science Learning

This issue can be very significant for science educators. For some students learning science, it may not be just that they have acquired lifeworld conceptions at odds with scientific ideas but that the scientific ideas presented in the curriculum are contrary to fundamental commitments about how the world must be. So many learners in some parts of the world may enter the science classroom believing that all the main types of living thing are the products of special creation, that is, descended from ancestors created as complete organisms, in their modern form, and that the heavens are unchanging. If these are not simply incidentally acquired ideas, but derive from worldview commitments that are intrinsically tied in with issues of culture, community, identity, self-worth, etc., then these students will tend to interpret teaching in terms of such commitments. Where this is not possible then they will often reject the teaching.

It is not only students meeting scientific ideas from certain theistic worldviews (e.g. some Christian and Islamic perspectives) that may find some aspects of school science incongruous with existing commitments. Students from many indigenous populations are likely to find the reductionist, analytical approach of modern Western science at odds with holistic ways of understanding the world applied in their culture (Kawagley et al., 1998). The metaphor of 'border crossing' has been used to describe the process of entering into the culture of the science classroom. Aikenhead and Jegede (1999, p. 269) acknowledge that barriers to border crossing may be most severe among students from developing countries who find 'that school science is like a foreign culture to them' due to 'fundamental differences between the culture of Western science and their indigenous cultures'. However, they also suggest that 'many students in industrialised countries share this feeling of foreignness as well'.

Worldview as Conceptual Habitat

If conceptual ecology comprises of various components such as conceptions, analogies and epistemological belief, then to posit worldview as a component of conceptual ecology could seem to assign it no more status than an alternative conception or familiar image. This would not do justice to the influence of worldview.

This raises the issue of the nature of the conceptual ecology notion: that is, is it more than just a metaphor? The notion of conceptual ecology can be seen as a pedagogic device – as a means of drawing attention to how learning takes place in a complex context, with many potentially interacting factors influencing the learning process. However, it can also be seen as a form of model. This would require us to move beyond the metaphor (a reference to a non-specific similarity) to consider a formal analogy between conceptual ecology and biological ecology. Analogy allows learning by mapping between two parallel structures, from the more familiar to the less familiar (see Chap. 7). Teachers commonly use analogies to 'make the unfamiliar, familiar' as a means of using learners' existing knowledge as a basis for learning new concepts by identifying the structural similarities (Harrison & Treagust, 2006). In teaching, the teacher suggests an analogy and shows how to map from analogue to target. However, individuals can also explore potentially useful analogies for themselves, and analogical processes have been seen to be extremely important in the way scientists form new ideas (Nersessian, 2008).

In terms of the conceptual ecology notion, to refer to worldview as *a component* of a conceptual ecology is probably too weak a suggestion – where different components of the ecology may thrive or fall into disuse, the worldview will only shift, if at all, through a slow process of succession. Rather, then, in terms of the ecological analogy, the learner's worldview is akin to the habitat in which the conceptual ecology develops.

Once we adopt ecology as an analogy, we are in effect using a model, and in principle, a testable model. We might posit component features of the model that could lead to testable hypotheses:

- Concepts exist in a kind of ecology: they can take root, thrive or whither according to the environmental conditions.
- Some conceptions may be much better established than others.
- Conceptions may be in competition for the same niche in the ecology.
- Conceptual 'fitness' can only be judged in the context of the ecological conditions.
- A new conception requires a niche in which to become established.
- Worldviews offer very different habitats, suitable for rather different conceptions to thrive.

It is clearly possible to continue to develop such an analogy:

- The neonate offers a new habitat for conceptual development. In the biological case, a new habitat would have geological and physical conditions established, but no biota yet. By analogy, in the conceptual ecology, the child will have genetic predispositions, etc. but will not have formed any conceptions about the world.
- Change of worldview is a rare and potentially a major disruption of conceptual ecology akin to a major traumatic event (e.g. earthquake, flood), which disrupts an ecosystem and may allow very different succession of species.

One might suggests that the development of a particular conceptual habitat (worldview) will reflect the local (cultural) climate, and sometimes, for some learners, science lessons may seem like short periods of bad weather – (intellectual) storms at odd with more familiar climatic conditions and with potential to wreak havoc with the fine balance of the (conceptual) ecosystem. Conceptual ecology seems to offer many such opportunities for thinking about conceptual development in terms of the analogy with ecosystems. However, being able to suggest analogical mappings does not in any way assure that the conceptual case *is* like the ecological situation in useful ways. Rather, the analogy can be used as a creative device to suggest possible avenues for testing in research. Conceptual ecology is a fertile metaphor, but the extent to which it should be adopted – for example, as a way for teachers to think about their work – is a matter for empirical testing through research.

Part V
Conclusion

This book began with a claim that all was not well in the way that many research reports in science education discuss the central themes of student knowledge, understanding, thinking and learning. Research studies commonly report how many students had a certain understanding or expressed certain ideas. We have also seen that some papers offer quite detailed accounts of aspects of student subject knowledge or learning progression. Yet it is also clear that these key foci of our research interests are non-observables and can only be inferred indirectly. We therefore use models of various forms to relate the phenomena that can be observed in classrooms, or in interviews with particular students, or in terms of the various productions students make during their studies or in research contexts (e.g. written work), to the constructs we claim to be researching and writing about: people's knowledge, understanding, thinking, beliefs, etc.

My argument has not been that this is to be in some way criticised, as it is unavoidable. However, my motivation in writing this book derives from the view that the way we often write about student knowledge, student understanding, student thinking, student learning, etc. *seems to imply* either that we can directly observe these things or at least that there is an unproblematic relationship between what we can observe and measure and the conclusions we can draw about these various mental properties. That is, often the modelling process is rendered invisible in our reports (and I suspect sometimes also in the thinking during the research which leads to those reports). This makes a very problematic area of research appear much more straightforward and often leads to research results appearing much more definitive than perhaps is justified. It seems this may also in part explain the high incidence of scholars disagreeing on what research does tell us (Taber, 2009b).

In the first chapter of the book I set out this position, offering examples of the kinds of often apparently definitive conclusions made in research reports. This was not to criticise those particular papers, which were just offered as examples of common trends, but to highlight the way such conclusions often seem to imply definitive new knowledge that may be inappropriate given the necessary caveats of work of this kind. Such caveats are barely acknowledged in many papers published in the literature. So the starting point for this volume was the suggestion that a major problem for research into learners' ideas and learning in science is the way that many studies offered accounts of research as a fairly unproblematic process of identifying aspects of students' thinking, understanding, knowledge and learning.

The rest of the book has presented an account of these very processes, but from a perspective of recognising this area of enquiry as heavily dependent upon *developing models* of cognitive properties and processes. This book offers an account based on current research and scholarship where it is made explicit just how research in this area is necessarily contingent upon commitment to (some set of) ontological models of the nature of human cognition and how this relates to our subjective mental experiences, and so commitment to an epistemological position on how much we can possibly know about these research foci, and how we can best go about that work. Some readers may not accept some of these commitments – but that is why it is so important to be explicit about the assumptions upon which our work is based.

If a research study assumes a model of cognition that we personally feel is invalid, then one necessary link in the logical chain of building up a case for the paper's findings is unconvincing – and so the findings themselves become suspect. Similarly, if the methodology adopted takes for granted an implicit model of how data collected (based on the public representations made by study participants) relates to a construct such as understanding, and the reader finds that model inadequate, then again the paper's knowledge claims will not seem well supported. It would be much better if such a model was explicit, making it easier for readers of research to appreciate and evaluate the epistemological assumptions of researchers. However, in many papers it seems to be assumed that these are things that can be taken for granted, which I suggest is due to our ubiquitous reliance on our everyday folk notions of the mind and the ability to communicate ideas to others.

The account offered here is intended to inform research. It will be clear that in some places the account offered is necessarily provisional and uncertain, given the current state of knowledge about cognition and the nature of conceptual change. The account in this volume may not ultimately be judged as 'right', as some of the models suggested may need to be revised or abandoned in the light of new research. However, the process of attempting to make explicit (i) how we should operationalise such notions as knowledge, understanding, etc. when undertaking research into them and (ii) the limitations and affordances of our processes of enquiring into these constructs through the kinds of data we can collect in educational research at least indicate just how tenuous and dependent upon modelling our research claims necessarily are.

This dependence upon modelling is the case whether the models we are drawing upon are taken-for-granted models shared in the lifeworld or explicit models set out as part of the technical basis for our research. My argument is that if one accepts that our research into student thinking and learning is inevitably a matter of modelling, then we can serve the research programme much better by at least making sure that the modelling process is explicit in our thinking and writing about our research. This final chapter will draw together the different key threads considered in the book, to summarise the argument, and highlight key points to take forward into further research.

Chapter 16
Implications for Research

This book is intended to frame a major area of science education, which explores students' ideas and their learning, from the perspective that research into this area must inherently be about constructing, testing and developing models. It might seem that this should be self-evident to science educators, as they are well aware that science itself is about constructing theories of the world, and that posing and testing models is a central part of that activity. If we wish to see science education research as scientific, then it should be no surprise that research in our field centrally involves similar model-building activities.

However, as the introduction suggested, research reports often do not make explicit that many of the key constructs science education researchers rely upon to design studies, to elicit and interpret data, and to form conclusions, are just that: constructs invented to model features of *the phenomena* of science education, that is, primarily the things that science learners, say, do and write. By contrast, what students know, think and learn are not phenomena at all (they are not directly observed features of the world): they are conjectured theoretical entities that form parts of our explanatory schemes. Yet, too often, research reports published in the literature are written as though student ideas, knowledge, understanding and learning are phenomena that can be unproblematically observed and reported.

It was suggested that this tendency to 'under-problematise' the process of research in our field (which sometimes impedes progress by leading to apparently definitive and yet inconsistent findings from different studies) is in part – and perhaps a very large part – intrinsically linked with the nature of the research foci and a key aspect of human cognition. As part of normal human development, we acquire a 'theory of mind' which allows us to posit mental qualities to others, and which becomes part of the common discourse of the lifeworld. It becomes 'natural' for us to take for granted that people think and know and believe and understand and recall things (even though the evidence is only ever indirect), and that their subjective experience of these processes is much as our own. Whilst these assumptions are likely very necessary for the normal functioning of society, they also seem to

K.S. Taber, *Modelling Learners and Learning in Science Education: Developing Representations of Concepts, Conceptual Structure and Conceptual Change to Inform Teaching and Research*, DOI 10.1007/978-94-007-7648-7_16,
© Springer Science+Business Media Dordrecht 2013

commonly undermine the usual scientific (critical) attitude when the constructs we develop as part of our lifeworld experience – the mental register as I have labelled it – are adopted as if they are technical terms that have been operationalised for systematic enquiry. Theory of mind means that in our everyday lives, we do treat the thoughts, knowledge and beliefs of others as if they are phenomena: that is, we interpret the behaviour of others such that we often think we know what they think or what they believe *without being aware* that we are actually developing models of their mental activity.

This has become second nature in the normal adult. This is similar to how once we have learnt a language we listen to talk without hearing the phonemes, but – due to the processing that occurs in perception – we actually 'hear' the words people say. Researchers in linguistics have to learn to hear through these natural interpretive processes to analyse speech, and researchers exploring the cognition of others similarly need to step back from and develop a critical attitude to the automatic processing that 'tells us' what another knows, thinks and believes. We cannot do without those processes, but they have evolved to give quick best-guess interpretations that allow us to act 'online' in normal social interactions: and we have to supplement them with more careful, reflective and justifiable analytical processes in our research.

If this book does no more than encourage some colleagues working in science education to develop their awareness about these issues, then I think that will have justified the effort expended in writing the book. Ideally, every time we read (or write) about student thinking or learning, about what someone knows or understands, we should be alerted that the claims refer to conjectured unobservables and so must be based on interpretations of indirect evidence. That is, they are the researcher's interpretations of the representations in the public space of the outcomes of internal cognitive processes, only some of which will ever be consciously experienced by the individual research participants themselves. Such interpretations involve a form of modelling, inevitably drawing upon choices of theoretical commitments (ontological, about the nature of the conjectured unobservables; epistemological, about how we could infer the properties of those conjectured unobservables). Given that, surely our research programme will benefit from both acknowledging how it is based on such modelling processes and being explicit about the models being adopted in particular studies.

In the present book, I have offered an account of one researcher's understanding of the models and modelling processes involved in this area of work: how we might understand mental experience and the cognitive processes underpinning it (Part II), how we might make sense of the key notion of knowledge (Part III) and how we might understand the nature of conceptual learning (Chap. 13). The models I have selected and presented (and in some cases constructed for this book) may be useful to others working in the research programme into student thinking and learning in science. Conversely, other researchers may feel these models are flawed, or the progress of time may reveal that some aspects of the account I have given are woefully inadequate. Yet by making the models explicit and – just as important – being explicit about their status as models, we open them up for inspection and critique. When the models researchers use are taken for granted and implicit, they are less available for critique, and the argument chain supporting knowledge claims in research reports is not readily available for interrogation and critique.

Challenges for Research

It should also be clear from this volume that acknowledging the degree to which research in this area is a matter of developing models, rather than simply observing straightforward phenomena, reveals some key challenges for research in science education.

The Challenge of the Persuasiveness of the Mental Register

I have argued that much of the tendency to present research findings as relatively unproblematic derives from the ways the themes of much of this research are closely tied to a 'register' of terms used in everyday life to communicate about mental phenomena. That is, we all know what it is to know, to understand, to believe, to think, to change our minds, to learn something new. We have all experienced internal mental states that we describe in these terms, and (assuming that others' mental experience must be much like our own) we have interpreted others using this kind of language (we have 'understood' them) by relating their reports of their own mental experiences to our own internal experiences. In effect we all have an informal 'theory of mind' and it generally works well in everyday discourse.

Yet because it derives from such a taken-for-granted human faculty, the everyday 'lifeworld' mental register is not subject to the safeguards put in place for technical terminology. We talk about a student's understanding of electrical current in much the same we might talk about the current itself: yet we do not have the equivalent of ammeters or indicator lamps in the former case. That is, we do not have reliable instrumentation based on well-accepted theory which allows us to give confident reports of student understanding in the way we might expect to be able to report a current of 0.35 (+/−0.02) amperes. Yet despite this, we often talk about and indeed write-up our work on student thinking and learning as if our instrumentation and the models underpinning it were just as robust and well established.

There is a major challenge for researchers here. It has been suggested that whereas teaching is about making the unfamiliar familiar, an important part of research is in making the familiar become unfamiliar. Here, researchers have to be alert to the mental register, and whenever referring to learners' ideas, thinking, knowledge, understanding, etc. pause to ask:

- Am I relying here on my informal notions of mental life?
- Am I sufficiently recognising that these terms refer to theoretical constructs that are components of a model of human cognition?

The former perspective is by far easier, but I have argued it does not tend to support robust research. The latter perspective is more appropriate in a research programme but admits a great many complications. In the appendix, 'Testing the logic of research', I have set out a simple set of formal questions that may be useful when considering the adequacy of research when dealing with components of the mental register. Some researchers, especially novice researchers, and readers of research may find this checklist helpful when interrogating research reports – including drafts of their own writing. The checklist is equally relevant to other research foci in science education (and beyond) but could perhaps be especially useful when the taken for grantedness of the mental register could lead us to 'drop our guard' in terms of research rigour.

Having established this starting point, the book then presented an account of what we think we know, about the nature of student thinking, understanding, knowing and learning in science. My aim here was not to present a definitive account (although I offered the best account I could) but rather to emphasise the extent to which any account necessarily involved a process of applying and/or developing models. So Part II set out the nature of the human cognitive system and how it relates to the public space to which we all have access through our senses. Whilst it is possible to offer supported models here, it is clear that such models are simplifications of what is actually very complex apparatus.

A key issue, though, was how each of us can only directly observe our own mental experience (which itself only reflects part of our cognitive activity). The rather obvious point that we cannot see another think, or know, or believe, or understand, or learn, is easily forgotten because we operate with the theory of mind as a core and taken-for-granted part of our cognitive apparatus. After all, we are socialised to metaphorically 'see the cogs turning' and recognise 'the moment when the penny drops' or when the 'light came on'. Part II set out the necessary limitations involved in interpreting what *is* observable to draw inferences about another's mental states. This part reinforced the argument that there are necessary implicit modelling processes involved in our everyday interactions with others, which need to be made explicit when we shift from the everyday discourse to the technical work of research.

A key point made was the need to adopt the approach often taken in cognitive sciences of not confusing discussion relating to the physical substrate of cognition (the physiology and neurology), the functional systems processing level (in terms of what cognition involves and so what the apparatus of cognition actually does) and the mental level (in terms of how we conceptualise and describe the mental experiences we have as a result of cognitive activity). Whilst advances in neuroscience bring increasing understanding of how brains work, in some sense, the biology of the nervous system is less relevant than understanding the functioning of a person's cognitive system: what functions are carried out and with what capabilities. Here cognitive science offers a great deal of useful knowledge about human cognition, which provides an objective basis for describing what is going on when we talk of knowing, thinking, remembering, believing, etc. Whilst it may be most 'natural' to talk about (and be interested in) the mental level, research focusing at

Fig. 16.1 An easier
configuration of numbers to
process

1914

1918

1939

1945

this level is primarily suitable for phenomenological approaches and does not offer the 'objectivity' needed to answer many of the research questions that arise when we enquire into student learning in science.

In this volume I have deliberately adopted the rather dry language of *the learner as a cognitive system that is processing information*. This choice is not in any way intended to undermine the importance of personal experience, or the humanity of learners, but simply to make explicit the way the researcher needs to approach the work of modelling another's cognition.

The Challenge of the Dynamic Nature of Memory

One key area where the folk psychology of the mental register can readily lead us astray is in using the notion of memory. In everyday conversation the memory is usually characterised as some kind of store of past experiences, from which – at least sometimes – those past experiences can be brought out of storage for inspection. Yet although experience often seems consistent with such a description, it is a misleading conception. We have seen that there appear to be two quite different types of memory within a person's cognitive system: working memory and long-term memory. So working memory seems to be a small capacity working 'space' that can access information represented in the system and mentipulate it in various ways. We have conscious experiences relating to this processing involving both perceptual information from current sensory input and representations from long-term memory.

Yet this working memory acts as a limiter on the amount of new information we can consciously process at any time, such that we are biased towards operating on data in ways that fits with existing thinking (we can recognise, and so later recall, the pattern of numbers in Fig. 16.1 more readily than the same numbers as arranged in Fig. 5.12).

Long-term memory seems to be quite unlike the careful storage space of folk psychology where we can later revisit earlier experiences. It seems that the apparatus that allows us to use representations of past experience to make sense of current experience is continuously modifying those representations both in view of current 'input' and, in the background, over time to provide more effective linkage between relatable representations. The outcome is cognitive apparatus that is very effective at maintaining a stable coherent view of the world by iteratively relating and updating our representations of that world, even if at the cost of blurring or even completely changing those representations so that they cease to be a reliable guide to our past experiences. Perhaps the price we pay for minds that can quickly come to a view of the world which we experience as stable and familiar is minds which take liberties in representing both the here and now and the past.

I have suggested that given the centrality of memory to cognition, and learning, it is perhaps surprising that more studies in science education have not focused specifically on this topic. We have many studies of students' ideas that show they are at odds with what we think they have been taught, but there has been very little attempt to explore how students' accounts of what they think they have seen and heard in science lessons may shift over time. Given the importance of memory to effective learning, and the complications that can result from its dynamic nature, constantly rewriting our representations of experience, there is perhaps a strong case for more work in this area.

One specific hypothesis of practical importance suggested in Chap. 5 was that the use of regular summative assessments as part of teaching is more likely to encourage the development and establishment of new alternative conceptions than formative approaches to assessment that lead to immediate feedback (i.e. teacher questioning, followed by examination and evaluation of responses during the same teaching episode). It would be difficult to test this suggestion by a formal 'experimental' study, both because of the threats to validity in carrying out experiments with human subjects (novelty effects, expectancy effects and so forth) and because of the ethical considerations in deliberately asking teachers to employ teaching approaches that we have good reason to think might disadvantage learners.

However, given that both

(a) It is well recognised that in many educational contexts much assessment of students seems to be primarily summative in nature
(b) There is a strong research base to suggest that learning is better supported when teachers undertake more formative assessment rather than primarily summative assessment (Black & Wiliam, 2003)

it may be possible to investigate this hypothesis through quasi-experimental methods complemented by case studies, in contexts where some teachers are modifying their classroom practice to include more formative assessment tasks.

Indeed given the central importance of memory in learning, it is surprising there are so few studies from within science education that explicitly focus on memory. This is especially so given how this theme links to the tenets of the constructivist perspective on learning. It was suggested in Chap. 5 that research which follows students' thinking over extended periods of time (months, years) could offer insights into the way that memory operates in learning science.

Certainly, given the common experience that learners' performances in demonstrating understanding and knowledge of taught scientific models do often regress after completing a course (module, topic) of study, further research into the nature of 'forgetting' science learning would seem to be indicated. This is particularly so given that 'forgetting' would seem not to be a matter of simply failing to access, but rather that over a period of time when representations are not being actively drawn upon in thinking, there is modification of those representations (Taber, 2003). It was suggested in Chap. 5 that teachers are often operating with incorrect lay ideas about how student memory operates and that research is needed to find out how to best fit teaching to the actual operating characteristics of the human cognitive system. That is, how should we best teach science when the learner's memory is understood as a dynamic and evolving system to model experience with built-in drives for ongoing and online updating of interpretive networks rather than being a store of records of past experiences.

The Challenge of the Limited Purview of Consciousness

Another key issue raised in the discussion in this volume is the question of the extent to which conscious mental experiences give access to the different processes of cognition that support science learning. Much research in science education has relied on what learners write and say – public representations of the outcomes of processes that are not directly accessible to researchers. It seems likely that consciousness reflects only one small part of thinking – perhaps the part most linked to metacognitive control. If much of importance in cognition is not directly accessible to learners themselves, then even when learners are willing to share their mental experiences and are able to describe them in ways researchers can make sense of, researchers' accounts only concern the conscious outcomes of a good deal of background processing that can at best only be inferred.

It is widely recognised that much creative work is undertaken without conscious awareness. Perhaps it is not so different with logical thinking as well. Just because the thinker can report steps in a process of, for example, completing a set exercise does not mean we can assume those steps fully reflect the preconscious thinking underpinning the task. Perhaps (like much scientific work) what is experienced and reported draws on the context of justification and is a post hoc reconstruction of thinking processes actually occurring at some inaccessible level of the mind. There are real issues here to tax the minds of researchers (consciously and preconsciously). Research suggests that many important knowledge elements supporting cognition are not open to direct access by consciousness. Conscious processes are important, but we probably do not really yet understand how much of what is important in learning science is inaccessible to introspection and needs to be investigated by approaches that do not rely on learners' self-reports. As the tip-of-the-tongue experience may suggest, what we consciously access is just the visible tip of the cognitive iceberg.

The Challenge of Identifying Public Knowledge

Another issue discussed in some depth in this volume is the nature of knowledge. In particular, the status of public knowledge systems, such as 'scientific knowledge', was raised. Whilst it was relatively clear how we might understand the knowledge of an individual scientist, it was less clear how public knowledge was to be understood. It was suggested that it could be considered as in some sense the knowledge residing within a community, where each individual has their own dynamic system of knowledge, and through interaction with others modifies their own and perhaps influences others' knowledge. This type of understanding leads to a notion of a complex and vast system of nodes (the individuals 'knowers'), each of which might be seen as the location of one of many partial versions of the distributed knowledge, but this does not support us in identifying 'the' public version of knowledge (e.g. 'the' scientific concept of energy).

An alternative is to consider public knowledge as located in formal records such as academic journals. However, here, even leaving aside the likely lack of consistency across the literature on many issues, we are dealing with public representations in a form that can only be understood by being reinterpreted in the minds of those who access these records. So if we consider knowledge to reside in such inscriptions, then it is knowledge that requires further interpretation before it can be applied.

So the conclusions drawn were that (i) public systems exist in networks of many individual minds, but that such public systems of knowledge do not offer definite versions of canonical knowledge (as judgements have to be made about which members of the community are currently holding the most current version of scientific knowledge or how a consensual version may be arrived at), and (ii) sources such as the research literature hold representations of some aspects of the personal knowledge of these individual minds (and sometimes negotiated representations that are considered to sufficiently represent the versions of knowledge in the minds of groups of co-workers), but cannot be considered to hold knowledge itself.

The conclusion implies that discussions of 'current scientific thinking', or 'scientific knowledge', are problematic. This does not exclude such notions from the discourse of research in science education, but does suggest that their use in accounts of research should be accompanied by an explanation of how such terms are understood and operationalised in those studies. Given that no two scientists (or two teachers or two researchers) are likely to have exactly the same understanding of any complex scientific concept, studies which make claims about whether or not learners have knowledge that matches scientific knowledge should be clear about the way scientific knowledge has been modelled in the research reported.

The Challenge of the Uncertain Nature of Personal Knowledge

Moreover, knowledge remains a problematic notion, even when discussion is limited to a single person. Our use of knowledge shows it is often uncertain, context-dependent, multifaceted and so forth. Individuals set a task intended to elicit their knowledge, such as answering an interview question, will formulate an answer which, assuming they are honest and motivated to complete the task, represents their conscious thinking as they access various internal resources – either directly (i.e. using working memory to work with recalled representations) or indirectly by reflecting on the task and reporting the outcomes of preconscious processing offered to consciousness. When research is undertaken carefully, we might hope that the researcher's reports are effective representations *of* valid interpretations *of* the representations made (e.g. the spoken answer given) by the research participant.

It might seem rather pedantic to refer to such reports as 'representations of interpretations of representations' of the learner's thinking, and perhaps it is understandable that – providing we are aware of these intermediate stages – we tend to consider that what is reported is an aspect of the learners' knowledge of some topic. Given the nature of much human knowledge, researchers need to consider whether we can be confident that on a different day, or in response to a rephrased question, the study participant would access the same cognitive resources, thinking about the focal topic in the same way, and produce an equivalent account of their knowledge. This certainly cannot be assumed to always be the case. Where studies have sought to explore student thinking in depth, across contexts and over time, it has usually become clear that knowledge is nuanced, multilayered and subject to wavering degrees of commitment. Yet many studies do not employ techniques that can provide such detailed accounts. This is a major challenge for researchers.

It was pointed out in Chap. 6 that student responses that may seem incoherent or inconsistent to a researcher could actually reflect a range of different situations. It was suggested that where students appear to hold manifold conceptions, this might sometimes reflect tenuous and fluid understanding or could reflect relatively stable conceptual structures that are slowly evolving towards greater coherence and integration or could even indicate a sophisticated epistemological stance in the face of apparently contrary and uncertain evidence. Such different interpretations can all be considered viable in terms of the literature and yet would lead us to make very different judgements about the student's current state of knowledge and appropriate pedagogy for supporting its development. Research is surely needed to explore such issues in a wider range of situations (different grade/age levels, concept areas, curriculum contexts). Such research should both usefully look at how to discriminate between different underlying causes of apparent fragmentation and incoherence in student thinking and how different situations respond to various teaching approaches.

The Challenge of Conceptual Development

Despite conceptual development/change being a central issue in education and the work of teachers, as well as having long been a key focus of research (Vosniadou, 2008a), it is arguably an area where the current state of scholarship offers limited basis for supporting classroom pedagogy. We have well-established ideas about types of conceptual change and possible mechanisms for, and impediments to, such change that offer general indications of the kinds of teaching approaches likely to be successful in bringing about different kinds of conceptual change.

These different theories can provide testable hypotheses for the kinds of teaching approaches likely to be most productive in relation to particular types of teaching context – when students at a certain level of development, holding particular conceptions, are to be taught particular knowledge. Yet there is less in the way of research testing out these specific ideas across a range of science topics for different ages of learners.

In this volume, I have explored the nature and implications of a Vygotskian model of concept development, but there is limited research to thoroughly test these ideas in authentic classroom contexts. I have argued, for example, that conceptual development in an area such as metals is likely to be different from in a concept area such as energy (Fig. 15.10 cf. Fig. 15.11). I have presented a hypothetical account of conceptual development in metals that suggests development may be largely incremental without requiring major ontological shifts, whereas lifeworld notions of energy make it more difficult to simply redirect student prior learning towards scientifically acceptable thinking. This is a viable account based on research but needs to be thoroughly tested.

It is also an account that invites some interesting questions that may be quite informative in understanding features of conceptual change in learning science. For example, how can we understand the way a learner develops an understanding of 'metal' in chemistry and in technology that allows them to discriminate between two partially overlapping and partially distinct concepts of metal? Do models of different 'layers' in conceptual structure (Fig. 15.14) offer more than a persuasive image and actually reflect features of structure that can reliably be found in research data from learners?

There need to be more studies to test the theories and the models we can build based on them. These studies need to explore enough concept areas and educational levels to help build a general model that be used to inform teachers and curriculum developers across science topics and ages.

The Challenge of the Idiosyncratic Nature of Cognition

Moreover, each individual person has a cognitive system primed by their (usually somewhat unique) genetic inheritance as a starting point for the iterative process of making sense of their own personal experience in interacting with their environment. The result is a mind that is unique, depending upon a vast representational network of resources to support cognition. So each person's understanding of the world is inevitably at least somewhat idiosyncratic.

This creates a challenge for research that explores foci such as student understanding (or thinking or knowledge or learning). One aspect of this challenge is deciding how detailed a model of an individual participant's understanding (or thinking or knowledge or learning) is required for it to be considered a valid model of what is necessarily going to be a nuanced and complex and somewhat unique focus. This will likely vary from study to study, but it is clear from the discussion in this volume that simple models will not do full justice to the nature of human understanding (or thinking or knowledge or learning).

The second, related, issue concerns the extent to which researchers need to offer generalised accounts to be useful to inform teaching. A key aspect of the modelling process for researchers in this field is constructing models that reflect commonalities across populations. That this is necessarily a process of simplification is not in itself problematic – useful models are often considerable simplifications of what is being modelled – but the tension, long recognised in the research programme, between (a) recognising diversity and detailing complexity and yet (b) finding valid and useful ways to summarise results in terms of small numbers of general patterns has only become more challenging as we have come to realise just how complicated the foci of our research are.

The Challenge of Sociocultural Perspectives on Learning

The challenges listed above are probably quite apparent from the account given earlier in the book. However, one other challenge is perhaps less obvious. My focus in this book has largely taken the individual mind as the unit of analysis. The book is very strongly influenced by the personal constructivist tradition. To *my mind*, it is not possible to meaningfully undertake research into student thinking, understanding, knowing and learning without focusing very clearly on individual learners and their own discrete cognitive systems.

Yet the social context has not been entirely missing from the book, and it is quite clear that throughout the analysis presented, a good deal of attention has been given to the issue of how minds work together. So I have puzzled over such matters as how we can best model:

• How we communicate our ideas to offers
• How we can claim to report on the minds of others
• How teaching and other social interaction can influence learning by feeding into the individual's cognitive system
• How communities can somehow be understood to hold a kind of distributed system of knowledge

Social and cultural perspectives certainly offer useful perspectives and insights related to such themes, but often this seems to be largely at the expense of treating individual cognition as something that can be assumed and taken for granted. Studying the individual mind in isolation is certainly not sufficient for informing teaching that is inherently a social activity embedded in a cultural context. However, studies of the social processes in the classroom are also inevitably limited unless a viable model of what is going on in the individual minds that are interacting in the social context informs them. Research with a social or cultural focus that takes into account what we are learning about how individuals can be said to know, understand and learn can be extremely valuable, but research with such foci that proceeds without regard to the nature – and affordances and limitations – of individual cognition is unlikely to inform the development of better science pedagogy. Certainly perspectives such as constructionism and ideas such as distributed cognition do not currently provide answers to key questions about the nature of human knowledge and how it may be developed that I find satisfactory.

One criticism of my previous book discussing this general area of research (Taber, 2009b) seemed to suggest that I was clinging to a stale personal constructivist perspective when the agenda had moved on to exciting new areas of research. I am all for new, exciting perspectives, and I certainly feel we need different approaches to illuminate topics as important and multifaceted as learning in science. I welcome sociocultural studies that can tell us much about how people form communities to construct knowledge together. However, to my reading, these studies offer answers at a different level of analysis to that which has been the prime concern of this book. Just as research at the physiological/neurological level, treating learners as cells and tissues, and research looking at learners as individual agents in the world driven by various motivations, can complement studies of learners modelled as cognitive systems, so can research at social/cultural level.

Research at such different levels helps us build up a fuller understanding of student learning, and this is especially so when we are able to make connections across levels. Research at the social/cultural levels complements work focused on aspects of classroom learning seen in terms of individual cognition. Such studies are exploring different aspects of the same complex events, so not only is research at the different levels needed, but ultimately convincing models and theories from the different perspectives need to be consistent if they are to lead to a coherent understanding of classroom teaching and learning.

So if we need to undertake and coordinate research at these different levels to develop a full picture of science learning, then it would only be sensible to move on from research informed by personal constructivist perspectives once we feel we had satisfactorily answered the research questions we have posed at that level. I do not think we have reached this point: mostly we have simply refined our ways of asking the questions to better appreciate the challenges for the work ahead (Taber, 2009b). We have done enough to appreciate how partial many of the answers in the existing literature actually are. When we appreciate fully the level of modelling inherent in this type of work, we can more clearly see how limited our current knowledge is. We can appreciate how apparently conflicting claims may each rest upon a series of modelling steps (each in turn underpinned by its own set of ontological and/or epistemological assumptions), and this can inform an inclusive approach to building a synthesis of apparently incongruent knowledge claims. Making explicit the modelling processes inherent in such research can both inform the adoption of findings as the basis for developing practice and contribute to the refinement of questions to guide future research. There may be other exciting perspectives on how students learn science, and some of them may even offer more immediate scope for making good progress, but the perspective explored in the present volume has certainly not been exhausted. Any 'grand theory' of teaching and learning science that we may aspire to will rely heavily on this area of research.

Constructing more authentic models of the learner and their learning remains a central task for research in science education. I hope the present analysis has made some contribution to clarifying just what such work may involve.

Appendix

Testing the Logic of Research

The assumption here is that research in science education has a strong logical component. Knowledge claims presented in research should logically follow from the analysis of data that is systematically collected. Yet, of course, to support research conclusions that data collected must be suitable to be used as evidence to answer research questions, so research design and methodology need to be fit for purpose. Research questions should themselves reflect the conceptual framework set out for the study and any theoretical perspective employed in the study (Taber, 2013a, In press).

The same considerations apply here whether the 'object' of study is something that can be directly observed – such as manipulations during a practical activity or the occurrence of keywords in student classroom dialogue – or something less accessible such as student thinking or understanding. An argument here has been that often when the research concerns entities from the mental register (e.g. understanding), there is a tendency to inappropriately take for granted shared understandings (sic) of such terms and to underestimate the epistemological challenges in learning about them.

Perhaps it would be wise if all those involved in science education (researchers and teachers and other users of research) developed an attitude of having a 'mental register radar' that flags up this potential problem whenever a research report is about such things as student knowledge and thinking. Seeing a study was about such foci might alert us to the need to check that we are treating the components of the mental register as technical terms, just as we would expect precision if the focus of a study was teacher qualifications, acid strength, identifying genes or student progression rates into higher education. In such cases we expect to know precisely what the research is about and how well the researchers are able to identify/characterise/count/measure qualifications, acids, genes or progression. The same should apply if the work is about student thinking, knowledge, understanding or learning.

K.S. Taber, *Modelling Learners and Learning in Science Education: Developing Representations of Concepts, Conceptual Structure and Conceptual Change to Inform Teaching and Research*, DOI 10.1007/978-94-007-7648-7,
© Springer Science+Business Media Dordrecht 2013

I present here a checklist of questions that might be used by those reading research papers, either as evaluators (editors, referees) or as practitioners or policymakers looking to see what implications research has for their own work. The checklist may also be useful to those writing research papers, especially those who are new to the research process.

The list could be applied to whatever the research is about – but may be especially useful when dealing with the kind of unobservables that are the foci of the present book.

The ontological question	*What is the object of the study?*
	Is the report clear about how the authors understand the nature of what they are researching?
The epistemological question	*What can be known about the object of the study?*
	Is the report clear about how the authors understand the limitations of knowing about what they are researching?
The methodological question	*How do the authors go about finding out about the object of the study?*
	Is there an explicit research design that employs methodology consistent with the authors' presentation of the nature of what they are researching and the limitations on knowing about it?
The reasoning question	*What knowledge claims about the object of the study are presented as findings?*
	Do the findings follow logically from the evidence presented and reflect the responses to the ontological, epistemological questions?
	Are the findings consistent with the nature of the object of the study presented in the paper?
	Are the findings consistent with the limitations on knowing about the object of study as presented in the paper?

References

Aaron, R. I. (1971). *Knowing and the function of reason*. Oxford, UK: Oxford University Press.

Abimbola, I. O. (1988). The problem of terminology in the study of student conceptions in science. *Science Education, 72*(2), 175–184.

Abrahams, I. (2011). *Practical work in school science: A minds-on approach*. London: Continuum.

Ahtee, M., & Varjola, I. (1998). Students' understanding of chemical reaction. *International Journal of Science Education, 20*(3), 305–316.

Aikenhead, G. S. (1996). Science education: Border crossing into the sub-culture of science. *Studies in Science Education, 27*(1), 1–52.

Aikenhead, G. S., & Jegede, O. J. (1999). Cross-cultural science education: A cognitive explanation of a cultural phenomenon. *Journal of Research in Science Teaching, 36*(3), 269–287. doi:10.1002/(sici)1098-2736(199903)36:3<269::aid-tea3>3.0.co;2-t.

Alvarez, P., & Squire, L. R. (1994). Memory consolidation and the medial temporal lobe: A simple network model [Neurobiology]. *Proceedings of the National Academy of Sciences, 91*, 7041–7045.

Ambusaidi, A., & Al-Shuaili, A. (2009). Science education development in the Sultanate of Oman. In S. BouJaoude & Z. R. Dagher (Eds.), *Arab States* (Vol. 3, pp. 205–219). Rotterdam, The Netherlands: Sense.

Anderberg, E. (2000). Word meaning and conceptions. An empirical study of relationships between students' thinking and use of language when reasoning about a problem. *Instructional Science, 28*, 89–113.

Anderson, J. R. (1995). *Learning and memory: An integrated approach*. New York: Wiley.

Andersson, B. (1986). The experiential gestalt of causation: A common core to pupils' preconceptions in science. *European Journal of Science Education, 8*(2), 155–171.

Andrés, P. (2003). Frontal cortex as the central executive of working memory: Time to revise our view. *Cortex, 39*(4–5), 871–895. doi:10.1016/s0010-9452(08)70868-2.

Arlin, P. K. (1975). Cognitive development in adulthood: A fifth stage? *Developmental Psychology, 11*(5), 602–606.

Ashby, F. G., & Maddox, W. T. (2005). Human category learning. *Annual Review of Psychology, 56*, 149–178.

Ashton-Jones, E., & Thomas, D. K. (1990). Composition, collaboration, and women's ways of knowing: A conversation with Mary Belen. *Journal of Advanced Composition, 10*(2), 275–292.

Ault, C. R., Novak, J. D., & Gowin, D. B. (1984). Constructing vee maps for clinical interviews on molecule concepts. *Science Education, 68*(4), 441–462.

Ausubel, D. P. (1968). *Educational psychology: A cognitive view*. New York: Holt, Rinehart & Winston.

Ausubel, D. P. (2000). *The acquisition and retention of knowledge: A cognitive view*. Dordrecht, The Netherlands: Kluwer Academic Publishers.

Ausubel, D. P., & Robinson, F. G. (1971). *School learning: An introduction to educational psychology*. London: Holt.

Baars, B. J., & McGovern, K. (1996). Cognitive views of consciousness: What are the facts? How can we explain them? In M. Velmans (Ed.), *The science of consciousness: Psychological, neuropsychological and clinical reviews* (pp. 63–95). London: Routledge.

Baddeley, A. D. (1986). *Working memory*. Oxford, UK: Clarendon Press.

Baddeley, A. D. (1990). *Human memory: Theory and practice*. Hove, UK: Lawrence Erlbaum Associates.

Baddeley, A. D. (2000). The episodic buffer: A new component of working memory? *Trends in Cognitive Sciences, 4*(11), 417–423. doi:10.1016/s1364-6613(00)01538-2.

Baddeley, A. D. (2003). Working memory: Looking back and looking forward. *Nature Reviews Neuroscience, 4*(10), 829–839. doi:10.1038/nrn1201.

Baddeley, A. D., Hitch, G. J., & Allen, R. J. (2009). Working memory and binding in sentence recall. *Journal of Memory and Language, 61*, 438–456.

Bailey, R. (2006). Science, normal science and science education – Thomas Kuhn and Education. *Learning for Democracy, 2*(2), 7–20.

Banerjee, A. C. (1991). Misconceptions of students and teachers in chemical equilibrium. *International Journal of Science Education, 13*(4), 487–494.

Barba, G. D. (1993). Confabulation: Knowledge and recollective experience. *Cognitive Neuropsychology, 10*(1), 1–20.

Barker, V., & Millar, R. (1999). Students' reasoning about chemical reactions: What changes occur during a context-based post-16 chemistry course? *International Journal of Science Education, 21*(6), 645–665.

Beatty, J., Rasmussen, N., & Roll-Hansen, N. (2002). Untangling the McClintock myths. *Metascience, 11*(3), 280–298.

Behar, R. (2001). Yellow marigolds for Ochun: An experiment in feminist ethnographic fiction. *International Journal of Qualitative Studies in Education, 14*(2), 107–116. doi:10.1080/09518390010023630.

Behe, M. J. (2007). *The edge of evolution: The search for the limits of Darwinism*. New York: Free Press.

Bhaskar, R. (1975/2008). *A realist theory of science*. Abingdon, VA: Routledge.

Bhaskar, R. (1981). Epistemology. In W. F. Bynum, E. J. Browne, & R. Porter (Eds.), *Macmillan dictionary of the history of science* (p. 128). London: The Macmillan Press.

Biesta, G. J. J., & Burbules, N. C. (2003). *Pragmatism and educational research*. Lanham, MD: Rowman & Littlefield Publishers.

Billingsley, B. (2004). *Ways of approaching the apparent contradictions between science and religion*. Ph.D. thesis, University of Tasmania.

Bivall, P., Ainsworth, S., & Tibell, L. A. E. (2011). Do haptic representations help complex molecular learning? *Science Education, 95*(4), 700–719. doi:10.1002/sce.20439.

Black, P., & Wiliam, D. (2003). In praise of educational research: Formative assessment. *British Educational Research Journal, 29*(5), 623–637.

Black, P. J., & Lucas, A. M. (1993a). Introduction. In P. J. Black & A. M. Lucas (Eds.), *Children's informal ideas in science* (pp. xi–xiii). London: Routledge.

Black, P. J., & Lucas, A. M. (Eds.). (1993b). *Children's informal ideas in science*. London: Routledge.

Bliss, J. (1995). Piaget and after: The case of learning science. *Studies in Science Education, 25*, 139–172.

Bodner, G. M. (1986). Constructivism: A theory of knowledge. *Journal of Chemical Education, 63*(10), 873–878.

Bonatti, L. (1994). Why should we abandon the mental logic hypothesis? *Cognition, 50*(1–3), 17–39. doi:10.1016/0010-0277(94)90019-1.

Bourtchouladze, R. (2002). *Memories are made of this*. London: Weidenfeld & Nicolson.

Box, G. E. P., & Draper, N. R. (1987). *Empirical model-building and response surfaces*. New York: Wiley.

Boyle, H. N. (2006). Memorization and learning in Islamic schools. *Comparative Education Review, 50*(3), 478–495.

Braaten, M., & Windschitl, M. (2011). Working toward a stronger conceptualization of scientific explanation for science education. *Science Education, 95*(4), 639–669. doi:10.1002/sce.20449.

Brahler, C. J., & Walker, D. (2008). Learning scientific and medical terminology with a mnemonic strategy using an illogical association technique. *Advances in Physiology Education, 32*, 219–224. doi:10.1152/advan.00083.2007.

British Educational Research Association. (2000). *Good practice in educational research writing*. Southwell, UK: British Educational Research Association.

Brockhampton. (1997). *Dictionary of ideas*. London: Brockhampton Press.

Brossard, D., Lewenstein, B., & Bonney, R. (2005). Scientific knowledge and attitude change: The impact of a citizen science project. *International Journal of Science Education, 27*(9), 1099–1121. doi:10.1080/09500690500069483.

Brown, H. I. (1995). Ideas. In T. Honderich (Ed.), *The Oxford companion to philosophy* (pp. 389–390). Oxford, UK: Oxford University Press.

Brown, J. R. (1991). *Laboratory of the mind: Thought experiments in the natural sciences*. London: Routledge.

Brown, R., & McNeill, D. (1966/1976). The 'tip-of-the-tongue' phenomenon. In J. M. Gardiner (Ed.), *Readings in human memory* (pp. 243–255). London: Methuen & Company.

Bruillard, E., & Baron, G.-L. (2000). Computer-based concept mapping: A review of a cognitive tool for students. In D. Benzie & D. Passey (Eds.), *Proceedings of conference on educational uses of information and communication technologies* (pp. 331–338). Beijing, China: Publishing House of Electronics Industry.

Bruner, J. S. (1960). *The process of education*. New York: Vintage Books.

Bruner, J. S. (1964). The course of cognitive growth. *American Psychologist, 19*(1), 1–15.

Bruner, J. S. (1967a). On cognitive growth. In J. S. Bruner, R. Oliver, R. P. M. Greenfield, J. R. Hornesby, H. J. Kenney, M. Maccoby, N. Modiano, F. A. Mosher, D. R. Oslon, M. C. Potter, L. M. Reich, & A. M. Sonstroem (Eds.), *Studies in cognitive growth: A collaboration at the Centre for Cognitive Studies* (pp. 1–29). New York: Wiley.

Bruner, J. S. (1967b). Preface. In J. S. Bruner, R. Oliver, R. P. M. Greenfield, J. R. Hornesby, H. J. Kenney, M. Maccoby, N. Modiano, F. A. Mosher, D. R. Oslon, M. C. Potter, L. C. Reich, & A. M. Sonstroem (Eds.), *Studies in cognitive growth: A collaboration at the Centre for Cognitive Studies* (pp. vii–xv). New York: Wiley.

Bruner, J. S. (1987). The transactional self. In J. Bruner & H. Haste (Eds.), *Making sense: The child's construction of the world* (pp. 81–96). London: Routledge.

Burgess, P. W. (1996). Confabulation and the control of recollection. *Memory, 4*(4), 359–412.

Butts, B., & Smith, R. (1987). HSC chemistry students' understanding of the structure and properties of molecular and ionic compounds. *Research in Science Education, 17*(1), 192–201.

Cajori, F. (1922). On the history of caloric. *Isis, 4*(3), 483–492.

Camacho, F. F., & Cazares, L. G. (1998). Partial possible models: An approach to interpret students' physical representations. *Science Education, 82*(1), 15–29.

Caravita, S., & Halldén, O. (1994). Re-framing the problem of conceptual change. *Learning and Instruction, 4*(1), 89–111.

Carey, S. (1986). Cognitive science and science education. *American Psychologist, 41*(10: Special issue: Psychological science and education), 1123–1130.

Carey, S., & Spelke, E. (1996). Science and core knowledge. *Philosophy of Science, 63*(4), 515–533.

Changeux, J.-P. (1983/1997). *Neuronal man: The biology of mind* (L. Garey, Trans.). Princeton, NJ: Princeton University Press.

Chapanis, N. P., & Chapanis, A. (1964). Cognitive dissonance. *Psychological Bulletin, 61*(1), 1–22. doi:10.1037/h0043457.

Cheng, M. M. W. (2011). *Students' visualization of scientific ideas: Case studies of a physical science and a biological science topic*. Ph.D., King's College, University of London, London.

Chevallard, Y. (2007). Readjusting didactics to a changing epistemology. *European Educational Research Journal, 6*(2), 131–134.

Chi, M. T. H. (1992). Conceptual change within and across ontological categories: Examples from learning and discovery in science. In R. N. Giere (Ed.), *Cognitive models in science* (Vol. XV, pp. 129–186). Minneapolis, MN: University of Minnesota Press.

Chi, M. T. H. (2008). Three types of conceptual change: Belief revision, mental model transformation, and categorical shift. In S. Vosniadou (Ed.), *International handbook of research on conceptual change* (pp. 61–82). New York: Routledge.

Chi, M. T. H., & Slotta, J. D. (1993). The ontological coherence of intuitive physics. *Cognition and Instruction, 10*(2&3), 249–260.

Chi, M. T. H., Slotta, J. D., & de Leeuw, N. (1994). From things to processes; A theory of conceptual change for learning science concepts. *Learning and Instruction, 4*, 27–43.

Chin, C., & Brown, D. E. (2000). Learning in science: A comparison of deep and surface approaches. *Journal of Research in Science Teaching, 37*(2), 109–138.

Chomsky, N. (1999). Form and meaning in natural languages. In M. Baghramian (Ed.), *Modern philosophy of language* (pp. 294–308). Washington, DC: Counterpoint.

Churchland, P. S. (1980). A perspective on mind-brain research. *The Journal of Philosophy, 77*(4), 185–207.

Claxton, G. (1986). The alternative conceivers' conceptions. *Studies in Science Education, 13*, 123–130. doi:10.1080/03057268608559934.

Claxton, G. (1993). Minitheories: A preliminary model for learning science. In P. J. Black & A. M. Lucas (Eds.), *Children's informal ideas in science* (pp. 45–61). London: Routledge.

Claxton, G. (2005). *The wayward mind: An intimate history of the unconscious*. London: Little Brown.

Clinchy, B., & Zimmerman, C. (1985). *Growing up intellectually: Issues for college women*. Wellesley, MA: Wellesley Centers for Women, Wellesley College.

Cobern, W. W. (1994). *Worldview theory and conceptual change in science education*. Paper presented at the National Association for Research in Science Teaching, Anaheim, CA.

Coll, R. K., Lay, M. C., & Taylor, N. (2008). Scientists and scientific thinking: Understanding scientific thinking through an investigation of scientists views about superstitions and religious beliefs. *Eurasia Journal of Mathematics, Science & Technology Education, 4*(3), 197–214.

Collins, H. (2010). *Tacit and explicit knowledge*. Chicago: The University of Chicago Press.

Commons, M. L., Richards, F. A., & Armon, C. (Eds.). (1984). *Beyond formal operations: Late adolescent and adult cognitive development*. New York: Praeger.

Cooper, J. (2007). *Cognitive dissonance: Fifty years of a classic theory*. London: Sage.

Cosgrove, M. (1995). A study of science-in-the-making as students generate an analogy for electricity. *International Journal of Science Education, 17*(3), 295–310.

Cosgrove, M., & Osborne, R. (1985). Lesson frameworks for changing children's ideas. In R. J. Osborne & P. Freyberg (Eds.), *Learning in science: The implications of children's science* (pp. 101–111). Auckland, New Zealand: Heinemann.

Cowan, N., Chen, Z., & Rouder, J. N. (2004). Constant capacity in an immediate serial-recall task: A logical sequel to Miller (1956). *Psychological Science, 15*(9), 634–640. doi:10.1111/j.0956-7976.2004.00732.x.

Crain, W. (1992). *Theories of development: Concepts and applications* (3rd ed.). London: Prentice-Hall International.

Cray, D., Dawkins, R., & Collins, F. (2006, November 5). God vs. science. *Time*. Retrieved from http://www.time.com/time/printout/0,8816,1555132,00.html

Crick, F., & Koch, C. (1990). Towards a neurobiological theory of consciousness. *Seminars in the Neurosciences, 2,* 263–275.

Csikszentmihalyi, M. (1988). The flow experience and its significance for human psychology. In M. C. Csikszentmihalyi & I. S. Csikszentmihalyi (Eds.), *Optimal experience: Psychological studies of flow in consciousness* (pp. 15–35). Cambridge, MA: Cambridge University Press.

Dagher, Z. R. (2009). Epistemology of science in curriculum standards of four Arab countries. In S. BouJaoude & Z. R. Dagher (Eds.), *Arab States* (Vol. 3, pp. 41–60). Rotterdam, The Netherlands: Sense.

Darwin, C. (1859/1968). *The origin of species by means of natural selection, or the preservation of favoured races in the struggle for life.* Harmondsworth, UK: Penguin.

Deese, J. (1963). The learning of concepts. In L. D. Crow & A. Crow (Eds.), *Readings in human learning* (pp. 400–402). New York: David McKay Company.

Demetriou, A., & Mouyi, A. (2011). Processing efficiency, representational capacity, and reasoning: Modelling their dynamic interactions. In P. Barrouillet & V. Gaillard (Eds.), *Cognitive development and working memory: A dialogue between neo-Piagetian theories and cognitive approaches* (pp. 69–103). Hove, UK: Psychology Press.

Dent, N. (1995). Normative. In T. Honderich (Ed.), *The Oxford companion to philosophy* (p. 626). Oxford, UK: Oxford University Press.

D'Esposito, M. (2007). From cognitive to neural models of working memory. In J. Driver, P. Haggard, & T. Shallice (Eds.), *Mental processes in the human brain* (pp. 7–25). Oxford, UK: Oxford University Press.

diSessa, A. A. (1993). Towards an epistemology of physics. *Cognition and Instruction, 10*(2&3), 105–225.

Donaldson, M. (1978). *Children's minds.* London: Fontana.

Drever, J., & Wallerstein, H. (Eds.). (1964). *The Penguin dictionary of psychology* (Rev. ed.). Harmondsworth, UK: Penguin Books.

Driver, J., Haggard, P., & Shallice, T. (Eds.). (2007). *Mental processes in the human brain.* Oxford, UK: Oxford University Press.

Driver, R. (1983). *The pupil as scientist?* Milton Keynes, UK: Open University Press.

Driver, R., & Erickson, G. (1983). Theories-in-action: Some theoretical and empirical issues in the study of students' conceptual frameworks in science. *Studies in Science Education, 10,* 37–60.

Driver, R., Leach, J., Millar, R., & Scott, P. (1996). *Young people's images of science.* Buckingham, UK: Open University Press.

Dudai, Y., & Eisenberg, M. (2004). Rites of passage of the engram: Reconsolidation and the lingering consolidation hypothesis. *Neuron, 44*(1), 93–100. doi:10.1016/j.neuron.2004.09.003.

Duit, R. (2009). *Bibliography – Students' and teachers' conceptions and science education.* Kiel: http://www.ipn.uni-kiel.de/aktuell/stcse/stcse.html

Dunbar, K. (2001). What scientific thinking reveals about the nature of cognition. In K. Crowley, C. D. Schunn, & T. Okada (Eds.), *Designing for science: Implications from everyday, classroom, and professional settings* (pp. 115–140). Mahwah, NJ: Lawrence Erlbaum Associates.

Eastwood, J. L., Schlegel, W. M., & Cook, K. L. (2011). Effects of an interdisciplinary program on students' reasoning with socioscientific issues and perceptions of their learning experiences. In T. D. Sadler (Ed.), *Socio-scientific issues in the classroom: Teaching, learning and research* (pp. 89–126). Dordrecht, The Netherlands: Springer.

Eickelman, D. F. (1978). The art of memory: Islamic education and its social reproduction. *Comparative Studies in Society and History, 20*(04), 485–516. doi:10.1017/S0010417500012536.

Elbert, T., & Schauer, M. (2002). Burnt into memory. *Nature, 419,* 883.

Elkind, D., & Flavell, J. H. (Eds.). (1969). *Studies in cognitive development: Essays in honor of Jean Piaget.* New York: Oxford University Press.

Ericsson, K. A., Chase, W. G., & Faloon, S. (1980). Acquisition of a memory skill. *Science, 208*(4448), 1181–1182.

Ezcurdia, M. (1998). The concept-conception distinction. *Philosophical Issues, 9,* 187–192. doi:10.2307/1522969.

Facione, P. A. (1990). *Critical thinking: A statement of expert consensus for purposes of educational assessment and instruction. Research findings and recommendations*. Newark, DE: American Philosophical Association.

Fensham, P. J. (2004). *Defining an identity: The evolution of science education as a field of research*. Dordrecht, The Netherlands: Kluwer Academic Publishers.

Fernandez-Duque, D., Baird, J. A., & Posner, M. I. (2000). Executive attention and metacognitive regulation. *Consciousness and Cognition, 9*, 288–307. doi:10.1006/ccog.2000.0447.

Feynman, R. P., Leighton, R. B., & Sands, M. (Eds.). (1963). *The Feynman lectures on physics* (Vol. 1). Reading, MA: Addison-Wesley Publishing Company.

Finster, D. C. (1991). Developmental instruction: Part 2. Application of Perry's model to general chemistry. *Journal of Chemical Education, 68*(9), 752–756.

Flavell, J. H. (1963). *The developmental psychology of Jean Piaget* (Students' paperback ed.). London: D Van Nostrand Company.

Fodor, J. A. (1983). *The modularity of mind*. Cambridge, MA: MIT Press.

Fodor, J. A. (1985). Précis of the modularity of mind. *The Behavioral and Brain Sciences, 8*(01), 1–5. doi:10.1017/S0140525X0001921X.

Frolov, I. T. (1991). *Philosophy and history of genetics: The inquiry and the debates*. London: McDonald & Co.

Fuster, J. M. (1995). *Memory in the cerebral cortex: An empirical approach to neural networks in the human and nonhuman primate*. Cambridge, MA: The MIT Press.

Gardner, H. (1993). *Frames of mind: The theory of multiple intelligences* (2nd ed.). London: Fontana.

Gardner, H. (1998). *Extraordinary minds*. London: Phoenix.

Gauld, C. (1986). Models, meters and memory. *Research in Science Education, 16*(1), 49–54. doi:10.1007/bf02356817.

Gauld, C. (1989). A study of pupils' responses to empirical evidence. In R. Millar (Ed.), *Doing science: Images of science in science education* (pp. 62–82). London: The Falmer Press.

Gazzaniga, M. S., Fendrich, R., & Wessinger, C. M. (1994). Blindsight reconsidered. *Current Directions in Psychological Science, 3*(3), 93–96.

Gelman, S. A. (2009). Learning from others: Children's construction of concepts. *Annual Review of Psychology, 60*(1), 115–140. doi:10.1146/annurev.psych.59.103006.093659.

Gentner, D. (1983). Structure-mapping: A theoretical framework for analogy. *Cognitive Science, 7*, 155–170.

Gilbert, J. K. (1995). Studies and fields: Directions of research in science education. *Studies in Science Education, 25*, 173–197. doi:10.1080/03057269508560053.

Gilbert, J. K., Osborne, R. J., & Fensham, P. J. (1982). Children's science and its consequences for teaching. *Science Education, 66*(4), 623–633.

Gilbert, J. K., & Swift, D. J. (1985). Towards a Lakatosian analysis of the Piagetian and alternative conceptions research programs. *Science Education, 69*(5), 681–696.

Gilbert, J. K., & Treagust, D. F. (Eds.). (2009). *Multiple representations in chemical education*. Dordrecht, The Netherlands: Springer.

Gilbert, J. K., & Watts, D. M. (1983). Concepts, misconceptions and alternative conceptions: Changing perspectives in science education. *Studies in Science Education, 10*(1), 61–98.

Gilbert, J. K., & Zylbersztajn, A. (1985). A conceptual framework for science education: The case study of force and movement. *European Journal of Science Education, 7*(2), 107–120.

Gilley, S., & Loades, A. (1981). Thomas Henry Huxley: The war between science and religion. *The Journal of Religion, 61*(3), 285–308.

Glasersfeld, E. v. (1983). *Learning as constructive activity*. Paper presented at the 5th Annual Meeting of the North American Chapter of the International Group for the Psychology of Mathematics Education, Monteréal. http://srri.umass.edu/vonGlasersfeld/onlinePapers/pdf/vonGlasersfeld_080.pdf

Glasersfeld, E. v. (1988). The reluctance to change a way of thinking. *Irish Journal of Psychology, 9*(1), 83–90. Retrieved from http://www.univie.ac.at/constructivism/EvG/papers/110.pdf

Glasersfeld, E. v. (1989). Cognition, construction of knowledge, and teaching. *Synthese, 80*(1), 121–140.

Glasersfeld, E. v. (1990). An exposition of constructivism: Why some like it radical. *Monographs of the Journal for Research in Mathematics Education, 4*, 19–29. Retrieved from http://www.univie.ac.at/constructivism/EvG/papers/125.pdf

Goldman, A. (1995). Knowledge. In T. Honderich (Ed.), *The Oxford companion to philosophy* (pp. 447–448). Oxford, UK: Oxford University Press.

Gould, S. J. (2001). *Rocks of ages: Science and religion in the fullness of life*. London: Jonathan Cape.

Gregory, R. L. (1987). Ideas. In R. L. Gregory (Ed.), *The Oxford companion to the mind* (p. 337). Oxford, UK: Oxford University Press.

Gunckel, K. L., Mohan, L., Covitt, B. A., & Anderson, C. W. (2012). Addressing challenges in developing learning progressions for environmental science literacy. In A. C. Alonzo & A. W. Gotwals (Eds.), *Learning progression in science: Current challenges and future directions* (pp. 39–75). Rotterdam, The Netherlands: Sense.

Hammer, D. (2000). Student resources for learning introductory physics. *American Journal of Physics, 68*(7-Physics Education Research Supplement), S52–S59.

Hammer, D., Elby, A., Scherr, R. E., & Redish, E. F. (2005). Resources, framing, and transfer. In J. P. Mestre (Ed.), *Transfer of learning: From a modern multidisciplinary perspective* (pp. 89–119). Greenwich, CT: Information Age Publishing.

Harrison, A. G., & Coll, R. K. (Eds.). (2008). *Using analogies in middle and secondary science classrooms*. Thousand Oaks, CA: Corwin Press.

Harrison, A. G., & Treagust, D. F. (2002). The particulate nature of matter: Challenges in understanding the submicroscopic world. In J. K. Gilbert, O. de Jong, R. Justi, D. F. Treagust, & J. H. Van Driel (Eds.), *Chemical education: Towards research-based practice* (pp. 189–212). Dordrecht, The Netherlands: Kluwer Academic Publishers.

Harrison, A. G., & Treagust, D. F. (2006). Teaching and learning with analogies: Friend or foe? In P. J. Aubusson, A. G. Harrison, & S. M. Ritchie (Eds.), *Metaphor and analogy in science education* (pp. 11–24). Dordrecht, The Netherlands: Springer.

Hart, B. (1910). The conception of the subconscious. *Journal of Abnormal Psychology, 4*(6), 351–371. doi:10.1037/h0074022.

Herron, J. D., Cantu, L., Ward, R., & Srinivasan, V. (1977). Problems associated with concept analysis. *Science Education, 61*(2), 185–199.

Hickey, R., & Schibeci, R. A. (1999). The attraction of magnetism. *Physics Education, 34*(6), 383.

Higgs, J., & Titchen, A. (1995). The nature, generation and verification of knowledge. *Physiotherapy, 81*(9), 521–530. doi:10.1016/s0031-9406(05)66683-7.

Hirschfeld, L., & Gelman, S. A. (1994a). Towards a topography of mind: An introduction to domain specificity. In L. Hirschfeld & S. A. Gelman (Eds.), *Mapping the mind: Domain specificity in cognition and culture* (pp. 3–35). Cambridge, MA: Cambridge University Press.

Hirschfeld, L., & Gelman, S. A. (Eds.). (1994b). *Mapping the mind: Domain specificity in cognition and culture*. Cambridge, MA: Cambridge University Press.

Hitch, G. J., & Baddeley, A. D. (1976). Verbal reasoning and working memory. *Quarterly Journal of Experimental Psychology, 28*(4), 603–621.

Hofer, B. K., & Pintrich, P. R. (1997). The development of epistemological theories: Beliefs about knowledge and knowing and their relation to learning. *Review of Educational Research, 67*(1), 88–140.

Holdsworth, L. (1998). Is it repressed memory with delayed recall or is it false memory syndrome-the controversy and its potential legal implications. *Law & Psychology Review, 22*, 103–129.

Inoue, S., & Matsuzawa, T. (2007). Working memory of numerals in chimpanzees. *Current Biology, 17*(23), R1004–R1005. doi:10.1016/j.cub.2007.10.027.

Ivić, I., Pešikan, A., & Antić, S. (2002). *Active learning* (2nd ed.). Belgrade, Serbia: Institute of Psychology.

Jaakkola, T., Nurmi, S., & Veermans, K. (2011). A comparison of students' conceptual understanding of electric circuits in simulation only and simulation-laboratory contexts. *Journal of Research in Science Teaching, 48*(1), 71–93. doi:10.1002/tea.20386.

James, W. (1890). *The principles of psychology*. Retrieved from http://psychclassics.yorku.ca/James/Principles/index.htm

Johnson-Laird, P. N. (1983). *Mental models: Towards a cognitive science of language, inference and consciousness*. Cambridge, MA: Cambridge University Press.

Johnson-Laird, P. N. (2003a). Illusions of understanding. In A. J. Sanford (Ed.), *The nature and limits of human understanding* (pp. 3–25). London: T&T Clark Ltd.

Johnson-Laird, P. N. (2003b). Models, causation, and explanation. In A. J. Sanford (Ed.), *The nature and limits of human understanding* (pp. 26–46). London: T&T Clark Ltd.

Johnstone, A. H. (2000). Teaching of chemistry – Logical or psychological? *Chemistry Education: Research and Practice in Europe, 1*(1), 9–15.

Jonassen, D. (2009). Reconciling a human cognitive architecture. In S. Tobias & T. M. Duffy (Eds.), *Constructivist instruction: Success or failure?* (pp. 13–33). New York: Routledge.

Jungmann, K. (2001). Muon physics possibilities at a Muon-Neutrino factory. *Hyperfine Interactions, 138*(1), 463–473. doi:10.1023/a:1020826729142.

Kaiser, M. K., McCloskey, M., & Proffitt, D. R. (1986). Development of intuitive theories of motion: Curvilinear motion in the absence of external forces. *Developmental Psychology, 22*(1), 67–71. doi:10.1037/0012-1649.22.1.67.

Karmiloff-Smith, A. (1994). Précis of beyond modularity: A developmental perspective on cognitive science. *The Behavioral and Brain Sciences, 17*(04), 693–707. doi:10.1017/S0140525X00036621.

Karmiloff-Smith, A. (1996). *Beyond modularity: A developmental perspective on cognitive science*. Cambridge, MA: MIT Press.

Kawagley, A. O., Norris-Tull, D., & Norris-Tull, R. A. (1998). The indigenous worldview of Yupiaq culture: Its scientific nature and relevance to the practice and teaching of science. *Journal of Research in Science Teaching, 35*(2), 133–144. doi:10.1002/(sici)1098-2736(199802)35:2<133::aid-tea4>3.0.co;2-t.

Keil, F. C. (1992). *Concepts, kinds and cognitive development*. Cambridge, MA: MIT Press.

Keller, E. F. (1983). *A feeling for the organism: The life and work of Barbara McClintock*. New York: W H Freeman and Company.

Kelly, G. (1963). *A theory of personality: The psychology of personal constructs*. New York: W W Norton & Company.

Kitchener, R. F. (1987). Genetic epistemology, equilibration and the rationality of scientific change. *Studies in History and Philosophy of Science Part A, 18*(3), 339–366. doi:10.1016/0039-3681(87)90024-0.

Knight, N., Sousa, P., Barrett, J. L., & Atran, S. (2004). Children's attributions of beliefs to humans and God: Cross-cultural evidence. *Cognitive Science, 28*, 117–126.

Koestler, A. (1978/1979). *Janus: A summing up*. London: Pan Books.

Koffka, K. (1967). Principles of Gestalt psychology. In J. A. Dyal (Ed.), *Readings in psychology: Understanding human behavior* (2nd ed., pp. 9–13). New York: McGraw-Hill Book Company.

Kohlberg, L. (1973). Stages and aging in moral development – Some speculations. *The Gerontologist, 13*(4), 497–502. doi:10.1093/geront/13.4.497.

Kohlberg, L., & Hersh, R. H. (1977). Moral development: A review of the theory. *Theory into Practice, 16*(2), 53–59. doi:10.1080/00405847709542675.

Koltko-Rivera, M. E. (2006). Rediscovering the later version of Maslow's hierarchy of needs: Self-transcendence and opportunities for theory, research, and unification. *Review of General Psychology, 10*(4), 302–317.

Kopelman, M. D. (1987). Two types of confabulation. *Journal of Neurology, Neurosurgery & Psychiatry, 50*(11), 1482–1487. doi:10.1136/jnnp.50.11.1482.

Kosso, P. (2010). And yet it moves: The observability of the rotation of the earth. *Foundations of Science, 15*(3), 213–225. doi:10.1007/s10699-010-9175-x.

Kouider, S., & Dehaene, S. (2007). Levels of processing during non-conscious perception: A critical review of visual masking. In J. Driver, P. Haggard, & T. Shallice (Eds.), *Mental processes in the human brain* (pp. 155–185). Oxford, UK: Oxford University Press.

Kramer, D. A. (1983). Post-formal operations? A need for further conceptualization. *Human Development, 26*, 91–105.

Kuhl, P. K. (2004). Early language acquisition: Cracking the speech code. *Nature Reviews Neuroscience, 5*(11), 831–843. doi:10.1038/nrn1533.

Kuhn, D. (1999). A developmental model of critical thinking. *Educational Researcher, 28*(2), 16–46.

Kuhn, T. S. (1970). *The structure of scientific revolutions* (2nd ed.). Chicago: University of Chicago.

Kuhn, T. S. (1974/1977). Second thoughts on paradigms. In T. S. Kuhn (Ed.), *The essential tension: Selected studies in scientific tradition and change* (pp. 293–319). Chicago: University of Chicago Press.

Kuhn, T. S. (Ed.). (1977). *The essential tension: Selected studies in scientific tradition and change*. Chicago: University of Chicago Press.

Kuhn, T. S. (1996). *The structure of scientific revolutions* (3rd ed.). Chicago: University of Chicago.

Lakatos, I. (1970). Falsification and the methodology of scientific research programmes. In I. Lakatos & A. Musgrove (Eds.), *Criticism and the growth of knowledge* (pp. 91–196). Cambridge, UK: Cambridge University Press.

Lakoff, G. (1973). Hedges: A study in meaning criteria and the logic of fuzzy concepts. *Journal of Philosophical Logic, 2*(4), 458–508. doi:10.1007/bf00262952.

Lakoff, G., & Johnson, M. (1980a). Conceptual metaphor in everyday language. *The Journal of Philosophy, 77*(8), 453–486.

Lakoff, G., & Johnson, M. (1980b). The metaphorical structure of the human conceptual system. *Cognitive Science, 4*(2), 195–208.

Latour, B., & Woolgar, S. (1986). *Laboratory life: The construction of scientific facts* (2nd ed.). Princeton, NJ: Princeton University Press.

Laugksch, R. C. (2000). Scientific literacy: A conceptual overview. *Science Education, 84*, 71–94.

Lawson, A. E. (1985). A review of research on formal reasoning and science teaching. *Journal of Research in Science Teaching, 22*(7), 569–617.

Lawson, A. E. (2010). *Teaching inquiry science in middle and secondary schools*. Thousand Oaks, CA: Sage.

Lawson, A. E., & Wollman, W. T. (1976). Encouraging the transition from concrete to formal cognitive functioning-an experiment. *Journal of Research in Science Teaching, 13*(5), 413–430. doi:10.1002/tea.3660130505.

Lehrer, R., & Schauble, L. (2006). Scientific thinking and science literacy. In W. Damon, R. M. Lerner, K. A. Renninger, & I. E. Sigel (Eds.), *Handbook of child psychology* (Child psychology in practice 6th ed., Vol. 4, pp. 153–196). New York: Wiley.

Lemke, J. L. (1990). *Talking science: Language, learning, and values*. Norwood, NJ: Ablex Publishing Corporation.

Leslie, A. M. (1984). ToMM, ToBy and agency: Core architecture and domain specificity. In L. Hirschfeld & S. A. Gelman (Eds.), *Mapping the mind: Domain specificity in cognition and culture* (pp. 119–148). Cambridge, UK: Cambridge University Press.

Levinson, R. (2007). Teaching controversial socio-scientific issues to gifted and talented students. In K. S. Taber (Ed.), *Science education for gifted learners* (pp. 128–141). London: Routledge.

Libet, B. (1996). Neural processes in the production of conscious experience. In M. Velmans (Ed.), *The science of consciousness: Psychological, neuropsychological and clinical reviews* (pp. 96–117). London: Routledge.

Lidskog, R. (1996). In science we trust? On the relation between scientific knowledge, risk consciousness and public trust. *Acta Sociologica, 39*(1), 31–56.

Lightman, B. (2002). Huxley and scientific agnosticism: The strange history of a failed rhetorical strategy. *The British Journal for the History of Science, 35*(03), 271–289. doi:10.1017/S0007087402004715.

Lindahl, M. G. (2010). Of pigs and men: Understanding students' reasoning about the use of pigs as donors for xenotransplantation. *Science Education, 19*(9), 867–894. doi:10.1007/s11191-010-9238-y.

Lobato, J. (2006). Alternative perspectives on the transfer of learning: History, issues, and challenges for future research. *The Journal of the Learning Sciences, 15*(4), 431–449. doi:10.1207/s15327809jls1504_1.

Long, D. E. (2011). *Evolution and religion in American education: An ethnography.* Dordrecht, The Netherlands: Springer.

Lubben, F., Sadeck, M., Scholtz, Z., & Braund, M. (2009). Gauging students' untutored ability in argumentation about experimental data: A South African case study. *International Journal of Science Education, 32*(16), 2143–2166. doi:10.1080/09500690903331886.

Luria, A. R. (1976). *Cognitive development: Its cultural and social foundations.* Cambridge, MA: Harvard University Press.

Luria, A. R. (1987). *The mind of a Mnemonist.* Cambridge, MA: Harvard University Press.

Mahowald, M. W., & Schenck, C. H. (2000). Parasomnias: Sleepwalking and the law. *Sleep Medicine Reviews, 4*(4), 321–339. doi:10.1053/smrv.1999.0078.

Martin, B., & Richards, E. (1995). Scientific knowledge, controversy, and public decision-making. In S. Jasanoff, G. E. Markle, J. C. Petersen, & T. Pinch (Eds.), *Handbook of science and technology studies* (pp. 506–526). Version accessed at http://www.bmartin.cc/pubs/595handbook.html. Newbury Park, CA.

Marton, F. (1981). Phenomenography – Describing conceptions of the world around us. *Instructional Science, 10,* 177–200.

Matthews, M. R. (1992). Old wine in new bottles: A problem with constructivist epistemology. *Philosophy of Education Yearbook 1992.* Retrieved from available at http://www.ed.uiuc.edu/eps/PES-Yearbook/92_docs/Matthews.HTM

Matthews, M. R. (2002). Constructivism and science education: A further appraisal. *Journal of Science Education and Technology, 11*(2), 121–134.

Mathy, F., & Feldman, J. (2012). What's magic about magic numbers? Chunking and data compression in short-term memory. *Cognition, 122*(3), 346–362. doi:10.1016/j.cognition.2011.11.003.

McClary, L., & Talanquer, V. (2011). College chemistry students' mental models of acids and acid strength. *Journal of Research in Science Teaching, 48*(4), 396–413. doi:10.1002/tea.20407.

McCloskey, M. (1983). Intuitive physics. *Scientific American, 248*(4), 114–122.

McInerney, C., Bird, N., & Nucci, M. (2004). The flow of scientific knowledge from lab to the lay public: The case of genetically modified food. *Science Communication, 26*(1), 44–74. doi:10.1177/1075547004267024.

Medin, D. L., & Scott, A. (Eds.). (1999). *Folkbiology.* Cambridge, MA: The MIT Press.

Merrill, M. D. (2000, December 4–6). *Knowledge objects and mental models.* Paper presented at the International Workshop on Advanced Learning Technologies 2000: Advanced Learning Technology: Design and Development Issues, Massey University, Palmerston North, New Zealand.

Miller, A. I. (1986). *Imagery in scientific thought.* Cambridge, MA: MIT Press.

Miller, G. A. (1968). The magical number seven, plus or minus two: Some limits on our capacity for processing information. In *The psychology of communication: Seven essays* (pp. 21–50). Harmondsworth, UK: Penguin.

Modgil, S. (1974). *Piagetian research: A handbook of recent studies.* Windsor, UK: NFER Publishing.

Morris, H. M. (2000). *The long war on God: The history and impact of the creation/evolution conflict.* Green Forest, AR: Master Books.

Morton, J. B., & Munakata, Y. (2002). Active versus latent representations: A neural network model of perseveration, dissociation, and decalage. *Developmental Psychobiology, 40*(3), 255–265. doi:10.1002/dev.10033.

Muldoon, C. A. (2006). *Shall I compare thee to a pressure wave?: Visualisation, analogy, insight and communication in physics.* Ph.D., University of Bath, Bath.

Mullis, I. V. S., Martin, M. O., Ruddock, G. J., O'Sullivan, C. Y., Arora, A., & Erberber, E. (2005). *TIMSS 2007 assessment frameworks.* Chestnut Hill, MA: International Association for the Evaluation of Educational Achievement/TIMSS & PIRLS International Study Center.

Nagel, T. (1974). What is it like to be a bat? *Philosophical Review, 83*(4), 435–450.

Nersessian, N. J. (2008). *Creating scientific concepts.* Cambridge, MA: The MIT Press.

Newton, D. P. (2000). *Teaching for understanding: What it is and how to do it.* London: RoutledgeFalmer.

Nickerson, R. S. (1985). Understanding. *American Journal of Education, 93*(2), 201–239.

Nickerson, R. S. (1998). Confirmation bias: A ubiquitous phenomenon in many guises. *Review of General Psychology, 2*(2), 175–220.

Norman, D. A. (1983). Some observations on mental models. In D. Gentner & A. L. Stevens (Eds.), *Mental models* (pp. 7–14). Hillsdale, NJ: Lawrence Erlbaum Associates.

Norum, K. E. (2000). School patterns: A sextet. *International Journal of Qualitative Studies in Education, 13*(3), 239–250. doi:10.1080/09518390050019659.

Novak, J. D. (1990a). Concept mapping: A useful tool for science education. *Journal of Research in Science Teaching, 27*(10), 937–949.

Novak, J. D. (1990b). Concept maps and Vee diagrams: Two metacognitive tools to facilitate meaningful learning. *Instructional Science, 19*(1), 29–52. doi:10.1007/bf00377984.

OECD. (2007). *PISA 2006 science competencies for tomorrow's world* (Analysis, Vol. 1). Paris: Organisation for Economic Cooperation and Development.

Osborne, J., Simon, S., & Collins, S. (2003). Attitudes towards science: A review of the literature and its implications. *International Journal of Science Education, 25*(9), 1049–1079. doi:10.1080/0950069032000032199.

Palmer, D. (1997). The effect of context on students' reasoning about forces. *International Journal of Science Education, 19*(16), 681–696. doi:10.1080/0950069970190605.

Park, J. (2013). Prototypes, exemplars, and theoretical & applied ethics. *Neuroethics, 6*(2), 237–247. doi:10.1007/s12152-011-9106-8.

Parkin, A. J. (1987). *Memory & amnesia: An introduction.* Oxford, UK: Basil Blackwell.

Parkin, A. J. (1993). *Memory: Phenomena, experiment and theory.* Hove, UK: Psychology Press.

Patomäki, H., & Wight, C. (2000). After postpositivism? The promises of critical realism. *International Studies Quarterly, 44*(2), 213–237. doi:10.1111/0020-8833.00156.

Payne, D. G. (1987). Hypermnesia and reminiscence in recall: A historical and empirical review. *Psychological Bulletin, 101*(1), 5–27.

Peitsch, D., Fietz, A., Hertel, H., Souza, J., Fix Ventura, D., & Menzel, R. (1992). The spectral input systems of hymenopteran insects and their receptor-based colour vision. *Journal of Comparative Physiology. A, Neuroethology, Sensory, Neural, and Behavioral Physiology, 170*(1), 23–40. doi:10.1007/BF00190398.

Perry, W. G. (1970). *Forms of intellectual and ethical development in the college years: A scheme.* New York: Holt, Rinehart & Winston.

Peterson, R., Treagust, D. F., & Garnett, P. (1986). Identification of secondary students' misconceptions of covalent bonding and structure concepts using a diagnostic instrument. *Research in Science Education, 16*, 40–48.

Petri, J., & Niedderer, H. (1998). A learning pathway in high-school level quantum atomic physics. *International Journal of Science Education, 20*(9), 1075–1088.

Petruccioli, S. (1993). *Atoms, metaphors and paradoxes: Niels Bohr and the construction of a new physics* (I. McGilvray, Trans.). Cambridge, UK: Cambridge University Press.

Phang, F. A. (2009). *The patterns of physics problem-solving from the perspective of metacognition.* Unpublished Ph.D. thesis. Cambridge, UK: Faculty of Education, University of Cambridge.

Piaget, J. (1932/1977). *The moral judgement of the child.* Harmondsworth, UK: Penguin Books.

Piaget, J. (1970/1972). *The principles of genetic epistemology* (W. Mays, Trans.). London: Routledge & Kegan Paul.

Piaget, J. (1972). *Psychology and epistemology: Towards a theory of knowledge* (P. A. Wells, Trans.). Harmondsworth, UK: Penguin.

Piaget, J. (1979). *Behaviour and evolution* (D. Nicholson-Smith, Trans.). London: Routledge & Kegan Paul.

Pintrich, P. R., Marx, R. W., & Boyle, R. A. (1993). Beyond cold conceptual change: The role of motivational beliefs and classroom contextual factors in the process of conceptual change. *Review of Educational Research, 63*(2), 167–199.

Polanyi, M. (1962). *Personal knowledge: Towards a post-critical philosophy* (Corrected version ed.). Chicago: University of Chicago Press.

Polanyi, M. (1970). The logic of tacit inference. In F. J. Crosson (Ed.), *Human and artificial intelligence* (pp. 219–240). New York: Appleton-Century-Crofts.

Pope, M. L., & Denicolo, P. (1986). Intuitive theories – A researcher's dilemma: Some practical methodological implications. *British Educational Research Journal, 12*(2), 153–166.

Pope, M. L., & Gilbert, J. K. (1983). Personal experience and the construction of knowledge in science. *Science Education, 67*(2), 193–203.

Popper, K. R. (1934/1959). *The logic of scientific discovery.* London: Hutchinson.

Popper, K. R. (1970). Normal science and its dangers. In I. Lakatos & A. Musgrove (Eds.), *Criticism and the growth of knowledge* (pp. 51–58). Cambridge, UK: Cambridge University Press.

Popper, K. R. (1979). *Objective knowledge: An evolutionary approach* (Rev. ed.). Oxford, UK: Oxford University Press.

Posner, G. J., Strike, K. A., Hewson, P. W., & Gertzog, W. A. (1982). Accommodation of a scientific conception: Towards a theory of conceptual change. *Science Education, 66*(2), 211–227.

Pring, R. (2000). *Philosophy of educational research.* London: Continuum.

Przełęcki, M. (1974). A set theoretic versus a model theoretic approach to the logical structure of physical theories. *Studia Logica, 33*(1), 91–105. doi:10.1007/bf02120870.

Pugh, K. J. (2004). Newton's laws beyond the classroom walls. *Science Education, 88*(2), 182–196. doi:10.1002/sce.10109.

Ramachandran, V. S. (2003). Foreword. In L. Pessoa & P. De Weerd (Eds.), *Filling-in: From perceptual completion to cortical reorganization* (pp. xi–xxii). Oxford, UK: Oxford University Press.

Ratinen, I. J. (2011). Primary student-teachers' conceptual understanding of the greenhouse effect: A mixed method study. *International Journal of Science Education,* 1–27. doi:10.1080/09500 693.2011.587845.

Rees, G. (2007). Neural correlates of the contents of visual awareness in humans. In J. Driver, P. Haggard, & T. Shallice (Eds.), *Mental processes in the human brain* (pp. 187–202). Oxford, UK: Oxford University Press.

Reiner, M., Slotta, J. D., Chi, M. T. H., & Resnick, L. B. (2000). Naive physics reasoning: A commitment to substance-based conceptions. *Cognition and Instruction, 18*(1), 1–34. doi:10.1207/ s1532690xci1801_01.

Roth, I. (1986). An introduction to object perception. In I. Roth & J. P. Frisby (Eds.), *Perception and representation: A cognitive approach* (pp. 79–132). Milton Keynes, UK: Open University Press.

Roth, W.-M., & Tobin, K. (2006). Editorial: Announcing cultural studies of science education. *Cultural Studies of Science Education, 1*(1), 1–5. doi:10.1007/s11422-005-9005-6.

Rothenberg, A. (1995). Creative cognitive processes in Kekulé's discovery of the structure of the benzene molecule. *The American Journal of Psychology, 108*(3), 419–438.

Sacks, O. (1986). *The man who mistook his wife for a hat.* London: Pan Books.

Sadler, T. D. (Ed.). (2011). *Socio-scientific issues in the classroom: Teaching, learning and research* (Vol. 39). Dordrecht, The Netherlands: Springer.

Sadler, T. D., Klosterman, M. L., & Topcu, M. S. (2011). Learning science content and socio-scientific reasoning through classroom explorations of global climate change. In T. D. Sadler (Ed.), *Socio-scientific issues in the classroom: Teaching, learning and research* (pp. 45–77). Dordrecht, The Netherlands: Springer.

Sagan, C. (1990). Why we need to understand science. *The Skeptical Inquirer, 14*(3), 263–269.

Scardamalia, M., & Bereiter, C. (2006). Knowledge building: Theory, pedagogy, and technology. In R. K. Sawyer (Ed.), *The Cambridge handbook of the learning sciences* (pp. 97–115). Cambridge, UK: Cambridge University Press.

Schacter, D. L., & Addis, D. R. (2007). The cognitive neuroscience of constructive memory: Remembering the past and imagining the future. In J. Driver, P. Haggard, & T. Shallice (Eds.), *Mental processes in the human brain* (pp. 27–47). Oxford, UK: Oxford University Press.

Schutz, A., & Luckmann, T. (1973). The structures of the life-world (R. M. Zaner & H. T. Engelhardt, Trans.). Evanston, IL: Northwest University Press.

Schwedes, H., & Schmidt, D. (1992). Conceptual change: A case study and theoretical comments. In R. Duit, F. Goldberg, & H. Niedderer (Eds.), *Research in physics learning: Theoretical issues and empirical studies* (pp. 188–292). Kiel, Germany: Institut für die Pädagogik der Naturwissenschaften.

Scott, P. (1998). Teacher talk and meaning making in science classrooms: A review of studies from a Vygotskian perspective. *Studies in Science Education, 32*, 45–80.

Seger, C. A., & Miller, E. K. (2010). Category learning in the brain. *Annual Review of Neuroscience, 33*(1), 203–219. doi:10.1146/annurev.neuro.051508.135546.

Shayer, M., & Adey, P. (1981). *Towards a science of science teaching: Cognitive development and curriculum demand*. Oxford, UK: Heinemann Educational Books.

Sheffield, C. (1981). *Earth watch: A survey of the world from space*. London: Sidgwick & Jackson.

Shepardson, D. P., Wee, B., Priddy, M., & Harbor, J. (2007). Students' mental models of the environment. *Journal of Research in Science Teaching, 44*(2), 327–348. doi:10.1002/tea.20161.

Sijuwade, P. O. (2007). Recent trends in the philosophy of science: Lessons for sociology. *Journal of Social Sciences, 14*(1), 53–64.

Simons, D. J., & Chabris, C. F. (1999). Gorillas in our midst: Sustained in attentional blindness for dynamic events. *Perception, 28*(9), 1059–1074.

Sjøberg, S. (2010). Constructivism and learning. In E. Baker, B. McGaw, & P. Peterson (Eds.), *International encyclopaedia of education* (3rd ed., pp. 485–490). Oxford, UK: Elsevier.

Smith, E. L. (1991). A conceptual change model of learning science. In S. M. Glynn, R. H. Yeany, & B. K. Britton (Eds.), *The psychology of learning science* (pp. 43–63). Hillsdale, NJ: Lawrence Erlbaum Associates.

Smith, J. P., diSessa, A. A., & Roschelle, J. (1993). Misconceptions reconceived: A constructivist analysis of knowledge in transition. *The Journal of the Learning Sciences, 3*(2), 115–163.

Solomon, J. (1983). Learning about energy: How pupils think in two domains. *European Journal of Science Education, 5*(1), 49–59. doi:10.1080/0140528830050105.

Solomon, J. (1992). *Getting to know about energy – In school and society*. London: Falmer Press.

Solomon, J. (1993). The social construction of children's scientific knowledge. In P. Black & A. M. Lucas (Eds.), *Children's informal ideas in science* (pp. 85–101). London: Routledge.

Sperber, D. (1994). The modularity of thought and the epidemiology of representations. In L. Hirschfeld & S. A. Gelman (Eds.), *Mapping the mind: Domain specificity in cognition and culture* (pp. 39–67). Cambridge, UK: Cambridge University Press.

Stavy, R., & Tirosh, D. (2000). *How students (mis)understand science and mathematics: Intuitive rules*. New York: Teachers College Press.

Sternberg, R. J. (1980). Factor theories of intelligence are all right almost. *Educational Researcher, 9*(8), 6–13+18.

Sternberg, R. J. (2009a). A balance theory of wisdom. In J. C. Kaufman & E. L. Grigorenko (Eds.), *The essential Sternberg: Essays on intelligence, psychology and education* (pp. 353–375). New York: Springer.

Sternberg, R. J. (2009b). Sketch of a componential subtheory of human intelligence. In J. C. Kaufman & E. L. Grigorenko (Eds.), *The essential Sternberg: Essays on intelligence, psychology and education* (pp. 3–31). New York: Springer.

Strong, K., & Hutchins, H. (2009). Connectivism: A theory for learning in a world of growing complexity Impact. *Journal of Applied Research in Workplace E-Learning, 1*(1), 53–67.

Sugarman, S. (1987). *Piaget's construction of the child's reality*. Cambridge, UK: Cambridge University Press.

Sutherland, P. (1992). *Cognitive development today: Piaget and his critics*. London: Paul Chapman Publishing.

Sweller, J. (2007). Evolutionary biology and educational psychology. In J. S. Carlson & J. R. Levin (Eds.), *Educating the evolved mind: Conceptual foundations for an evolutionary educational psychology* (pp. 165–175). Charlotte, NC: Information Age Publishing.

Taber, K. S. (1994). Student reaction on being introduced to concept mapping. *Physics Education, 29*(5), 276–281.

Taber, K. S. (1997). Student understanding of ionic bonding: Molecular versus electrostatic thinking? *School Science Review, 78*(285), 85–95.

Taber, K. S. (1998a). An alternative conceptual framework from chemistry education. *International Journal of Science Education, 20*(5), 597–608.

Taber, K. S. (1998b). The sharing-out of nuclear attraction: Or I can't think about physics in chemistry. *International Journal of Science Education, 20*(8), 1001–1014.

Taber, K. S. (2000a). Finding the optimum level of simplification: The case of teaching about heat and temperature. *Physics Education, 35*(5), 320–325.

Taber, K. S. (2000b). Multiple frameworks?: Evidence of manifold conceptions in individual cognitive structure. *International Journal of Science Education, 22*(4), 399–417.

Taber, K. S. (2001a). Building the structural concepts of chemistry: Some considerations from educational research. *Chemistry Education: Research and Practice in Europe, 2*(2), 123–158.

Taber, K. S. (2001b). Shifting sands: A case study of conceptual development as competition between alternative conceptions. *International Journal of Science Education, 23*(7), 731–753.

Taber, K. S. (2002a). *Chemical misconceptions – Prevention, diagnosis and cure: theoretical background* (Vol. 1). London: Royal Society of Chemistry.

Taber, K. S. (2002b). Conceptualizing quanta – Illuminating the ground state of student understanding of atomic orbitals. *Chemistry Education: Research and Practice in Europe, 3*(2), 145–158.

Taber, K. S. (2003). Lost without trace or not brought to mind? – A case study of remembering and forgetting of college science. *Chemistry Education: Research and Practice, 4*(3), 249–277.

Taber, K. S. (2006a). Beyond constructivism: The progressive research programme into learning science. *Studies in Science Education, 42*, 125–184.

Taber, K. S. (2006b). Conceptual integration: A demarcation criterion for science education? *Physics Education, 41*(4), 286–287.

Taber, K. S. (2006c). Constructivism's new clothes: The trivial, the contingent, and a progressive research programme into the learning of science. *Foundations of Chemistry, 8*(2), 189–219.

Taber, K. S. (2007). *Classroom-based research and evidence-based practice: A guide for teachers.* London: Sage.

Taber, K. S. (2008a). Exploring conceptual integration in student thinking: Evidence from a case study. *International Journal of Science Education, 30*(14), 1915–1943. doi:10.1080/09500690701589404.

Taber, K. S. (2008b). Towards a curricular model of the nature of science. *Science Education, 17*(2–3), 179–218. doi:10.1007/s11191-006-9056-4.

Taber, K. S. (2009a). Learning at the symbolic level. In J. K. Gilbert & D. F. Treagust (Eds.), *Multiple representations in chemical education* (pp. 75–108). Dordrecht, The Netherlands: Springer.

Taber, K. S. (2009b). *Progressing science education: Constructing the scientific research programme into the contingent nature of learning science.* Dordrecht, The Netherlands: Springer.

Taber, K. S. (2011). The natures of scientific thinking: Creativity as the handmaiden to logic in the development of public and personal knowledge. In M. S. Khine (Ed.), *Advances in the nature of science research – Concepts and methodologies* (pp. 51–74). Dordrecht, The Netherlands: Springer.

Taber, K. S. (2012). Recognising quality in reports of chemistry education research and practice. *Chemistry Education Research and Practice, 13*(1), 4–7.

Taber, K. S. (2013a). *Classroom-based research and evidence-based practice: An introduction* (2nd ed.). London: Sage.

Taber, K. S. (2013b). Personal or collective knowledge: Harry Collins' notions of tacit knowledge and of the individual as an epistemic parasite. *Chemistry: Bulgarian Journal of Science Education, 22*(1), 114–135.

Taber, K. S. (2013c). Revisiting the chemistry triplet: Drawing upon the nature of chemical knowledge and the psychology of learning to inform chemistry education. *Chemistry Education Research and Practice, 14*(2), 156–168. doi:10.1039/C3RP00012E.

Taber, K. S. (2013d). A common core to chemical conceptions: Learners' conceptions of chemical stability, change and bonding. In G. Tsaparlis & H. Sevian (Eds.), *Concepts of matter in science education* (pp. 391–418). Dordrecht, The Netherlands: Springer.

Taber, K. S. (2013e). Conceptual frameworks, metaphysical commitments and worldviews: The challenge of reflecting the relationships between science and religion in science education. In N. Mansour & R. Wegerif (Eds.), *Science education for diversity: Theory and practice* (pp. 151–177). Dordrecht, The Netherlands: Springer.

Taber, K. S. (2013f). The relationship between science and religion – A contentious and complex issue facing science education. In B. Akpan (Ed.), *Science education: A global perspective.* Abuja, Nigeria: Next Generation Education Ltd.

Taber, K. S. (2013g). Representing evolution in science education: The challenge of teaching about natural selection. In B. Akpan (Ed.), *Science education: A global perspective.* Abuja, Nigeria: Next Generation Education Ltd.

Taber, K. S. (In press). Methodological issues in science education research: A perspective from the philosophy of science. In M. R. Matthews (Ed.), *International handbook of research in history and philosophy for science and mathematics education*. Springer.

Taber, K. S., Billingsley, B., Riga, F., & Newdick, H. (2011). Secondary students' responses to perceptions of the relationship between science and religion: Stances identified from an interview study. *Science Education, 95*(6), 1000–1025. doi:10.1002/sce.20459.

Taber, K. S., & Bricheno, P. A. (2009). Coordinating procedural and conceptual knowledge to make sense of word equations: Understanding the complexity of a 'simple' completion task at the learner's resolution. *International Journal of Science Education, 31*(15), 2021–2055. doi:10.1080/09500690802326243.

Taber, K. S., & García Franco, A. (2009). Intuitive thinking and learning chemistry. *Education in Chemistry, 46*(2), 57–60.

Taber, K. S., & García Franco, A. (2010). Learning processes in chemistry: Drawing upon cognitive resources to learn about the particulate structure of matter. *The Journal of the Learning Sciences, 19*(1), 99–142.

Taber, K. S., & Tan, K. C. D. (2011). The insidious nature of 'hard core' alternative conceptions: Implications for the constructivist research programme of patterns in high school students' and pre-service teachers' thinking about ionisation energy. *International Journal of Science Education, 33*(2), 259–297. doi:10.1080/09500691003709880.

Taber, K. S., Tsaparlis, G., & Nakiboğlu, C. (2012). Student conceptions of ionic bonding: Patterns of thinking across three European contexts. *International Journal of Science Education*, 1–31. doi:10.1080/09500693.2012.656150.

Tavakol, M., & Dennick, R. (2010). Are Asian international medical students just rote learners? *Advances in Health Sciences Education, 15*(3), 369–377. doi:10.1007/s10459-009-9203-1.

Thagard, P. (1992). *Conceptual revolutions*. Oxford, UK: Princeton University Press.

Thomson, R. (1959). *The psychology of thinking*. Harmondsworth, UK: Penguin Books.

Toulmin, S. (1972). *Human understanding: The collective use and evolution of concepts*. Princeton, NJ: Princeton University Press.

Toulmin, S., & Goodfield, J. (1962/1999). *The fabric of the heavens: The development of astronomy and dynamics*. Chicago: University of Chicago Press.

Trainor, A. (2003). Poem. *International Journal of Qualitative Studies in Education, 16*(5), 727–727. doi:10.1080/0951839032000143039.

Treagust, D. F. (1988). Development and use of diagnostic tests to evaluate students' misconceptions in science. *International Journal of Science Education, 10*(2), 159–169. doi:10.1080/0950069880100204.

Tsaparlis, G. (1994). Blocking mechanisms in problem solving from the Pascual-Leone's M-space perspective. In H.-J. Schmidt (Ed.), *Problem solving and misconceptions in chemistry and physics* (pp. 211–226). Dortmund, Germany: International Council of Associations for Science Education.

Van Dyck, D., & Op de Beeck, M. (1996). A simple intuitive theory for electron diffraction. *Ultramicroscopy, 64*(1–4), 99–107. doi:10.1016/0304-3991(96)00008-3.

Van Eijck, M. (2009). Scientific literacy: Past research, present conceptions, and future developments. In W.-M. Roth & K. Tobin (Eds.), *Handbook of research in North America* (Vol. 1, pp. 245–258). Rotterdam, The Netherlands: Sense.

Vertes, R. P. (2004). Memory consolidation in sleep: Dream or reality. *Neuron, 44*(1), 135–148. doi:10.1016/j.neuron.2004.08.034.

Vosniadou, S. (Ed.). (2008a). *International handbook of research on conceptual change*. New York: Routledge.

Vosniadou, S. (Ed.). (2008b). *International handbook of research on conceptual change*. London: Routledge.

Vygotsky, L. S. (1934/1986). *Thought and language*. London: MIT Press.

Vygotsky, L. S. (1934/1994). The development of academic concepts in school aged children. In R. van der Veer & J. Valsiner (Eds.), *The Vygotsky reader* (pp. 355–370). Oxford, UK: Blackwell.

Vygotsky, L. S. (1978). *Mind in society: The development of higher psychological processes*. Cambridge, MA: Harvard University Press.

Walker, M. P., & Stickgold, R. (2004). Sleep-dependent learning and memory consolidation. *Neuron, 44*(1), 121–133. doi:10.1016/j.neuron.2004.08.031.

Watson, J. B. (1924/1998). *Behaviorism*. New Brunswick, NJ: Transaction.

Watson, J. B. (1967). What is behaviourism? In J. A. Dyal (Ed.), *Readings in psychology: Understanding human behavior* (2nd ed., pp. 7–9). New York: McGraw-Hill Book Company.

Watson, O. (Ed.). (1968). *Longman modern English dictionary*. Harlow, UK: Longman Group Limited.

Watts, M., & Gilbert, J. K. (1983). Enigmas in school science: Students' conceptions for scientifically associated words. *Research in Science and Technological Education, 1*(2), 161–171.

Watts, M., & Pope, M. L. (1982). *A Lakatosian view of the young personal scientist*. Paper presented at the British Conference on Personal Construct Psychology, University of Manchester Institute of Science & Technology, Manchester.

Watts, M., & Taber, K. S. (1996). An explanatory gestalt of essence: Students' conceptions of the 'natural' in physical phenomena. *International Journal of Science Education, 18*(8), 939–954.

Watts, M., & Zylbersztajn, A. (1981). A survey of some children's ideas about force. *Physics Education, 16*(6), 360–365.

White, R. T. (1998). Research, theories of learning, principles of teaching and classroom practice: Examples and issues. *Studies in Science Education, 31*, 55–70. doi:10.1080/03057269808560112.

White, R. T., & Gunstone, R. F. (1992). *Probing understanding*. London: Falmer Press.

White, R. T., & Mitchell, I. J. (1994). Metacognition and the quality of learning. *Studies in Science Education, 23*, 21–37. doi:10.1080/03057269408560028.

Wiltgen, B. J., Brown, R. A. M., Talton, L. E., & Silva, A. J. (2004). New circuits for old memories: The role of the Neocortex in consolidation. *Neuron, 44*(1), 101–108. doi:10.1016/j.neuron.2004.09.015.

Witzig, S. B., Halverson, K. L., Siegel, M. A., & Freyermuth, S. K. (2011). The interface of opinion, understanding and evaluation while learning about a socioscientific issue. *International Journal of Science Education*, 1–25. doi:10.1080/09500693.2011.600351.

Wolpert, L., & Richards, A. (1988). *A passion for science*. Oxford, UK: Oxford University Press.

Wong, E. D. (1993). Understanding the generative capacity of analogies as a tool for explanation. *Journal of Research in Science Teaching, 30*(10), 1259–1272. doi:10.1002/tea.3660301008.

Yenilmez, A., & Tekkaya, C. (2006). Enhancing students' understanding of photosynthesis and respiration in plant through conceptual change approach. *Journal of Science Education and Technology, 15*(1), 81–87. doi:10.1007/s10956-006-0358-8.

Ziman, J. (1978/1991). *Reliable knowledge: An exploration of the grounds for belief in science*. Cambridge, UK: Cambridge University Press.

Index

K.S. Taber, *Modelling Learners and Learning in Science Education: Developing Representations of Concepts, Conceptual Structure and Conceptual Change to Inform Teaching and Research*, DOI 10.1007/978-94-007-7648-7, © Springer Science+Business Media Dordrecht 2013